D0722605

MEASURING COLOUR
Second Edition

ELLIS HORWOOD SERIES IN APPLIED SCIENCE AND INDUSTRIAL TECHNOLOGY

Series Editor: Dr D. H. SHARP, OBE, former General Secretary, Society of Chemical Industry; formerly General Secretary, Institution of Chemical Engineers; and former Technical Director, Confederation of British Industry.

This collection of books is designed to meet the needs of technologists already working in the fields to be covered, and for those new to the industries concerned. The series comprises valuable works of reference for scientists and engineers in many fields, with special usefulness to technologists and entrepreneurs in developing countries.

Students of chemical engineering, industrial and applied chemistry, and related fields, will also find these books of great use, with their emphasis on the practical technology as well as theory. The authors are highly qualified chemical engineers and industrial chemists with extensive experience, who write with the authority gained from their years in industry.

Published and in active publication

PRACTICAL USES OF DIAMONDS
A. BAKON, Research Centre of Geological Technique, Warsaw, and A. SZYMANSKI, Institute of Electronic Materials Technology, Warsaw

NATURAL GLASSES
V. BOUSKA *et al.*, Czechoslovak Society for Mineralogy & Geology, Czechoslovakia

POTTERY SCIENCE: Materials, Processes and Products
A. DINSDALE, lately Director of Research, British Ceramic Research Association

MATCHMAKING: Science, Technology and Manufacture
C. A. FINCH, Managing Director, Pentafin Associates, Chemical, Technical and Media Consultants, Stoke Mandeville, and S. RAMACHANDRAN, Senior Consultant, United Nations Industrial Development Organisation for the Match Industry

THE HOSPITAL LABORATORY: Strategy, Equipment, Management and Economics
T. B. HALES, Arrowe Park Hospital, Merseyside

OFFSHORE PETROLEUM TECHNOLOGY AND DRILLING EQUIPMENT
R. HOSIE, formerly of Robert Gordon's Institute of Technology, Aberdeen

MEASURING COLOUR: Second Edition
R. W. G. HUNT, Visiting Professor, The City University, London

MODERN APPLIED ENERGY CONSERVATION
Editor: K. JACQUES, University of Stirling, Scotland

CHARACTERIZATION OF FOSSIL FUEL LIQUIDS
D. W. JONES, University of Bristol

PAINT AND SURFACE COATINGS: Theory and Practice
Editor: R. LAMBOURNE, Technical Manager, INDCOLLAG (Industrial Colloid Advisory Group), Department of Physical Chemistry, University of Bristol

CROP PROTECTION CHEMICALS
B. G. LEVER, International Research and Development Planning Manager, ICI Agrochemicals

HANDBOOK OF MATERIALS HANDLING
Translated by R. G. T. LINDKVIST, MTG, Translation Editor: R. ROBINSON, Editor, *Materials Handling News*. Technical Editor: G. LUNDESJO, Rolatruc Limited

FERTILIZER TECHNOLOGY
G. C. LOWRISON, Consultant, Bradford

NON-WOVEN BONDED FABRICS
Editor: J. LUNENSCHLOSS, Institute of Textile Technology of the Rhenish-Westphalian Technical University, Aachen, and W. ALBRECHT, Wuppertal

REPROCESSING OF TYRES AND RUBBER WASTES: Recycling from the Rubber Products Industry
V. M. MAKAROV, Head of General Chemical Engineering, Labour Protection, and Nature Conservation Department, Yaroslavl Polytechnic Institute, USSR, and V. F. DROZDOVSKI, Head of the Rubber Reclaiming Laboratory, Research Institute of the Tyre Industry, Moscow, USSR

PROFIT BY QUALITY: The Essentials of Industrial Survival
P. W. MOIR, Consultant, West Sussex

EFFICIENT BEYOND IMAGINING: CIM and its Applications for Today's Industry
P. W. MOIR, Consultant, West Sussex

TRANSIENT SIMULATION METHODS FOR GAS NETWORKS
A. J. OSIADACZ, UMIST, Manchester

Series continued at back of book

MEASURING COLOUR
Second Edition

R. W. G. HUNT
Colour Consultant and Visiting Professor of Physiological Optics
The City University, London

ELLIS HORWOOD
NEW YORK LONDON TORONTO SYDNEY TOKYO SINGAPORE

First published in 1991, reprinted in 1992 by
ELLIS HORWOOD LIMITED
Market Cross House, Cooper Street,
Chichester, West Sussex, PO19 1EB, England

A division of
Simon & Schuster International Group
A Paramount Communications Company

Typeset in Times by Ellis Horwood
Printed and bound in Great Britain
by Hartnolls, Bodmin, Cornwall

British Library Cataloguing-in-Publication Data

Hunt, R. W. G.
Measuring colour, second edition. —
(Ellis Horwood series in applied science and industrial technology)
I. Title II. Series
535.6
ISBN 0–13–567686–X

Library of Congress Cataloging-in-Publication Data

Hunt, R. W. G. (Robert William Gainer), 1923–
Measuring colour / R. W. G. Hunt. — 2nd ed.
p. cm. — (Ellis Horwood series in applied science and industrial technology)
Includes bibliographical references and index.
ISBN 0–13–567686–X
1. Colorimetry. I. Title. II. Series.
QC495.H84 1991
535.6′028′7–dc20 91–14380
CIP

Contents

9 Fluorescent colours

Prologue

This is the story of Mister Chrome
      who started out to paint his home.
The paint ran out when half way through
      so to the store he quickly flew
to buy some more of matching hue,
      a delicate shade of egg-shell blue.
But when he tried this latest batch,
      he found it simply didn't match.
No wonder he was in a fix,
      for of the colours we can mix,
the major shades and those between,
      ten million different can be seen.

You foolish man, said Missis Chrome,
      you should have taken from the home
a sample of the colour done;
      you can't remember every one.
Taking care that she had got
      a sample from the early pot,
she went and bought her husband more
      of better colour from the store.
Before she paid, she checked the shade,
      and found a perfect match it made.
In triumph now she took it home,
      and gave it straight to Mister Chrome.
He put in on without delay,
      and found the colour now okay.

But, after dark, in tungsten light,
they found the colour still not right.
So to the store they both went now,
with samples clear, and asked them how
a paint that matched in daylight bright
could fail to match in tungsten light.
The man's reply to their complaint
was that the pigments in the paint
had been exchanged, since they had bought,
for others of a different sort.
To solve the problem on their wall,
he gave them paint to do it all
from just one batch of constant shade,
and then at last success was made.

To compensate them for their trouble,
the store sent to them curtains double.
They hung them up with great delight;
they matched in tungsten and daylight.
A neighbour then did make a call
and fixed his eye upon the wall;
the paint, he said was all one colour,
but clearly saw the curtains duller!

Though colours strange at times appear,
the moral of this tale is clear:
to understand just what we see,
object, light, and eye, all three,
must colour all our thinking through
of chromic problems, old or new!

Preface to the first edition

This book is intended to provide the reader with the basic facts needed to measure colour. It is a book about principles, rather than a guide to instruments. With the continual advances in technology, instruments are being improved all the time, so that any description of particular colorimeters, spectroradiometers, or spectrophotometers is likely to become out of date very quickly. For such information, manufacturers' catalogues are a better source of information than books. But the principles of measuring colour are not subject to rapid change, and are therefore appropriate for treatment in the more permanent format offered by books.

Recommendations about the precise way in which the basic principles of colour measurement should be applied have for over 50 years been the province of the International Commission on Illumination (CIE). The second edition of its Publication No. 15, *Colorimetry* includes several new practices, and it is therefore timely to restate the principles of colorimetry together with these latest international recommendations on their application; this is the aim of *Measuring Colour*.

Colour is, of course, primarily a sensation experienced by the individual. For this reason, the material has been set in the context of the colour vision properties of the human observer: the first chapter is a review of our current knowledge of colour vision; and the last chapter provides a description of a model of colour vision that can be used to extend colour measurement, beyond the territory covered by the CIE at present, to the field of colour appearance.

Preface to the second edition

The second edition contains all the material of the first edition, together with four new chapters. Two of these chapters provide entirely new material: one is on light sources, and the other is on precision and accuracy in colorimetry. The other two new chapters provide expanded treatments of metamerism and of the colorimetry of fluorescent materials. Extensive revisions have been made to the chapter on the model of colour vision, so as to present it in its latest version. Finally, minor revisions have been made to the rest of the book to improve the treatment in various respects.

Acknowledgements for the first edition

I am most grateful to Dr M. R. Pointer of Kodak Limited for kindly making many helpful comments on the text, for providing some of the numerical data, and for help with the proof reading. My grateful thanks for help are also due to Dr A. Hård in connection with the section on the NCS, to Dr H. Terstiege with that on the DIN system, and to Dr A. Nemcsics with that on the Coloroid system. For permission to reproduce figures, my thanks are due to the Institute of Physics for Fig. 3.5; to John Wiley & Sons for Figs 7.5, 7.12, 7.19 and 8.1; to Dr A. Hård for Fig. 7.12; to Dr H. Terstiege for Fig. 7.19; and to Academic Press for Figs 9.1, 9.2, and 9.3. I would also like to thank Dr J. Schanda for kindly supplying me with copies of recent CIE documents.

With regard to the colour plates, my thanks are due to the following for kindly supplying the originals: Dr A. A. Clarke and Dr M. R. Luo, of Loughborough University of Technology, for Plates 2 and 3; Dr A. Hård for Plate 5; Dr H. Terstiege for Plate 8; and Mr R. Ingalls for Plates 1, 6 and 7. I would also like to thank the Munsell Corporation for permission to reproduce Plate 4.

I am also most grateful to Mr A. J. Johnson, and some of his colleagues, of Crosfield Electronics Limited, for kindly supplying the separations for the colour illustrations.

Finally my grateful thanks are due to my wife for editorial assistance and for help with the proof reading.

Acknowledgements for the second edition

I am very grateful for help that has kindly been given to me by experts on the subject matter of the new material in this second edition. Dr F. W. Billmeyer has made many suggestions for improving the new chapters on metamerism, on precision and accuracy in colorimetry, and on the colorimetry of fluorescent materials. Miss M. B. Halstead, Mr D. O. Wharmby, and Dr M. G. Abeywickrama have helped with the new chapter on light sources. Dr R. F. Berns has helped with the section on correcting for errors in spectral data, and Dr W. H. Venable with the section on the computation of tristimulus values. I am indebted to Mr J. K. C. Kempster for the data on which Fig. 6.2 is based. Once again, I am most grateful to Dr M. R. Pointer for general comments, for help with computations, and for proof reading, and to my wife for editorial help and for proof reading.

1

Colour vision

1.1 INTRODUCTION

Ten million! That is the number of different colours that we can distinguish, according to one reliable estimate (Judd & Wyszecki 1975). It is, therefore, no wonder that we cannot remember colours well enough to identify a particular shade. Ladies are thus well advised to take samples of their dress colours with them, when purchasing accessories that are intended to match. They are also usually well aware that it is not enough to examine the colour match in just one type of light in a store, but to see it in daylight as well as in artificial light. Finally, a second opinion about the match, expressed by a friend or a shop assistant, is often wisely sought.

The above activity involves the three basic components of colour: sources of light, objects illuminated by them, and observers. Colour, therefore, involves not only material sciences, such as physics and chemistry, but also biological sciences such as physiology and psychology; and, in its applications, colour involves various applied sciences, such as architecture, dyeing, paint technology, and illuminating engineering. Measuring colour is, therefore, a subject that has to be broadly based and widely applied.

Without observers possessing the faculty of sight, there would be no colour. Hence it is appropriate to start by considering the nature of the colour vision provided by the human eye and brain. Before doing this, however, a brief description must be given of the way in which it is necessary to characterize the nature of the light which stimulates the visual system.

1.2 THE SPECTRUM

It is fair to say that understanding colour finds its foundations in the famous experiments performed by Isaac Newton in 1666. Before this date, opinions on the nature of colours and the relationships between them were most vague and of very little use, but, after Newton's work became known, the road was open for progress based on experimental facts.

The historic experiments were performed in Trinity College, Cambridge, when Newton made a small hole, a third of an inch in diameter, in the wall of an otherwise entirely dark room; through this hole, the direct rays of the sun could shine and form an image of the sun's disc on the opposite wall of the room, like a pin-hole camera. Then, taking a prism of glass, and placing it close to the hole, he observed that the light was spread out fan-wise into what he, first, called a *spectrum*: a strip of light, in this case about ten inches long, and coloured red, orange, yellow, green, blue, indigo, and violet along its length. The natural conclusion, which Newton was quick to draw, was that white light was not the simple homogeneous entity which it was natural to expect it to be, but was composed of a mixture of all the colours of the spectrum.

The next question which arose was whether these spectral colours themselves, red, green, etc., were also mixtures and could be spread out into further constituent colours. A further experiment was performed to test this suggestion. A card with a slit in it was used to obscure all the light of the spectrum, except for one narrow band. This band of light, say a yellow or a green, was then made to pass through a second prism, but the light was then seen not to be spread out any further, remaining exactly the same colour as when it emerged from the slit in the card. It was, therefore, established that the spectral colours were in fact the basic components of white light.

The inclusion by Newton of indigo in the list of spectral colours is rather puzzling since, to most people, there appears to be a gradual transition between blue and violet with no distinct colour between them, as there is in the case of orange between red and yellow. Several explanations of the inclusion of indigo have been suggested, but the most likely is that Newton tried to fit the colours into a scale of tones in a way analogous to the eight-tone musical scale; to do this he needed seven different colours to correspond to the seven different notes of the scale (McLaren 1985).

In Fig. 1.1, the main bands of colour in the spectrum are shown against a scale of

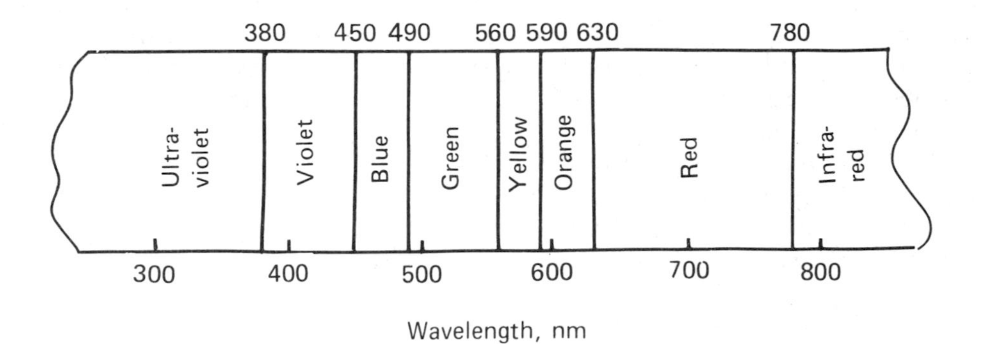

Fig. 1.1 — The colour names usually given to the main regions of the spectrum.

the *wavelength* of the light. Light is a form of electro-magnetic radiation, as is also the case for X-rays, radar, and radio waves, for instance, and the property of these radiations that gives them their particular characteristics is their wavelength. Radio waves have quite long wavelengths, typically in the range from about a metre to several kilometres, whereas X-rays have extremely short wavelengths, typically about a millionth of a millimetre or shorter. Light waves have wavelengths in between, ranging from slightly above to slightly below a half a millionth of a metre. To obtain convenient numbers for the wavelengths of light, the unit used for expressing them is the *nanometre* (abbreviation, *nm*), which is a millionth of a millimetre, or 10^{-9} of a metre; this is the unit used in Fig. 1.1. It must be emphasized that the colour names and wavelength boundaries given in Fig. 1.1 are only intended as a rough guide; each colour gradually merges into the next so that there is really no exact boundary; moreover, the colour appearance of light of a given wavelength depends on the viewing conditions, and is also liable to be slightly different from one observer to another. Even so, the names given in Fig. 1.1 are useful to bear in mind when considering data that are presented as functions of wavelength. The infra-red and ultra-violet regions are outside the *visible* spectrum; they can provide radiant energy that tans the skin or warms the body, for instance, but they cannot normally be seen as light. In colour science, although it is the long-established practice to identify different parts of the spectrum by using wavelength, it would be more fundamental to use *frequency*. This is because, for light from any part of the spectrum, as it passes through a medium, its wavelength decreases by being divided by the refractive index of that medium; however, the velocity also decreases in the same proportion, so that the frequency (the velocity divided by the wavelength) remains constant. The wavelengths quoted are usually in air, and, although those in vacuum would be more fundamental, they differ by only about 3 parts in 10 000. (The velocity of light in vacuum is about 2.998×10^8 metres per second.)

1.3 CONSTRUCTION OF THE EYE

A diagrammatic representation of a cross-section of the human eye is given in Fig. 1.2. Most of the optical power is provided by the curved surface of the *cornea*, and the main function of the *lens* is to alter that power by changing its shape, being thinner for viewing distant objects and thicker for near objects. The cornea and lens acting together form a small inverted image of the outside world on the *retina*, the light-sensitive surface of the eye. The *iris*, the annular shaped coloured part of the eye that we see from the outside, changes its shape, having a central hole that is only about 2 mm in diameter in bright light, but which is larger in dim light, having a maximum diameter of about 8 mm. The central hole referred to is the *pupil*, the optical aperture through which the light passes. The iris, by changing its shape, provides some compensation for changes in the level of illumination under which objects are seen; however, this compensation only amounts to a factor of about 8 to 1, rather than the 16 to 1 to be expected from the ratio of the squares of the diameters, because rays that pass through the edge of the pupil are less effective in stimulating the retina than those that pass through the centre, a property known as the *Stiles-Crawford effect*.

The retina lines most of the interior of the approximately spherically-shaped eye-ball, and this provides the eye with a very wide field of view. However, the retina is

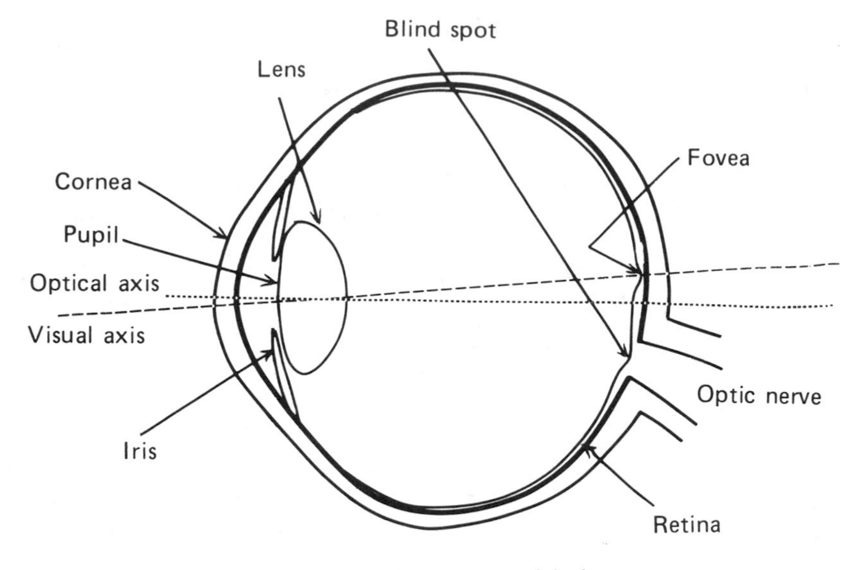

Fig. 1.2 — Cross-sectional diagram of the human eye.

far from being uniform in sensitivity over its area. Colour vision is limited to stimuli seen within about 40° of the visual axis (Hurvich 1981), and outside this area vision is virtually monochromatic and used mainly for the detection of movement. Within the 40° on either side of the eye's axis, the ability to see both colour and fine detail gradually increases as the eye's axis is approached, the area of sharpest vision being termed the *fovea*, which comprises approximately the central 1½° diameter of the visual field. An area within this, termed the *foveola*, corresponds to a field of about 1°. A curious feature of the fovea and foveola is that they are not centred on the *optical axis* of the eye, but lie about 4° to one side as shown in Fig. 1.2, thus resulting in the *visual axis* being offset by this amount. About 10° to the other side of the optical axis (equivalent to about 14° from the fovea) is the *blind spot*, where the nerve fibres connecting the retina to the brain pass through the surface of the eye-ball, and this area has no sensitivity to light at all. There is also an area covering part of the fovea, called the *yellow spot* or *macula lutea*, containing a yellowish pigment. In addition to these spatial variations in the retina, there are changes in the types of light receptors present in different areas. In the foveola, the receptors are all of one type, called *cones*; outside this area, there is, in addition, another type, called *rods*. The ratio of cones to rods varies continuously from all cones and no rods in the foveola to nearly all rods and very few cones beyond about 40° from the visual axis. Finally, the individual cones and rods are connected to the brain by nerve fibres in very different ways, depending on their position: in the foveola, there are about the same number of nerve fibres as cones; but, as the angle from the visual axis increases, the number of nerve fibres decreases continuously until as many as several hundred rods and cones may be served by each nerve fibre.

1.4 THE RETINAL RECEPTORS

The function of the rods in the retina is to give monochromatic vision under low levels of illumination. This *scotopic* form of vision operates when the stimuli have luminances of less than some hundredths of a candela per square metre (cd/m^2; for a summary of photometric terms and units, see Appendix 1).

The function of the cones in the retina is to give colour vision at normal levels of illumination. This *photopic* form of vision operates when stimuli have luminances of several cd/m^2 or more.

There is a gradual change from photopic to scotopic vision as the illumination level is lowered, and in this *mesopic* form of vision both cones and rods make significant contributions to the visual response. The wavelengths to which the rods are most sensitive are shorter than in the case of the cones, and, as a result, as the illumination level falls through the mesopic range, the relative brightnesses of red and blue colours change. This can often be seen in a garden at the end of the day; red flowers that look lighter than blue flowers in full daylight, look darker than the blue ones as the light fades. This is known as the *Purkinje phenomenon*.

The rods and cones are so named because of their shapes, but they are all very small, being typically of about a five-hundredth of a millimetre in diameter, with a length of around a twenty-fifth of a millimetre. They are packed parallel to one another and facing end on towards the pupil of the eye so that the light is absorbed by them as it travels along their length. They are connected to the nerve fibres via an extremely complicated network of cells situated immediately on the pupil-side of their ends. The nerve fibres then travel across the pupil-side of the retina to the blind spot where they are collected together to form the *optic nerve* which connects the eye to the brain. In each eye, there are about 6 million cones, 100 million rods, and 1 million nerve fibres

1.5 SPECTRAL SENSITIVITIES OF THE RETINAL RECEPTORS

The rods and the cones are not equally sensitive to all wavelengths of light. In the case of the rods, the initial step in the visual process is the absorption of light in a photosensitive pigment called *rhodopsin*. This pigment absorbs light most strongly in the blue-green part of the spectrum, and decreasingly as the wavelength of the light becomes either longer or shorter. As a result, the spectral sensitivity of the scotopic vision of the eye is as shown by the broken curve of Fig. 1.3. This curve is obtained by having observers adjust the strength of a beam of light of one wavelength until the sensation it produces has the same intensity as a beam of fixed strength of a reference wavelength. If the strength of the variable beam had to be, for example, twice that of the fixed beam, then the scotopic sensitivity at the wavelength of the variable beam would be regarded as a half of that at the wavelength of the fixed beam. These relative sensitivities are then plotted against wavelength to obtain the broken curve of Fig. 1.3, the maximum value being made equal to 1.0 by convention. To obtain a sensitivity curve representing scotopic vision, it is necessary to use beams of sufficiently low intensity to be entirely in the scotopic range, and the curve of Fig. 1.3 was obtained in this way. It is based on results obtained from about 70 observers (22 in a study by Wald 1945; and 50 in a study by Crawford 1949) and represents scotopic

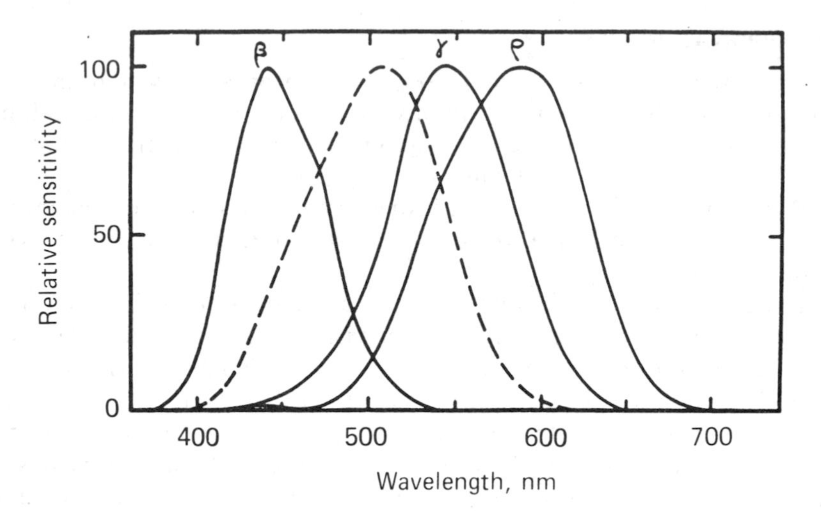

Fig. 1.3 — Broken line: the spectral sensitivity of the eye for scotopic (rod) vision. Full lines: spectral sensitivity curves representative of those believed to be typical of the three different types of cones, ρ, γ, and β, of the retina that provide the basis of photopic vision.

vision of observers under 30 years of age; above this age, progressive yellowing of the lens of the eye makes the results rather variable. The curve represents the scotopic sensitivity for light incident on the cornea, and thus the effects of any absorptions in the ocular media are included. The strengths of the beams can be evaluated in various ways, but the convention has been adopted to use the amount of power (energy per unit time) per small constant-width wavelength interval. If the beams used have the same small width of wavelength throughout the spectrum, then all that is required is to know the relative power in each beam. But, if the beams have different widths of wavelength, then the relative powers per unit wavelength interval have to be determined for each beam.

A system having a single spectral sensitivity function, such as that shown by the broken line in Fig. 1.3, cannot, on its own, provide a basis for colour vision. Thus, although, for example, light of wavelength 500 nm would result in a response about 30 times as great as the same strength of light of wavelength 600 nm, the two responses could be made equal simply by increasing the strength of the 600 nm beam by a factor of about 30 times. The system is thus not able to distinguish between changes in wavelength and changes in intensity, and this is what is needed to provide a basis for colour vision.

In the case of the cones of the human retina, it has not yet been possible to isolate and identify any photosensitive pigments, and our knowledge of them has had to be obtained by indirect means. These include very careful measurements of the light absorbed at each wavelength of the spectrum by individual cones removed from eyes that have become available for study (Dartnall, Bowmaker & Mollon 1983), and deductions from experiments on colour matching together with data on colour defective (colour blind, to be discussed in section 1.10) vision (Estévez 1979). As a result of these studies, sets of curves typified by those shown by the full lines of Fig. 1.3 have been obtained.

The exact shapes of the curves that best typify the spectral sensitivities of the cones are still a matter of some debate, but the set shown in Fig. 1.3 shows all the important features of any reasonably plausible set, and is entirely adequate for our present descriptive purposes. These curves represent the spectral sensitivities for light incident on the cornea, and allowance has thus again been made for any absorptions in the ocular media; these types of curve are often referred to as *action spectra*. For convenience, they have been plotted so that their maxima are all equal.

The full curves of Fig. 1.3 have been labelled ρ, γ, and β, to distinguish them. If Fig. 1.3 is compared to Fig. 1.1, it is clear that the ρ curve has a maximum sensitivity in the yellow–orange part of the spectrum, the γ curve in the green part, and the β curve in the blue–violet part. Various designations have been used by different authors for the three types of cone to which the three curves refer, including L, M, S (Long, Medium, and Short wavelength), π_5, π_2, π_1 (after the work of Stiles), and R, G, B (sensitive mainly to the Reddish, Greenish, and Bluish thirds of the spectrum). The R, G, B designation is perhaps the most widely used, but it is convenient to keep these symbols to represent red, green, and blue lights and colours, and hence the similar, but distinctive, Greek symbols, ρ, γ, β, have been adopted instead.

It is clear that there are *three* different curves for the cones, and they correspond to three different types of cone, each containing a different photosensitive pigment. We now have a basis for colour vision. For, if we consider again lights of wavelengths 500 and 600 nm, it is clear that the 500 nm light will produce about twice as much γ response as ρ response, whereas the 600 nm light will produce about twice as much ρ response as γ response. If the strengths of the beams are altered, the ρ and γ responses will also alter, but their ratios will always remain typical of those for their respective wavelengths. Hence, in this case, the strengths of the signals can indicate the intensities of the lights, and the ratios the wavelengths of the lights. We can thus distinguish between changes in the intensity, and changes in the spectral composition, of the lights, and hence a basis for colour vision exists. Most colours consist of a mixture of many wavelengths of the spectrum, not just a single wavelength as considered so far; but the above argument is quite general, and changes in spectral composition of such spectrally complex colours will cause changes in the ratios of the cone responses, and changes in the amount of light will cause changes in the strengths of the responses. Of course, in general, there will also be changes in the ratios of the β to γ and β to ρ responses as the spectral composition is changed, and these further assist in the discrimination of colours.

The different types of cone, ρ, γ, and β, are, as far as is known, distributed more or less randomly in the retinal mosaic of receptors on which the light falls, but there are many fewer β cones than ρ and γ; one estimate of their relative abundances is that they are in the ratios of 40 to 20 to 1 for the ρ, γ, and β cones, respectively (Walraven & Bouman 1966). This rather asymmetrical arrangement is, in fact, very understandable. Because the eye is not corrected for chromatic aberration, it cannot simultaneously focus sharply the three regions of the spectrum in which the ρ, γ, and β cones are most sensitive, that is, wavelengths of around 580, 540, and 440 nm, respectively. The ρ and γ peak wavelengths are closer together than is the case for the γ and β peaks; hence, if the eye focuses light of wavelength about 560 nm, both the ρ and γ responses will correspond to images that are reasonably sharp. The β cones then have to receive an image that is much less sharp, and hence it is unnecessary to provide such a fine network of β cones to detect it.

1.6 VISUAL SIGNAL TRANSMISSION

When light is absorbed in a receptor in the retina, the molecules of the photosensitive pigment are excited, and, as a result, a change in electrical potential is produced. This change then travels through a series of relay cells, and eventually results in a series of voltage pulses being transmitted along a nerve fibre to the brain. The rate at which these pulses are produced provides the signal modulation, a higher rate indicating a stronger signal, and a lower rate a weaker signal. However, zero signal may be indicated by a *resting rate*, and rates lower than this can then indicate an opposite signal. The pulses themselves are all of the same amplitude, and it is only their frequency that carries information to the brain. The frequencies involved are typically from a few per second to around 400 per second.

It might be thought that, as there are four different types of receptor, the rods and the three different types of cone, there would be four different types of signal, transmitted along four different types of nerve fibre, each indicating the strength of the response from one of the four receptor types. However, there is overwhelming evidence that this is not what happens (Mollon 1982). While much still remains unknown about the way in which the signals are encoded for transmission, the simple scheme shown in Fig. 1.4 can be regarded as a plausible framework for incorporating

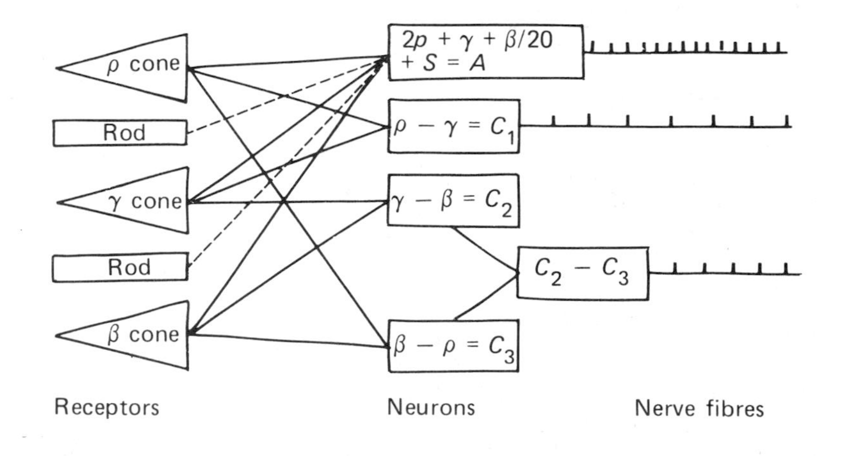

Fig. 1.4 — Greatly oversimplified and hypothetical diagrammatic representation of possible types of connections between some retinal receptors and some nerve fibres.

some of the salient features of what is believed to take place. The strengths of the signals from the cones is represented by the symbols, ρ, γ, and β. These strengths will depend on the amount of radiation usefully absorbed by the three different types of cone, and on various other factors (as will be discussed in more detail in Chapter 12).

The rod and cone receptors are shown to be connected to *neurons* (nerve cells) that eventually result in just three, not four, different types of signal in the nerve fibres. One of these signals is usually referred to as an *achromatic signal*; its neurons collect inputs from both rods and all three types of cone. Because of the different abundances of the ρ, γ, and β cones, the cone part of its signal is represented as:

$$2\rho + \gamma + \beta/20$$

If the scotopic contribution from the rods is represented by S, then the total achromatic signal is:

$$2\rho + \gamma + \beta/20 + S = A$$

The other two signals in the nerve fibres are usually referred to as *colour difference* signals. Three basic difference signals are possible between cones:

$$\rho - \gamma = C_1$$

$$\gamma - \beta = C_2$$

$$\beta - \rho = C_3 \ .$$

To transmit all three of these signals would be slightly redundant, since, if two of them are known, the third can be deduced from the fact that $C_1 + C_2 + C_3 = 0$. In fact, there is various evidence to suggest that the signals transmitted resemble:

$$C_1 = \rho - \gamma$$

and

$$C_2 - C_3 = \gamma - \beta - (\beta - \rho) = \rho + \gamma - 2\beta \ .$$

Behavioural studies have shown the presence of well developed colour vision in various species; in many cases, physiological experiments on these species have revealed signals of three general types that are broadly similar to the signals, A, C_1, and $C_2 - C_3$, proposed above.

1.7 BASIC PERCEPTUAL ATTRIBUTES OF COLOUR

We shall now consider some of the perceptual attributes of colour, in the context of these visual signals. There are three basic attributes, brightness, hue, and colourfulness; they are defined as follows:

Brightness
Attribute of a visual sensation according to which an area appears to exhibit more or less light. (Adjectives: *bright* and *dim*.)

Hue
Attribute of a visual sensation according to which an area appears to be similar to one, or to proportions of two, of the perceived colours red, yellow, green, and blue.

Colourfulness
Attribute of a visual sensation according to which an area appears to exhibit more or less of its hue.

The achromatic channel could be largely responsible for providing a basis for the attribute of brightness: all colours have a brightness, and, as this channel collects responses from all types of receptor, it could indicate an overall magnitude of response for all colours. Hence we could have:

A	large	Bright colours
A	small	Dim colours

If we assume that, for white, grey, and black colours, $\rho = \gamma = \beta$, then the colour difference signals, C_1, C_2, and C_3, would be zero for these colours. The hues of colours could then perhaps be indicated thus:

C_1	positive	Reddish colours
C_1	negative	Greenish colours
$C_2 - C_3$	positive	Yellowish colours
$C_2 - C_3$	negative	Bluish colours

The particular hue of any colour could then perhaps be indicated by the ratio of C_1 to $C_2 - C_3$, which corresponds to C_1, C_2, and C_3 being in constant ratios to one another; and the colourfulness of colours could then perhaps be indicated by the strengths of the signals C_1 and $C_2 - C_3$, zero indicating zero colourfulness (that is, white, grey, or black, the *achromatic colours*), and increasing signals indicating the degree to which the hue is exhibited in colours possessing a hue (the *chromatic colours*).

1.8 COLOUR CONSTANCY

One of the most important practical uses of colour is as an aid to the recognition of objects. But objects can be illuminated under a very wide range of conditions; in particular, the level and colour of the illumination can vary very considerably. Thus bright sunlight represents a level of illumination that is typically about a thousand times that inside a living room; and electric tungsten filament lighting is much yellower than daylight. However, the human visual system is extremely good at compensating for changes in both the level and the colour of the lighting; as a result of this *adaptation*, objects tend to be recognized as having nearly the same colour in very many conditions, a phenomenon known as *colour constancy*. Colour constancy is only approximate, and considerable changes in colour appearance can sometimes occur, as in the tendency for colours that appear purple in daylight to appear distinctly redder in tungsten light; but colour constancy is, nevertheless, an extremely powerful and important effect in colour perception.

We can, for the moment, regard colour constancy as corresponding to the ρ, γ, and β responses being approximately equal for whites, greys, and blacks, no matter what the level and colour of the illuminant. (This will be discussed more fully in section 3.13 and in Chapter 12.)

1.9 RELATIVE PERCEPTUAL ATTRIBUTES OF COLOUR

Let us now consider, as an example, a white and a grey patch seen side by side on a piece of paper. If we observe the patches in bright sunlight they will look very bright, and if we take them into the shade, or indoors, they will look less bright. But the white will still look white, and the grey will still look grey. By means which are not fully understood, the eye and brain subconsciously allow for the fact that the lower brightnesses are not caused by changes in the objects, but by changes in the illumination. The same is also true for changes in the colour of the illuminant over the range of typical 'white light' illuminants. Thus, in tungsten light, the patches still look approximately white and grey, and certainly not yellow and brown. (It is to explain these phenomena that the *Retinex theory* was produced by Edwin Land; see, for example, Land & McCann 1971.) This is such an important phenomenon that certain relative perceptual attributes of colours are given separate names.

The term *lightness* is used to describe the brightness of objects relative to that of a similarly illuminated white. Thus, whereas brightness could depend on the magnitude of an achromatic signal such as A, lightness could depend on a signal such as A/A_n, where the suffix n indicates that the responses are for an appropriately chosen reference white. Changes in the level of illumination would tend to change A and A_n in the same proportions, thus tending to keep A/A_n constant; hence lightness would tend to remain constant for a given colour. Whites and greys could then be recognized as such by their lightnesses, independent of their brightnesses.

Just as it is possible to judge brightness relative to that of a white, so it is also possible to judge colourfulness in proportion to the brightness of a white, and the relative colourfulness then perceived is called *chroma*. It is well known that, as the illumination level falls, the colourfulness of objects decreases. Thus, in bright daylight, a scene may look very colourful, but it will look less so under dark clouds; and as the light fails in the evening the colourfulness gradually reduces to zero when scotopic levels are reached. But, over most of the photopic range of illumination levels, the colours of objects are recognized as approximately constant. Let us take, as an example, a red tomato on a white plate. In bright daylight, the red tomato looks very colourful: its red hue is exhibited very strongly. But, if we then view it at a much lower level of illumination indoors it will look less colourful (its hue will not be exhibited so strongly); but the white plate will also look less bright, and the visual system then subconsciously judges that the lower colourfulness in the dimmer light is caused by the lower level of illumination characterized by the lower brightness of the white. The colourfulness judged in proportion to the brightness of the white is then seen to be unchanged, and this relative colourfulness is the chroma. Thus, whereas colourfulness could depend on the magnitudes of signals such as C_1 and $C_2 - C_3$, chroma could depend on the magnitudes of signals such as C_1/A_n and $(C_2 - C_3)/A_n$ where the suffix n again indicates that the responses are for the white.

Lightness and chroma are therefore defined as follows:

Lightness

The brightness of an area judged relative to the brightness of a similarly illuminated area that appears to be white or highly transmitting. (Adjectives: *light* and *dark*.)

Chroma

The colourfulness of an area judged in proportion to the brightness of a similarly illuminated area that appears to be white or highly transmitting. (Adjectives: *strong* and *weak*.)

Because these two attributes are defined with reference to a 'similarly illuminated area' they do not apply to *unrelated colours*, that is, colours perceived to belong to areas seen in isolation from other colours, but only to *related colours*, that is, colours perceived to belong to areas seen in relation to other colours. Self-luminous colours, such as light sources, are usually perceived as unrelated colours; colours produced by objects reflecting light in ordinary viewing conditions are usually perceived as related colours. A television display is self-luminous, but, within the picture area, related colours can be seen if the portrayal is of illuminated objects. A transmitting colour can be perceived as a related colour if seen in suitable relationship to other areas, as in a stained glass window in a church; it is for this reason that the words 'highly transmitting' are included in the above definitions.

If we consider the case of the tomato on the plate again, because the tomato is a solid object, the level of illumination will vary considerably over its surface, being high where the light falls on it perpendicularly, low where it falls on it at a glancing angle, and even lower for those parts in shadow. The brightness of a similarly illuminated white can then be readily judged only for a few areas, and hence lightness and chroma can only be evaluated in these areas. But it is also possible to judge colourfulness relative to the brightness of the same area, instead of relative to that of a similarly illuminated white; when this is done the attribute is called *saturation*. This attribute can be readily judged at all parts of the tomato; hence, saturation, together with hue, can then be used to judge the uniformity of colour over the surface of the tomato. Saturation could depend on signals such as C_1/A and $(C_2 - C_3)/A$. Saturation is then defined as follows:

Saturation

The colourfulness of an area judged in proportion to its brightness.

Because the judgement of saturation does not require the concept of a similarly illuminated white, it is applicable to both related and unrelated colours. Consider, as an example of an unrelated colour, a red traffic-light signal seen first directly, and then reflected in a piece of plane glass, such as a shop window. When seen directly, the red signal will usually look quite bright and colourful. But its reflection will look both less bright and less colourful. However, it will still look red, not pink, because its lower colourfulness will be judged in proportion to its lower brightness and it will be perceived to have the same saturation. Thus recognition of the colour of the traffic-light will depend on its hue and saturation, not on its hue and colourfulness.

In photopic conditions, the rod response, S, is often negligibly small; in this case, when ρ, γ, and β are in constant ratios to one another, $C_1/(C_2 - C_3)$, C_1/A, and $(C_2 - C_3)/A$ are all constant, and this implies constant hue and saturation, but not constant brightness (which is in accord with the discussion of the 500 nm and 600 nm lights in section 1.5).

1.10 COLOUR BLINDNESS

It has long been known that some observers have colour vision that is markedly different from the average of most observers. Such people are popularly known as 'colour blind', but a more appropriate term is *colour defective*, because, in most cases, what is involved is a reduction in colour discrimination, not a complete loss of it.

There are various types of colour deficiency, and their exact causes are still a matter of some debate (Ruddock & Naghshineh 1974). However, the most likely causes for the various categories are given below in brackets:

Protanopia

Severely reduced discrimination of the reddish and greenish contents of colours, with reddish colours appearing dimmer than normal. (ρ cones missing).

Deuteranopia

Severely reduced discrimination of the reddish and greenish contents of colours, without any colours appearing appreciably dimmer than normal. (γ cones missing; the similarity of the shapes of the ρ and γ spectral sensitivity curves on the short wavelength side of their peaks could preserve approximately normal brightness of colours. Alternatively ρ and γ cones present, but $\rho - \gamma$ colour-difference signals missing.)

Tritanopia

Severely reduced discrimination of the bluish and yellowish contents of colours. (β cones missing; the very small contribution of the β cones to the achromatic signal would preserve approximately normal brightness of colours.)

Cone monochromatism

No colour discrimination, but approximately normal brightnesses of colours. (γ and β cones missing; that is, a combination of deuteranopia and tritanopia. Alternatively, all three types of cone present, but no colour-difference signals present.)

Rod monochromatism

No colour discrimination, and brightnesses typical of scotopic vision. (No cones present.)

Protanomaly

Some reduction in the discrimination of the reddish and greenish contents of colours, with reddish colours appearing dimmer than normal. (ρ cones having a spectral sensitivity curve shifted along the wavelength axis towards that of the normal γ cones; perhaps some γ pigment in the ρ cones.)

Deuteranomaly

Some reduction in the discrimination of the reddish and greenish contents of colours, without any colours appearing abnormally dim. (γ cones having a spectral sensitivity curve shifted along the wavelength axis towards that of the normal ρ cones; perhaps some ρ pigment in the γ cones.)

Tritanomaly

Some reduction in the discrimination of the bluish and yellowish contents of colours. (β cones having a spectral sensitivity curve shifted along the wavelength axis towards that of the normal γ cones; perhaps some γ pigments in the β cones.)

The degree of reduction in colour discrimination in the cases of protanomaly, deuteranomaly, and tritanomaly (referred to as *anomalous trichromatism*) varies from only slight differences from normal observers, to nearly as great a loss as occurs in protanopia, deuteranopia, and tritanopia (referred to as *dichromatism*). Colour matches made by normal observers are accepted by dichromats, but not always by anomalous trichromats (these mismatches could be caused by the filtering effects of one pigment on another if mixed pigments are present in the cones).

The occurrence of these different types of defective colour vision varies enormously, and is different for men and for women. The following figures are estimates for western races, based on various surveys.

Type	*Men %*	*Women %*
Protanopia	1.0	0.02
Deuteranopia	1.1	0.01
Tritanopia	0.002	0.001
Cone monochromatism	Very rare	Very rare
Rod monochromatism	0.003	0.002
Protanomaly	1.0	0.02
Deuteranomaly	4.9	0.38
Tritanomaly	Unknown	Unknown

The nature of the colour confusions likely to be made by colour defective observers will be described in terms of colorimetry in section 3.6. Detection of colour deficiency is usually carried out by means of confusion charts, such as the widely used *Ishihara* charts; but accurate classification of the type of deficiency usually requires the use of other test methods (see, for instance, Fletcher & Voke 1985). Most cases of colour deficiency are hereditary, but progressive tritanomaly can sometimes be acquired with certain diseases.

1.11 COLOUR PSEUDO-STEREOPSIS

When saturated red and blue lettering is viewed on a dark or black background, some observers perceive the red letters as standing out in front of the plane of the paper, and the blue letters lying behind it, even though the letters are all actually in the plane of the paper. However, although a majority of observers see this phenomenon, there is a minority for whom the reverse effect occurs, with the red letters receding and the blue letters advancing; and there is a third, and still smaller group, who see all the letters in the same plane as the paper. This effect is caused by a combination of the chromatic aberration of the eye and the fact that the pupils of the eyes are not always central with respect to their optical axes. This is illustrated in Fig. 1.5. In the left hand

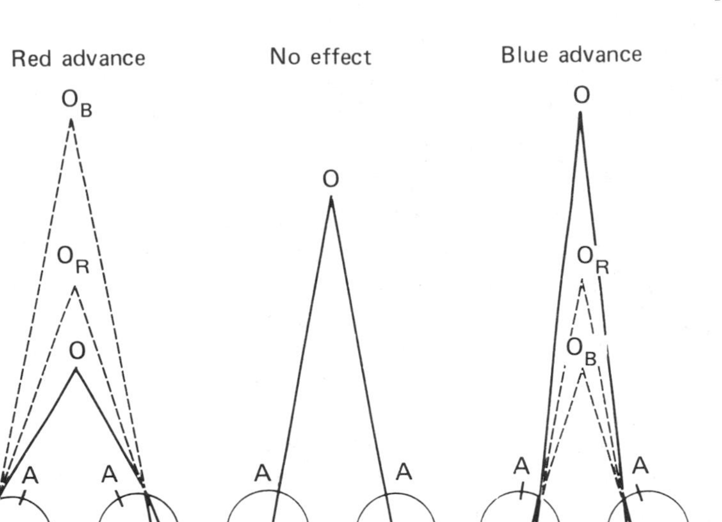

Fig. 1.5 — Illustration (not to scale) of the basis for colour pseudo-stereopsis.

diagram, the pupils are displaced outwards relative to the optical axes, A, so that the rays from an object at O are dispersed by refraction with images of blue (B) light in the two eyes being closer to one another than in the case of red (R) light. The red light therefore appears to emanate from an object that is closer than an object from which blue light appears to emanate. In the right hand diagram, the pupils are displaced inwards relative to the optical axes, A, with the result that a blue object now appears nearer than a red object. In the centre diagram, the pupils are concentric with the optical axis, A, and no effect occurs.

REFERENCES

Crawford, B. H. *Proc. Phys. Soc. B*, **62**, 321 (1949).

Dartnall, H. J. A., Bowmaker, J. K. & Mollon, J. D. *Proc. Roy. Soc. Lond. B.*, **220**, 115 (1983).

Estévez, O. PhD Thesis, University of Amsterdam (1979).

Fletcher, R. & Voke, J. *Defective Colour Vision*, Hilger, Bristol (1985).

Hurvich, L. M. *Color Vision*, p.21, Sinauer Associates, Sunderland, Mass., U.S.A. (1981).

Judd, D. B. & Wyszecki, G. *Color in Business Science and Industry*, 3rd ed., p. 388, Wiley, New York (1975).

Land, E. H. & McCann, J. J. *J. Opt. Soc. Amer.*, **6**1, 1 (1971).

McLaren, K. *Color Res. Appl.*, **10**, 225 (1985).

Mollon, J. D. *Ann. Rev. Psychol.* **33**, 41 (1982).

Ruddock, K. H. & Naghshineh, S. *Mod. Prob. Ophth.*, **13**, 210 (1974).
Wald, G. *Science*, **101**, 653 (1945).
Walraven, P. L. & Bouman, M. A. *Vision Research*, **6**, 567 (1966).

GENERAL REFERENCES

Boynton, R. M. *Human Color Vision*, Holt and Rinehart-Winston, New York (1979).
Hunt, R. W. G. *Colour Terminology*, *Color Res. Appl.* **3**, 79-87 (1978).

2

Spectral weighting functions

2.1 INTRODUCTION

When white light is passed through a prism, or some other suitable device, to form a spectrum, it normally appears to be brightest somewhere near the middle, around the green part. As the part considered is increasingly displaced from the brightest part, the brightness decreases continually until, at the two ends, it merges into complete darkness. These changes in brightness along the length of the spectrum are not usually caused mainly by changes in the power present, but by changes in the sensitivity of the eye to different wavelengths.

2.2 SCOTOPIC SPECTRAL LUMINOUS EFFICIENCY

For the scotopic type of vision that occurs at low light levels, we saw, in section 1.5, that the photosensitive pigment, rhodopsin, on which the rods depend, absorbs light most strongly in the blue-green part of the spectrum and less strongly in the other parts. As a result, the sensitivity throughout the spectrum of the eye in scotopic vision is as shown by the broken line in Fig. 2.1 (which was also shown as the broken line in Fig. 1.3). This function was obtained by having observers adjust the power in the beam of each wavelength of the spectrum until its brightness matched that of a beam of fixed strength of a reference wavelength (see section 1.5).

But what is the situation if light of a second wavelength is present in the same stimulus? At each wavelength, the power of the light can be multiplied by the height, at that wavelength, of the broken curve in Fig. 2.1, and the resulting products added together; then, if this sum is equal for any two stimuli, experiment shows that they look equally intense if viewed under the same scotopic conditions. The broken curve of Fig. 2.1 can, therefore, be thought of as a function representing the different efficiencies with which different parts of the spectrum excite the scotopic visual system. By convention, the maximum efficiency is given a value of 1.0, and for scotopic vision this occurs at a wavelength of approximately 510 nm. It is clear that, for wavelengths above and below this, the efficiency falls until it reaches approximately zero for wavelengths greater than about 620 nm, and less than about 400 nm. The broken curve of Fig. 2.1 has been standardized internationally by the CIE

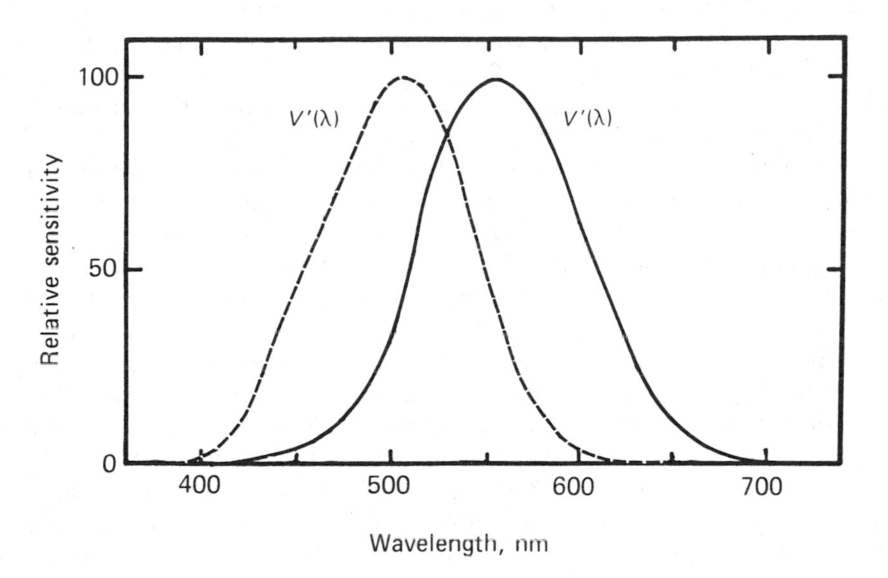

Fig. 2.1 — The spectral luminous efficiency curves, $V'(\lambda)$, for low level (scotopic) vision, and $V(\lambda)$, for high level (photopic) vision.

(Commission Internationale de l'Éclairage),and is known as the *spectral luminous efficiency for scotopic vision*, and is given the symbol $V'(\lambda)$. Its values are given in full in Appendix 2. An observer having a relative spectral sensitivity function that is the same as the $V'(\lambda)$ function is known as the *CIE standard scotopic photometric observer*.

The $V'(\lambda)$ function can be used as a weighting function to determine which of any two lights, whatever their spectral composition, will appear, under the same scotopic viewing conditions, to have the greater intensity, or whether they have the same intensity. This is done by using the following formula:

$$L' = K'_m(P_1V'_1 + P_2V'_2 + P_3V'_3 + \dots\dots)$$

In this formula K'_m is a constant, P_1, P_2, P_3, etc., are the amounts of power per small constant-width wavelength interval throughout the visible spectrum, and V'_1, V'_2, V'_3, etc., are the values of the $V'(\lambda)$ function at the central wavelength of each interval. The effect of this formula is to evaluate the radiation by giving most weight to its components having wavelengths near 510 nm, and progressively less and less weight as the wavelength differs more and more from 510 nm, until at the two ends of the spectrum no weight is given at all. The result is an evaluation of the radiation, not in terms of its total power, but in terms of its ability to stimulate the eye (at low levels of illumination). The use of weighting functions in this way is a very important procedure in photometry and colorimetry, and different weighting functions are used to represent different visual properties. The $V'(\lambda)$ function is used to represent the relative spectral sensitivity of the eye in scotopic conditions.

In the above formula, if K'_m is put equal to 1700, and P is the radiance in watts per steradian and per square metre, L' is the *scotopic luminance* expressed in scotopic candelas per square metre. (A steradian is a unit of solid angle defined as: the solid angle that, having its vertex in the middle of a sphere, cuts off an area on the surface of the sphere equal to that of a square with side of length equal to that of the radius of the sphere.) If P is in some other radiant measure, then L' will be in the corresponding scotopic photometric measure; for example, if P is the irradiance in watts per square metre, then L' will be the scotopic illuminance in scotopic lux. (For details of these corresponding photometric and radiometric measures, see Appendix 1, Table A1.1).

It is important to realize the limitations of photometric measures. They can tell us whether two lights are equally intense; or, if one is more intense, they can tell us by what factor the other must be increased in radiant power (keeping the relative spectral composition the same) for the two intensities to appear equal. But photometric measures do not tell us how bright a given light will look, because this depends on the conditions of viewing. For example, a motor car headlight that appears dazzlingly bright at night, appears to have a much more modest brightness when seen in bright sunlight. Moreover, even for a fixed set of viewing conditions, the relationship between photometric measures and brightness is not one of simple proportionality. (This topic will be discussed more fully in Chapter 12.)

Let us now consider, as an example, two lights, B and R, whose spectral radiances, P_B and P_R, per small constant-width wavelength interval throughout the spectrum, are as shown in Table 2.1. (To obtain convenient numbers, the powers are expressed in microwatts, μW.) The table shows how the $V'(\lambda)$ function is used to obtain the weighted spectral power, PV', at each wavelength. These products are then summed, and multiplied by 1700 (and by 10^{-6} to allow for the use of microwatts) to obtain the scotopic luminance, L', in scotopic candelas per square metre. It is seen that the scotopic luminances are 0.00806 for light B and 0.00680 for light R; we therefore conclude that the scotopic luminance of light B is greater than that of light R by a factor of 806/680, that is by 1.18 times. Hence the powers in light R would have to be increased by a factor of 1.18 to ensure that light R appeared as intense as light B when seen under the same scotopic viewing conditions. In Table 2.1, data at every 10 nm have been used; this is usually a sufficiently close sampling of the spectral data for determining scotopic measures.

2.3 PHOTOPIC SPECTRAL LUMINOUS EFFICIENCY

If we wish to compare the intensities of stimuli seen under photopic conditions of vision, we have to use, not the $V'(\lambda)$ function shown by the broken line of Fig. 2.1,

Table 2.1 — Calculation of scotopic values using the CIE $V'(\lambda)$ function

Wavelength, λ	$V'(\lambda)$	Power in microwatts, μW, in interval $\lambda-5$ to $\lambda+5$ nm P_B	P_R	Scotopic light P_BV'	P_RV'
380 nm	0.001	1.1	0	0.001	0
390	0.002	1.05	0	0.002	0
400	0.009	1.0	0	0.009	0
410	0.035	0.95	0	0.033	0
420	0.097	0.9	0	0.087	0
430	0.200	0.85	0.05	0.170	0.010
440	0.328	0.8	0.1	0.262	0.033
450	0.455	0.75	0.15	0.341	0.068
460	0.567	0.7	0.2	0.397	0.113
470	0.676	0.65	0.25	0.439	0.169
480	0.793	0.6	0.3	0.476	0.238
490	0.904	0.55	0.35	0.497	0.316
500	0.982	0.5	0.4	0.491	0.393
510	0.997	0.45	0.45	0.449	0.449
520	0.935	0.4	0.5	0.374	0.468
530	0.811	0.35	0.55	0.284	0.446
540	0.650	0.3	0.6	0.195	0.390
550	0.481	0.25	0.65	0.120	0.313
560	0.329	0.2	0.7	0.066	0.230
570	0.208	0.15	0.75	0.031	0.156
580	0.121	0.1	0.8	0.012	0.097
590	0.066	0.05	0.85	0.003	0.056
600	0.033	0	0.9	0	0.030
610	0.016	0	0.95	0	0.015
620	0.007	0	1.0	0	0.007
630	0.003	0	1.05	0	0.003
640	0.001	0	1.1	0	0.001
650	0.001	0	1.15	0	0.001
Totals				4.739	4.002
$\times 1700$				8056	6803
$\times 10^{-6}$				0.00806	0.00680

but the *photopic spectral luminous efficiency* function, $V(\lambda)$, shown by the full line in Fig. 2.1. Its values are given in full in Appendix 2. It is clear that the $V(\lambda)$ function, although of similar general shape as the $V'(\lambda)$ function, peaks at a wavelength of about 555 nm instead of at about 510 nm. This represents a change in relative spectral sensitivity towards an increase for reddish colours and a decrease for bluish colours, which, as mentioned in Section 1.4, is referred to as the *Purkinje phenomenon*.

The experimental basis for obtaining the $V(\lambda)$ function is not as simple as that used for the scotopic $V'(\lambda)$ function. Because the different colours of the spectrum are not seen at scotopic levels, the observer's task of adjusting the power at each wavelength to obtain equality of intensity with a reference beam is a comparatively simple one. But, if the same technique is used at photopic levels, the large difference in colour between the light of most of the wavelengths and that of the reference beam makes the task very much more difficult. However, if the two beams are seen, not side by side, but alternately, it has been found that a rate of alternation can be set that is too fast for colour differences to be detected, but slow enough for intensity differences still to be visible. Using this rate of alternation, it is then a somewhat easier task to set the intensities of the two beams so that no flicker can be seen, and this then corresponds to equality of intensity. This technique is known as *flicker photometry*, and it was used to obtain the $V(\lambda)$ function shown by the full line of Fig. 2.1. An observer having a relative spectral sensitivity function that is the same as the $V(\lambda)$ function is known as the *CIE standard photometric observer*. (Flicker photometry is not the only experimental method of obtaining the $V(\lambda)$ function, as will be discussed in section 2.8, where other methods, and other photopic functions, will be briefly reviewed.)

The $V(\lambda)$ function is used in the same way as the $V'(\lambda)$ function for evaluating the relative intensities of stimuli, but using the formula:

$$L = K_m(P_1V_1 + P_2V_2 + P_3V_3 + \ldots\ldots.) \ ,$$

where K_m is another constant, P_1, P_2, P_3, etc., are the amounts of power per small constant-width wavelength interval throughout the visible spectrum, and V_1, V_2, V_3, etc., are the values of the $V(\lambda)$ function at the central wavelength of each interval. If K_m is put equal to 683 and P is the radiance in watts per steradian and per square metre, L is the *luminance* expressed in candelas per square metre (cd/m^2). As before, if other radiant measures are used for P, the corresponding photometric measures are obtained (see Appendix 1, Table A1.1, for details).

In Table 2.2, two stimuli CB and CR are evaluated using $V(\lambda)$ as a weighting function; they have the same relative spectral power distributions as the stimuli B and R, respectively, of Table 2.1, but the powers are high enough to be in the photopic range (the units are watts instead of microwatts). It is seen that now the phototopic luminances are 1595 for light CB and 5116 for light CR; we therefore conclude that the photopic luminance for light CR is greater than that for light CB by a factor of 5116/1594, that is by 3.21 times. Hence the powers in light CB would have to be increased by a factor of 3.21 to ensure that it appeared as intense as light CR when seen in the same viewing conditions and judged by flicker photometry; but in Table 2.1 light B had a higher scotopic luminance than light A. This illustrates the *Purkinje phenomenon* described in section 1.4, lights B and CB being bluish, and lights R and CR being reddish. In Table 2.2, data at every 10 nm has been used, and, as before, this is usually a sufficiently close sampling of the spectral data.

Once again, these photometric measures do not tell us how bright a stimulus will look. As with scotopic measures, the brightness is greatly affected by the viewing conditions, and, in any one set of viewing conditions, brightness and photometric measures are not proportional. But there is an additional factor operating in the case of photopic vision. For colours of the same luminance, there is a tendency for the

Table 2.2 —Calculation of photopic values using the CIE $V(\lambda)$ function

Wavelength, λ	$V(\lambda)$	Power in watts, W, in interval $\lambda-5$ to $\lambda+5$ nm P_{CB}	P_{CR}	Photopic light $P_{CB}V$	$P_{CR}V$
400 nm	0.000	1.0	0	0	0
410	0.001	0.95	0	0.001	0
420	0.004	0.9	0	0.004	0
430	0.012	0.85	0.05	0.010	0.001
440	0.023	0.8	0.1	0.018	0.002
450	0.038	0.75	0.15	0.028	0.006
460	0.060	0.7	0.2	0.042	0.012
470	0.091	0.65	0.25	0.059	0.023
480	0.139	0.6	0.3	0.083	0.042
490	0.208	0.55	0.35	0.114	0.073
500	0.323	0.5	0.4	0.162	0.129
510	0.503	0.45	0.45	0.226	0.226
520	0.710	0.4	0.5	0.284	0.355
530	0.862	0.35	0.55	0.302	0.474
540	0.954	0.3	0.6	0.286	0.572
550	0.995	0.25	0.65	0.249	0.647
560	0.995	0.2	0.7	0.199	0.696
570	0.952	0.15	0.75	0.143	0.714
580	0.870	0.1	0.8	0.087	0.696
590	0.757	0.05	0.85	0.038	0.643
600	0.631	0	0.9	0	0.568
610	0.503	0	0.95	0	0.478
620	0.381	0	1.0	0	0.381
630	0.265	0	1.05	0	0.278
640	0.175	0	1.1	0	0.192
650	0.107	0	1.15	0	0.123
660	0.061	0	1.2	0	0.073
670	0.032	0	1.25	0	0.040
680	0.017	0	1.3	0	0.022
690	0.008	0	1.35	0	0.011
700	0.004	0	1.4	0	0.006
710	0.002	0	1.45	0	0.003
720	0.001	0	1.5	0	0.002
730	0.001	0	1.55	0	0.002
Totals				2.335	7.490
× 683				1595	5116

brightness to increase gradually as the colourfulness of the colour increases (this phenomenon is known as the *Helmholtz-Kohlrausch effect*). Thus, if a white and a colourful red of the same luminance are compared side by side, the red usually looks brighter than the white. Similarly, if a colourful blue-green and a white have the same luminance, then the blue–green usually looks brighter. But if we add the red and blue–green lights together, and compare this mixture with that of the two white lights added together, we find that the brightnesses of the two mixtures are now similar. This arises because the mixture of the red and blue–green results in a whitish colour, and the additional brightnesses associated with the high colourfulnesses of the red and the blue–green colours have disappeared. This state of affairs is sometimes described by saying that luminances are additive, but brightnesses are not. (The effects of these factors will be discussed more fully in Chapter 12.)

The type of mixing of colours referred to above is called *additive* to distinguish it from *subtractive* mixing, in which the light passes through colorants which subtract part of the light in different parts of the spectrum by absorption. Additive mixture can be either by superimposing beams of light on a diffuser as shown in Fig. 2.2 (or on

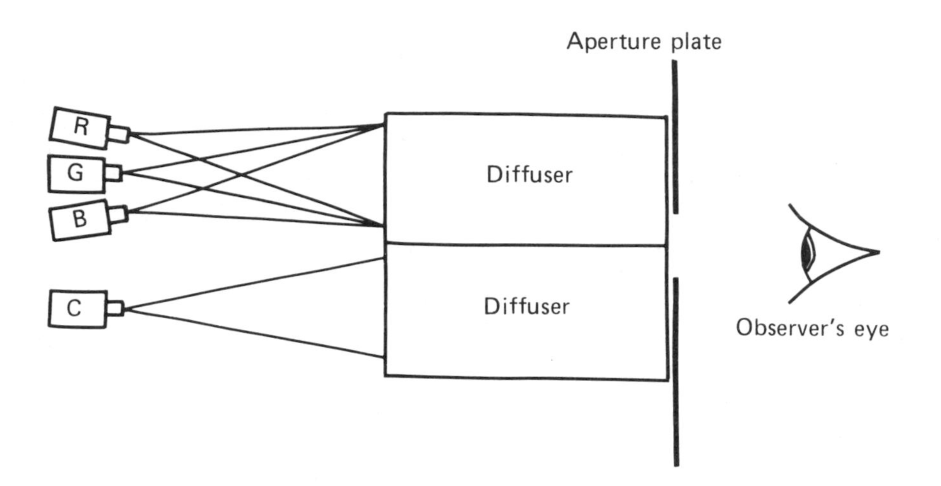

Fig. 2.2 — Principle of trichromatic colour matching by additive mixing of lights. R, G, and B are sources of red, green, and blue light, whose intensities can be adjusted; C is the light whose colour is to be matched. The diffusers result in uniform fields for viewing by the observer.

other types of beam combiners), or by viewing beams in succession at a frequency high enough to remove all sense of flicker, or by viewing them in adjacent areas that are too small to resolve (as in typical television display tubes). Subtractive mixture can be either by mixing pigments together in a suitable vehicle (as artists often do) or by having dyes in successive thin layers (as in colour photography).

The equations given for evaluating L and L' imply that the contributions of light of all the various wavelengths are exactly additive after weighting by the $V(\lambda)$, or $V'(\lambda)$, function, as appropriate. In the case of the $V'(\lambda)$ function, this is to be expected because only a single photosensitive pigment is involved. But, in the case of the $V(\lambda)$ function, the photopic achromatic signal, on which it presumably depends, is composed of contributions from the three different photosensitive pigments in the three types of cone. Moreover, there is considerable evidence that the output signals from the cones are not proportional to the light absorbed, but more nearly to its square-root (or some similar function, as will be discussed in Chapter 12). However, the contribution of the β cones is very small, and the spectral sensitivities of the γ and ρ cones overlap considerably (see Fig. 1.3); the result of this is that a single function, such as the $V(\lambda)$ function, can represent the effective spectral sensitivity of the achromatic signal quite closely. This is borne out by experiments that show that luminances are additive. By this is meant that, if two colour stimuli, A and B, are perceived to be equally intense, as judged by flicker photometry, and two other stimuli, C and D, are similarly perceived to be equally intense, then the additive mixtures of A with C and B with D will also be similarly perceived to be equally intense. If the criterion is equality of brightness in side by side comparisons, as already explained, additivity does not hold. However, the discrepancies from brightness additivity can be small for many stimuli, especially those of low colourfulness. When brightness additivity occurs it is sometimes referred to as *Abney's Law*. (Methods of allowing for the contribution of colourfulness towards brightness will be discussed in sections 11.4 and 12.21.)

For most practical applications, photopic levels of illumination apply, and the convention has been adopted that, unless otherwise indicated, all photometric measures are assumed to be based on the $V(\lambda)$ function and to represent photopic vision. In those cases where the $V'(\lambda)$ function has been used, the adjective *scotopic* precedes the photometric term (for example, scotopic luminance, scotopic cd/m^2), and a prime is added to the symbol (for example, L').

2.4 COLOUR-MATCHING FUNCTIONS

Colour vision is basically a function of three variables. There are three different types of cone, and, although the rods provide a fourth spectrally different receptor, there is overwhelming evidence that, at some later stage in the visual system, the number of variables is reduced once again to three, as indicated in Fig. 1.4. Hence, it is to be expected that the evaluation of colour from spectral power data should require the use of three different spectral weighting functions. At levels of illumination that are high enough for colour vision to be operating properly, there is evidence that the output from the rods is in some way rendered ineffective. At levels where both cones and rods are operating together, colour vision must be based on the four different spectral sensitivities of the cones and the rods, but, at these levels, colour discrimination is not very good, and any disturbing influence of the rods is usually not of appreciable practicable importance (although rod activity can be detected by special experimental techniques (Trezona 1973; Palmer 1981), and its effects will be considered in Chapter 12).

In view of the small effects of rod intrusion, it might be thought that a good basis for evaluating colour from spectral power data would be the use of the three cone spectral sensitivity curves shown by the full lines in Fig. 1.3. However, this is not what was done in establishing the internationally accepted method of evaluating colour, and the reason is that the curves of the type shown in Fig. 1.3 are not known with sufficient precision. Instead, the basis chosen was three-colour matching or *trichromatic matching*, as it is usually called.

In Fig. 2.2, the basic experimental arrangement for trichromatic matching is shown diagrammatically. The test colour to be matched is seen in one half of the field of view, and, in the other half, the observer sees an additive mixture of beams of red, green, and blue light. The amounts of red, green, and blue light are then adjusted until the mixture matches the test colour in brightness, hue, and colourfulness. The fact that the colours can be matched in this way, using just three *matching stimuli* in the mixture is a consequence of the fact that there are only three spectrally different types of cone in the retina.

For trichromatic matches to form a proper basis for a system of colour measurement, various parts of the experimental system must be precisely specified. Because, as we saw in section 1.3, the retina varies considerably in its properties from one part to another, it is necessary to specify the angular size of the matching field. Standards now exist for two field sizes, 2° and 10°, but the original system was based on the 2° field size, so this will be considered first. Then the precise colours of the red, green, and blue matching stimuli need to be specified. In the original work that led to the 2° data, two separate experimental arrangements were used. J.Guild, at the National Physical Laboratory at Teddington, used a tungsten lamp and coloured filters; W.D.Wright, at Imperial College, Kensington, used monochromatic bands of light isolated from a spectrum formed by a system of prisms. In order to combine the two sets of results, each set of data was transformed mathematically to what would have been obtained if the following monochromatic matching stimuli had been used:

Red	700 nm
Green	546.1 nm
Blue	435.8 nm

The green and blue stimuli were chosen to coincide with two prominent lines in the mercury discharge spectrum, because this facilitated wavelength calibration; the red stimulus was chosen to be in a part of the spectrum where hue changes vary slowly with wavelength, so as to reduce the effects of any errors in wavelength calibration. The results were also transformed to what would have been obtained if the amounts of red, green, and blue had been measured using, not units of luminance, but units that result in a white being matched by equal amounts of the three matching stimuli. Compared with the unit used for the amount of red, the photometric value of the unit used for the green was rather larger, and that for the blue was very much smaller. The reason for the very considerable change in the case of the blue is that, because, as we saw in section 1.5, there are many fewer β cones in the retina than γ or ρ cones, a typical beam of white light is matched by a red, green, and blue mixture in which the luminance of the blue is much less than that of the green or red. For example, to match 5.6508 cd/m^2 of a white light consisting of equal amounts of power per small constant-width wavelength interval throughout the spectrum (a stimulus known as the *equi-energy stimulus*, S_E), the amounts of these red, green, and blue matching stimuli are:

Red	1.0000 cd/m^2
Green	4.5907 cd/m^2
Blue	0.0601 cd/m^2
Mixture	5.6508 cd/m^2

But, because white is a colour that is not perceptually biased to either red, green, or blue, it was decided to measure the amount of green in a unit that was 4.5907 cd/m^2, and the amount of blue in a unit that was 0.0601 cd/m^2. The match of 5.6508 cd/m^2 of S_E is then represented by:

Red	1.0000 cd/m^2
Green	1.0000 new green units
Blue	1.0000 new blue units

Of course, changing the units does not change the amount of matching stimulus in the match; it merely expresses it by using a more convenient number, just as, if an article cost 10 000 cents, it is more convenient to change the unit, and to speak of it as costing 100 dollars. The amounts of the three matching stimuli, expressed in the units adopted for them, are known as *tristimulus values*.

In Fig. 2.3 the probable sensitivity curves of the three types of cone are

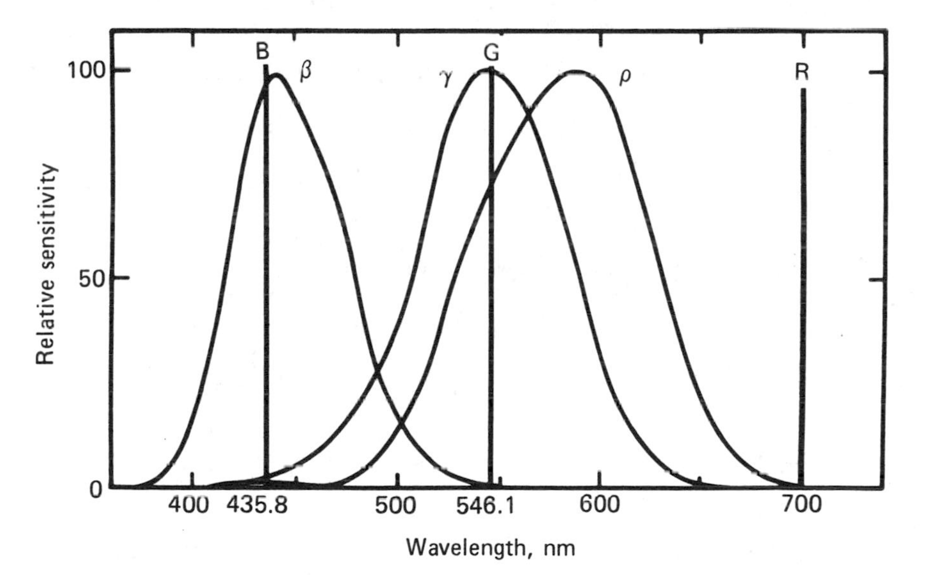

Fig. 2.3 — Representative spectral sensitivity curves of the three different types, ρ, γ, and β, of cone in the retina, together with the wavelengths, R, G, and B used for defining the CIE 1931 Standard Colorimetric Observer, 700, 546.1, and 435.8 nm, respectively.

reproduced from Fig. 1.3. Superimposed on this diagram are shown the wavelengths of the three matching stimuli, R, G, B. Although the ρ curve is very low at 700 nm, we can take it that the R stimulus will excite the ρ cones and not the γ or β; the B stimulus will excite mainly the β cones, but with some very small excitations of the γ and ρ cones; but the G stimulus, although exciting the γ cones most, and the β cones very little, clearly excites the ρ cones quite strongly. From the point of view of trichromatic colour matching, the R stimulus can give us all the ρ excitation that we need, so the excitation of the ρ cones by the G stimulus is unwanted, and in fact leads to a complication in colour matching.

For simplicity, let us assume that the ordinate scale in Fig. 2.3 shows the cone responses per unit power at each wavelength. It is then clear from Fig. 2.3 that one power unit of light of wavelength 500 nm will produce responses of approximately the following magnitudes:

$$\rho = 20 \qquad \gamma = 40 \qquad \beta = 20$$

For simplicity, let us also assume that 100 units of B produce cone responses as shown by the ordinates on Fig. 2.3, that 100 units of G also produce cone responses as shown by the ordinates, and that 100 units of R produce cone responses equal to 100 times the ordinates; the reason why this large factor appears in the case of R is that, as can be seen from Fig. 2.3, whereas G and B are near the peaks of the curves for the γ and β cones, respectively, R is displaced well away from the peak of that for the ρ cones. Fig. 2.3 then shows that:

100 units of B produce	$\beta = 100$	$\gamma = 5$	$\rho = 4$
100 units of G produce	$\beta = 0$	$\gamma = 100$	$\rho = 75$
100 units of R produce	$\beta = 0$	$\gamma = 0$	$\rho = 100$

To match 1 power unit of 500 nm light we must produce:

$$\beta = 20 \quad \gamma = 40 \quad \rho = 20$$

We proceed as follows:

20 of B produces	$\beta = 20$	$\gamma = 1$	$\rho = 0.8$
39 of G produces	$\beta = 0$	$\gamma = 39$	$\rho = 29.2$

So together:

20 of B plus 39 of G produce	$\beta = 20$	$\gamma = 40$	$\rho = 30$

We have thus produced a ρ response of 30, without having added any R stimulus. The only way in which a match can be made is to add some of the R stimulus to the 500 nm light, thus:

1 unit of 500 nm light produces	$\beta = 20$	$\gamma = 40$	$\rho = 20$
10 of R produces	$\beta = 0$	$\gamma = 0$	$\rho = 10$

So together:

1 of 500 nm plus 10 of R produce	$\beta = 20$	$\gamma = 40$	$\rho = 30$

We thus have that:

1 of 500 nm plus 10 of R is matched by 39 of G plus 20 of B

and by convention this is written as:

1 of 500 nm is matched by − 10 of R + 39 of G + 20 of B

The use of the negative sign does not, of course, mean that there is any such thing as negative light! It is only a convenient way of expressing the experimental fact that one of the matching stimuli, in this case the red, had to be added to the colour being matched instead of to the mixture.

If light of 500 nm was then additively mixed with light of say 600 nm, we might have:

1 of 500 nm is matched by	− 10 of R + 39 of G + 20 of B
1 of 600 nm is matched by	95 of R + 30 of G + 0 of B

So together:

1 of 500 nm + 1 of 600 nm is matched by

85 of R + 69 of G + 20 of B

The negative amount of R in the match on 500 nm has been subtracted from the positive amount in the match on 600 nm; and experiment shows that this correctly predicts the amounts of R, G, and B needed to match this mixture. The result is quite general, and the amounts of matching stimuli required in matches can be added and subtracted by using the ordinary rules of algebra. This is sometimes referred to as a consequence of *Grassmann's Laws* (see Appendix 7, entry for *Grassmann's Laws*, for details of these laws).

In Fig. 2.4 the amounts of R, G, and B needed to match a constant amount of

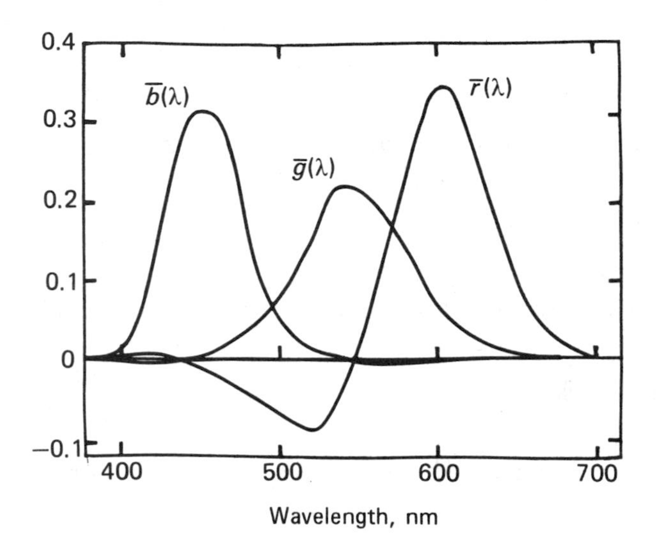

Fig. 2.4 — The colour-matching functions for the CIE 1931 Standard Colorimetric Observer, expressed in terms of matching stimuli, R, G, and B, consisting of monochromatic stimuli of wavelengths 700, 546.1, and 435.8 nm, respectively.

power per small constant-width wavelength interval at each wavelength of the spectrum are shown. Curves of this type are called *colour-matching functions*, and are designated by the symbols $\bar{r}(\lambda)$, $\bar{g}(\lambda)$, and $\bar{b}(\lambda)$. The $\bar{r}(\lambda)$ curve shows the amount of R needed for a match to be made for each wavelength of the spectrum; and the $\bar{g}(\lambda)$ and $\bar{b}(\lambda)$ curves similarly show the amounts of G and B, respectively. The $\bar{r}(\lambda)$ curve is strongly positive in the orange part of the spectrum, and negative in the blue–green part, with a small positive part in the violet. The $\bar{g}(\lambda)$ curve is strongly positive in the greenish part of the spectrum, with a small negative part in the violet; this negative part arises because these spectral colours have to be matched by R and B only with a small amount of G added to the test colour. The $\bar{b}(\lambda)$ curve is strongly positive in the bluish part of the spectrum with a small negative part in the yellow; this negative part arises because these spectral colours have to be matched by R and G only, with a small amount of B added to the test colour. As is to be expected, light of 700 nm is matched by R only, of 546.1 nm by G only, and of 435.8 nm by B only, the other two curves being zero at each of these wavelengths.

If the matching stimuli had consisted of monochromatic lights of different wavelengths, then the positions at which two of the three curves were zero would move to these different wavelengths, and the sizes and positions of the positive and negative parts of the curves would also have been different. But all sets of matching stimuli, no matter what their colours, give rise to colour-matching functions having some negative parts to their curves. The areas under the three curves of Fig. 2.4 are equal (subtracting the area of the negative part from the sum of the areas of the positive parts in the case of each curve), and this arises from the choice of the units used being such that equal amounts are required to match the equi-energy stimulus, S_E.

The amounts of R, G, and B needed to match the colours of the spectrum, represented by the curves of Fig. 2.4, were based on the experimental results obtained by 10 observers in Wright's investigation, and by 7 in Guild's. The two investigations produced similar results, and the average obtained, which is what is shown in Fig. 2.4, therefore represents the average colour matching properties of the eyes of 17 British observers. Subsequent investigations have shown that this average result represents normal human colour vision in 2° fields adequately (see, for instance, Stiles 1955), apart from the values being lower than they should be below 450 nm because of the use of the $V(\lambda)$ function which has this same feature; but the effect of this discrepancy on the colour-matching functions is negligible in most practical situations.

As has already been mentioned, experiment shows that colour matches are additive. Hence, if 1 power unit of one wavelength, λ_1, is matched by

$$\bar{r}_1 \text{ of R} + \bar{g}_1 \text{ of G} + \bar{b}_1 \text{ of B} \ ,$$

and 1 power unit of light of wavelength λ_2 by

$$\bar{r}_2 \text{ of R} + \bar{g}_2 \text{ of G} + \bar{b}_2 \text{ of B} \ ,$$

then the additive mixture of the two lights, λ_1 and λ_2 is matched by

$$(\bar{r}_1 + \bar{r}_2) \text{ of R} + (\bar{g}_1 + \bar{g}_2) \text{ of G} + (\bar{b}_1 + (\bar{b}_2) \text{ of B} \ .$$

This means that the colour-matching functions of Fig. 2.4 can be used as weighting functions to determine the amounts of R, G, and B needed to match any colour, if the amount of power per small constant-width wavelength interval is known for that colour throughout the spectrum. These amounts R, G, and B, are evaluated using the formulae:

$$R = k(P_1\bar{r}_1 + P_2\bar{r}_2 + P_3\bar{r}_3 + \dots\dots\dots)$$
$$G = k(P_1\bar{g}_1 + P_2\bar{g}_2 + P_3\bar{g}_3 + \dots\dots\dots)$$
$$B = k(P_1\bar{b}_1 + P_2\bar{b}_2 + P_3\bar{b}_3 + \dots\dots\dots) \ ,$$

where k is a constant; the values of P are the amounts of power per small constant-width wavelength interval throughout the spectrum, and $\bar{r}$, $\bar{g}$, and $\bar{b}$, are the heights of the colour-matching functions, at the central wavelength of each interval.

The luminance of a colour matched by those amounts, R, G, and B, of the three matching stimuli is given by:

$$L = 1.0000R + 4.5907G + 0.0601B \ ,$$

the expressions on the right being the amounts of the matching stimuli expressed in units of luminance again. The constant k can be chosen so that, if P is in watts per steradian and per square metre, L is in candelas per square metre (cd/m^2); but if P is in some other radiant measure, then L will be in the corresponding photometric measure (see Appendix 1, Table A1.1 for details).

2.5 TRANSFORMATION FROM R, G, B, TO X, Y, Z

The values of R, G, and B, calculated in the above way, could provide a system of colour specification which is precise, is based on representative human colour vision, and could be used to calculate colour specifications from spectral power data. But it is not used in the above form. In 1931, the CIE drew up the system of colour specification that has been adopted internationally; it was felt at that time that the presence of negative values in the colour-matching functions, and the fact that, for some colours, the value of one of the tristimulus values could be negative, would militate against the system being used without error and adopted without misgivings.

Therefore, to avoid negative numbers in colour specifications, the CIE recommended that the tristimulus values, R, G, B, should be replaced by a new set of tristimulus values, X, Y, Z, that are obtained by means of the equations:

$$\begin{aligned} X &= 0.49R \quad + 0.31G \quad + 0.20B \\ Y &= 0.17697R + 0.81240G + 0.01063B \\ Z &= 0.00R \quad + 0.01G \quad + 0.99B \end{aligned}$$

The numbers in these equations were carefully chosen so that X, Y, and Z would always be all positive for all colours. That this is possible can be illustrated by considering a blue–green spectral colour. Such a colour will have a negative value of R, but positive values of G, and B; in fact for light of wavelength 500 nm, $R = -0.072$, $G = 0.085$, and $B = 0.048$. The corresponding value of X is given by:

$$\begin{aligned} & 0.49 \times (-0.072) + 0.31 \times (0.085) + 0.20 \times (0.048) \\ &= -0.035 \quad + 0.026 \quad + 0.010 \\ &= +0.001 \ . \end{aligned}$$

The values of Y, and Z are also both positive for this colour. Hence the use of X, Y, and Z instead of R, G, and B has eliminated the presence of negative numbers in this colour specification, and in fact eliminates them for all colours.

There is, moreover, another advantage in using X, Y, and Z. It is clear from the equations used to evaluate them that fairly simple numbers have been used for the coefficients in the case of X and Z; but, in the case of Y, the coefficients are given to five places of decimals. The reason for this is that these coefficients have been carefully chosen so that they are in the same ratios as those of the luminances of the units used for measuring the amounts of the matching stimuli R, G, and B. That is to say:

$$0.17697,\ 0.81240,\ \text{and}\ 0.01063$$

are in the same ratios as

$$1.0000,\ 4.5907,\ \text{and}\ 0.0601\ .$$

This means that the value of Y is proportional to L, the luminance of the colour being specified. Hence, the ratio of the values of Y for any two colours, Y_1 and Y_2, is the same as the ratio of their luminances, L_1 and L_2:

$$Y_1/Y_2 = L_1/L_2\ .$$

In the equations given above relating X, Y, Z and R, G, B, the coefficients in the equation for X sum to unity, as is also the case for the equations for Y and Z. This means that, when $R=G=B$, it is the case that $X=Y=Z$. Hence, for the equi-energy stimulus, S_E, since $R=G=B$, it is also the case that $X=Y=Z$, and the equations were designed to achieve this result.

2.6 CIE COLOUR-MATCHING FUNCTIONS

The colour-matching functions, $\bar{r}(\lambda)$, $\bar{g}(\lambda)$, and $\bar{b}(\lambda)$, represent the amounts of R, G, and B needed to match a constant amount of power per small constant-width wavelength interval throughout the spectrum. The equivalent values of X, Y, and Z, which are denoted by the symbols $\bar{x}(\lambda)$, $\bar{y}(\lambda)$, and $\bar{z}(\lambda)$ can be calculated thus:

$$\bar{x}(\lambda) = 0.49\bar{r}(\lambda) + 0.31\bar{g}(\lambda) + 0.20\bar{b}(\lambda)$$
$$\bar{y}(\lambda) = 0.17697\bar{r}(\lambda) + 0.81240\bar{g}(\lambda) + 0.01063\bar{b}(\lambda)$$
$$\bar{z}(\lambda) = 0.00\bar{r}(\lambda) + 0.01\bar{g}(\lambda) + 0.99\bar{b}(\lambda)\ .$$

These values are called the *CIE colour-matching functions* and they define the colour matching properties of the *CIE 1931 Standard Colorimetric Observer*, often referred to as the *2° Observer*. These $\bar{x}(\lambda)$, $\bar{y}(\lambda)$, and $\bar{z}(\lambda)$ functions are shown in Fig.

2.5 by the full lines. It is clear that, as is to be expected, these functions have no negative parts; and, because, for S_E, $X = Y = Z$, the areas under the three curves are equal. The figure also illustrates another advantage of these functions: the $\bar{y}(\lambda)$ function is exactly the same shape as the $V(\lambda)$ function of Fig. 2.1, and this means that, for photopic vision, instead of having four spectral functions, $\bar{r}(\lambda)$, $\bar{g}(\lambda)$, $\bar{b}(\lambda)$, and $V(\lambda)$, only three are required, $\bar{x}(\lambda)$, $\bar{y}(\lambda) = V(\lambda)$, and $\bar{z}(\lambda)$. The $\bar{y}(\lambda)$ function has to be the same shape as the $V(\lambda)$ function because the values of Y are proportional to luminance; by making it have a maximum value of 1.0, the $\bar{y}(\lambda)$ function is made, not only the same shape as, but also identical to, the $V(\lambda)$ function.

The $\bar{x}(\lambda)$, $\bar{y}(\lambda)$, and $\bar{z}(\lambda)$ functions can be used as weighting functions to enable X, Y, and Z to be evaluated from spectral power data by using the formulae:

$$X = K(P_1\bar{x}_1 + P_2\bar{x}_2 + P_3\bar{x}_3 \,........)$$
$$Y = K(P_1\bar{y}_1 + P_2\bar{y}_2 + P_3\bar{y}_3 \,........)$$
$$Z = K(P_1\bar{z}_1 + P_2\bar{z}_2 + P_3\bar{z}_3 \,........)\ ,$$

where K is a constant; the values of P are the amounts of power per small constant-width wavelength interval throughout the spectrum, and the values of $\bar{x}$, $\bar{y}$, and $\bar{z}$ are the heights of the CIE colour-matching functions at the central wavelength of each interval. If P is in watts per steradian and per square metre and K is put equal to 683, then Y is the luminance in candelas per square metre (cd/m^2); when this is the case it is convenient to use the symbols X_L, Y_L, and Z_L for these luminance-related tristimulus values. If other radiometric measures are used for P, the corresponding photometric measures are obtained (see Appendix 1, Table A1.1 for details).

It is much more customary, however, to express the luminance separately from the colour specification, and to choose the constant K so that $Y = 100$ for a *perfect reflecting* (or *transmitting*) *diffuser* similarly illuminated and viewed.

A perfect diffuser reflects (or transmits), equally strongly in all directions, all the light that is incident on it at every wavelength throughout the visible spectrum. The value of Y then gives the percentage luminous reflection or transmission of the colour. (For non-diffusing samples a perfectly reflecting mirror, or a perfectly transmitting specimen, may be used instead.)

Specific terms are defined to denote different ways in which the light from the sample and the perfect diffuser may be received and evaluated. *Reflectance factor* and *transmittance factor* are used when the light reflected or transmitted by the sample and by the perfect diffuser lie within a defined cone. If this cone is a hemisphere, the terms *reflectance* and *transmittance* are used. If the light is monochromatic, the adjective *spectral* is used; for example, *spectral reflectance factor*. If the light is evaluated according to its radiant power, the adjective *radiant* is used; for example, *radiant reflectance factor*. If the light is evaluated by using the $V(\lambda)$ or $\bar{y}(\lambda)$ function as a weighting function, the adjective *luminous* is used; for example, *luminous reflectance factor*. If the cone in which light is reflected or transmitted is very small, the radiant term used is *radiance factor*, and the luminous term used is *luminance factor*. Y is then equal to whichever is the appropriate one of these luminous measures. (See Appendix 1, section A1.6.)

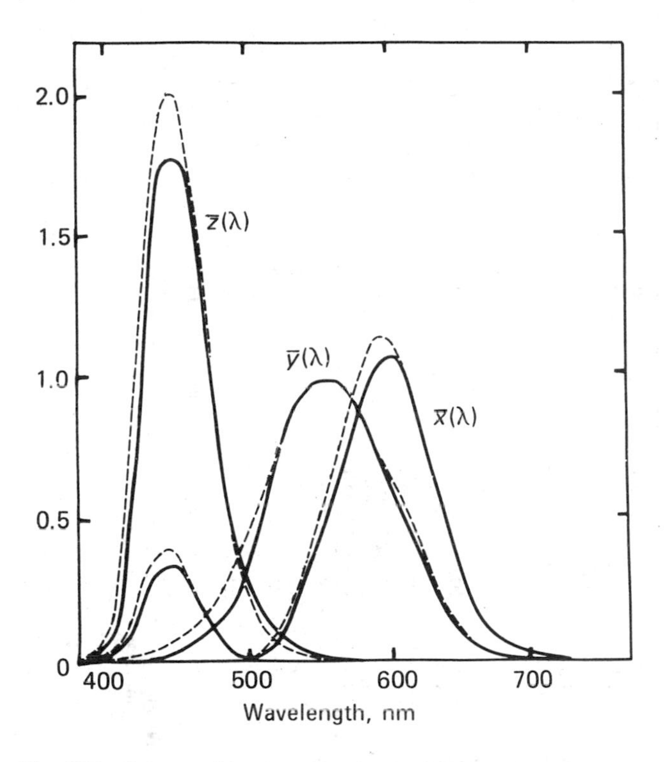

Fig. 2.5 — The CIE colour-matching functions for the 1931 Standard Colorimetric Observer (full lines), and for the 1964 Supplementary Standard Colorimetric Observer (broken lines).

The CIE colour-matching functions illustrated in Fig. 2.5 are the most important spectral functions in colorimetry. They are used to obtain X, Y, and Z tristimulus values from spectral power data. If two colour stimuli have the same tristimulus values, they will look alike, when viewed under the same photopic conditions, by an observer whose colour vision is not significantly different from that of the CIE 1931 Standard Colorimetric Observer; conversely, if the tristimulus values are different, the colours may be expected to look different in these circumstances.

In Table 2.3, an example is given of how the tristimulus values, X, Y, and Z, are obtained for two samples, A and B, having different spectral power distributions. The samples, in this case, are both reflecting surfaces, so that their spectral data are recorded as their spectral reflectance factors, $R_A(\lambda)$ and $R_B(\lambda)$. They are assumed to be viewed under the same light source, whose spectral power distribution per small constant-width wavelength interval, $S(\lambda)$, is given in the Table. In colorimetry, it is usual to work with *relative spectral power distributions*, and in this case $S(\lambda)$ is normalized by having an arbitrary value of 100 at 560nm, the values at all the other wavelengths having the correct ratios relative to this value. Also listed are the values for the CIE colour-matching functions $\bar{x}(\lambda)$, $\bar{y}(\lambda)$, and $\bar{z}(\lambda)$. The relative spectral power distributions of the two samples under this light source are then obtained by multiplying, at each wavelength, the value of R_A by S and R_B by S; such spectral

Table 2.3 — Calculation of tristimulus values from CIE colour-matching functions

Wavelength nm	$\bar{x}(\lambda)$	$\bar{y}(\lambda)$	$\bar{z}(\lambda)$	$S(\lambda)$	$S(\lambda)$x $\bar{y}(\lambda)$	$R_A(\lambda)$	$R_B(\lambda)$	X_A/K	Y_A/K	Z_A/K	X_B/K	Y_B/K	Z_B/K
400	0.0143	0.0004	0.0679	15	0.0	0.20	0.20	0.0	0.0	0.2	0.0	0.0	0.2
420	0.1344	0.0040	0.6456	21	0.1	0.40	0.10	1.1	0.0	5.4	0.3	0.0	1.4
440	0.3483	0.0230	1.7471	29	0.7	0.40	0.11	4.0	0.3	20.3	1.1	0.1	5.6
460	0.2908	0.0600	1.6692	38	2.3	0.10	0.32	1.1	0.2	6.3	3.5	0.7	20.3
480	0.0956	0.1390	0.8130	48	6.7	0.10	0.27	0.5	0.7	3.9	1.2	1.8	10.5
500	0.0049	0.3230	0.2720	60	19.4	0.20	0.20	0.1	3.9	3.3	0.1	3.9	3.3
520	0.0633	0.7100	0.0782	72	51.1	0.40	0.10	1.8	20.4	2.3	0.5	5.1	0.6
540	0.2904	0.9540	0.0203	86	82.0	0.40	0.10	10.0	32.8	0.7	2.5	8.2	0.2
560	0.5945	0.9950	0.0039	100	99.5	0.10	0.39	5.9	10.0	0.0	23.2	38.8	0.2
580	0.9163	0.8700	0.0017	114	99.2	0.10	0.36	10.4	9.9	0.0	37.6	35.7	0.1
600	1.0622	0.6310	0.0008	129	81.4	0.20	0.20	27.4	16.3	0.0	27.4	16.3	0.0
620	0.8544	0.3810	0.0002	144	54.9	0.40	0.11	49.2	21.9	0.0	13.5	6.0	0.0
640	0.4479	0.1750	0.0000	158	27.6	0.40	0.10	28.3	11.1	0.0	7.1	2.8	0.0
660	0.1649	0.0610	0.0000	172	10.5	0.10	0.75	2.8	1.0	0.0	21.3	7.9	0.0
680	0.0468	0.0170	0.0000	185	3.1	0.10	0.49	0.9	0.3	0.0	4.2	1.5	0.0
700	0.0114	0.0041	0.0000	198	0.8	0.20	0.20	0.5	0.2	0.0	0.5	0.2	0.0
Totals					539.3			144.0	129.0	42.4	144.0	129.0	42.4

For sample A,

$X_A = (100/539.3)144.0 = 26.7$
$Y_A = (100/539.3)129.0 = 23.9$
$Z_Z = (100/539.3)42.4 = 7.9$

For sample B,

$X_B = (100/539.3)144.0 = 26.7$
$Y_B = (100/539.3)129.0 = 23.9$
$Z_B = (100/539.3)42.4 = 7.9$

In this example, only values at 20 nm intervals, over the range of wavelengths from 400 to 700 nm, are included in the computation; this is the minimum number of values that can be used, and was done only to keep the example as simple as possible. For most applications it is necessary to take values either at every 10 nm, or preferably at every 5 nm, and to use a range of wavelengths from 380 to 780 nm.

products are referred to as *relative colour stimulus functions*. Each of these relative spectral power distributions is then weighted by each of the CIE colour-matching functions, and these weighted products summed to obtain:

$$X = K(R_1S_1\bar{x}_1 + R_2S_2\bar{x}_2 + R_3S_3\bar{x}_3 + ...)$$
$$Y = K(R_1S_1\bar{y}_1 + R_2S_2\bar{y}_2 + R_3S_3\bar{y}_3 + ...)$$
$$Z = K(R_1S_1\bar{z}_1 + R_2S_2\bar{z}_2 + R_3S_3\bar{z}_3 + ...) \ .$$

In Table 2.3, these products and summations are listed as X/K, Y/K, Z/K.

The value of the constant K that results in Y being equal to 100 for the perfect diffuser is given by:

$$K = 100/(S_1\bar{y}_1 + S_2\bar{y}_2 + S_3\bar{y}_3 + ...) \ .$$

This arises because, for the perfect diffuser, the reflectance factor $R(\lambda)$, is equal to 1 at all wavelengths, so that its Y tristimulus value is given by:

$$Y = 100 = K(S_1\bar{y}_1 + S_2\bar{y}_2 + S_3\bar{y}_3 + ...) \ .$$

The products $S\bar{y}$, and their summation, are therefore also evaluated in Table 2.3.

The tristimulus values for the two samples, X_A, Y_A, Z_A, and X_B, Y_B, Z_B are then obtained, and it is seen from the Table that $X_A = X_B$, $Y_A = Y_B$, and $Z_A = Z_B$. The samples will therefore look alike, when seen under the same photopic viewing conditions, by an observer whose colour vision is not significantly different from the CIE 1931 Standard Colorimetric Observer, for the particular light source, $S(\lambda)$, used.

In Table 2.3, the summation is carried out using data at every 20 nm, from 400 to 700 nm. This is the minimum number of values that can be used. For most applications it is necessary to take values either at every 10 nm, or preferably at every 5 nm, and to use a range of wavelengths from 380 to 780 nm. In Appendix 3, the values of the $\bar{x}(\lambda)$, $\bar{y}(\lambda)$, and $\bar{z}(\lambda)$ functions are given in full at every 5 nm (CIE 1986a); values at every 1 nm are also available (CIE 1971; CIE 1986b; Grum & Bartleson 1980; Wyszecki & Stiles 1982).

As mentioned in section 1.3, the retina varies considerably in its properties from one point to another, and matches made with a 2° field may not remain matches if the field size is altered. If the field size is reduced, the ability to discriminate one colour from another becomes less marked, and hence 2° matches usually appear to remain matches; but, if the field size is increased, colour discrimination becomes more pronounced, and 2° matches tend to break down. For this reason, in 1964, the CIE recommended a different set of colour-matching functions for samples having field sizes greater than 4°. These supplementary colour-matching functions, $\bar{x}_{10}(\lambda)$, $\bar{y}_{10}(\lambda)$, and $\bar{z}_{10}(\lambda)$, are shown in Fig. 2.5 by the broken lines, the full lines showing the $\bar{x}(\lambda)$, $\bar{y}(\lambda)$, and $\bar{z}(\lambda)$ functions for comparison. It can be seen from the figure that the two sets of functions are similar, but the differences are large enough to be significant. Values for these functions are given in Appendix 3 at every 5 nm (CIE

1986a); 1 nm values are also available (CIE 1971; CIE 1986b; Grum & Bartleson 1980; Wyszecki & Stiles 1982). The $\bar{x}_{10}(\lambda)$, $\bar{y}_{10}(\lambda)$, and $\bar{z}_{10}(\lambda)$ functions define the colour-matching properties of the *CIE 1964 Supplementary Standard Colorimetric Observer*, often referred to as the *10° Observer*. These functions can be used as weighting functions to obtain tristimulus values, X_{10}, Y_{10}, and Z_{10}, by means of procedures analogous to those adopted to obtain X, Y, and Z. All measures obtained using the $\bar{x}_{10}(\lambda)$, $\bar{y}_{10}(\lambda)$, and $\bar{z}_{10}(\lambda)$ colour-matching functions as a basis are distinguished by the presence of a subscript 10. It has been arranged that, for the equi-energy stimulus, S_E, just as $X = Y = Z$, so also $X_{10} = Y_{10} = Z_{10}$.

Unlike the $\bar{y}(\lambda)$ function, the $\bar{y}_{10}(\lambda)$ function has no photometric significance. In fact, photometric measures are not additive in 10° fields in the way that they are in 2° fields.

If the spectral data available are in terms of an absolute measure of a radiant quantity per small constant-width wavelength interval throughout the spectrum, it is referred to, not as the relative colour stimulus function, but as the *colour stimulus function*. As already mentioned, in this case, tristimulus values, X_L, Y_L, Z_L, can be obtained for which Y_L is equal to the absolute value of a photometric quantity; the constant K must then be put equal to 683 lumens per watt, and the radiometric quantity must correspond to the photometric quantity required (see Appendix 1, Table A1.1). However, this is seldom done, even for light sources; these are usually evaluated using convenient arbitrary values for K.

2.7 METAMERISM

When two colours having different spectral compositions match one another, they are said to be *metameric*, or *metamers*, and the phenomenon is called *metamerism*. A characteristic of metamerism is that, in general, metameric matches are upset if a different observer is used, or, in the case of reflecting or transmitting samples, if a different illuminant is used. Thus, for example, a change of field size from 2° to 10°, necessitating a change from the CIE 1931 Standard Colorimetric Observer to the CIE 1964 Supplementary Standard Colorimetry Observer, will usually indicate that metameric matches have been upset. Real observers will also often see that metameric matches are upset when field sizes are changed. In the case of illuminants, if a change is made from daylight to electric tungsten filament lighting, for example, then again, in general, both colorimetric computations and real observers will indicate that metameric matches have been upset. These effects are of great practical importance, and will be discussed further in Chapter 6.

2.8 SPECTRAL LUMINOUS EFFICIENCY FUNCTIONS FOR PHOTOPIC VISION

As mentioned in section 2.4, the values of the $V(\lambda)$ function at wavelengths below about 450 nm are lower than they should be, but the resulting errors in deriving photopic photometric measures are negligible in most practical cases. The $V(\lambda)$ function has therefore been retained both for photometric purposes, and in the form of the $\bar{y}(\lambda)$ function in the CIE X,Y,Z system of colorimetry. A set of values that is more correct was proposed by Deane B. Judd in 1951 (CIE 1951), and has been referred to as the *Judd correction*; in 1988 the CIE (CIE 1988b) provided a set of values known as the *CIE 1988 modified two degree spectral luminous efficiency function for photopic vision*, $V_M(\lambda)$; these values are given in Appendix 2.

The $V(\lambda)$ function was originally obtained by the method of flicker photometry. It has been found since that similar results for photopic efficiency can be obtained by three other methods:

(a) Step-by-step comparisons involving pairs of intermediate stimuli having small colour differences.
(b) Minimum-distinct-border evaluation, in which the intensities of two stimuli are judged equal when the border between them is minimally distinct when they are viewed with a negligible gap between them.
(c) Visual acuity evaluation, in which the intensities of two stimuli are judged equal to one another when the visibility of fine detail in each is judged equally clear.

When the intensities are judged in other ways, such as by equality when seen side by side in the presence of large colour differences, or by subjective scaling of apparent brightness, or by threshold intensities for detection in a given set of viewing conditions, then the functions obtained are usually broader than the $V(\lambda)$ function. Sets of values for such functions have been provided by the CIE (CIE 1988a) for both 2° and 10° field sizes, known as the *spectral luminous efficiency functions based upon brightness matching for monochromatic 2° and 10° fields*, $V_{b,2}(\lambda)$ and $V_{b,10}(\lambda)$, which are given in Appendix 2. It was found that the $V_M(\lambda)$ function could be used as the corresponding function to represent *point sources*, by which is meant those of less than 10 minutes of arc in angular size. It is important to note that the $V_{b,2}(\lambda)$ and $V_{b,10}(\lambda)$ functions, like the $\bar{y}_{10}(\lambda)$ function, cannot be used as weighting functions for stimuli consisting of a variety of wavelengths, because additivity for lights of different wavelengths does not generally hold in the case of brightness.

It must be emphasized that, in colorimetry, none of the above functions is used, except the $\bar{y}(\lambda)$ and $\bar{y}_{10}(\lambda)$ functions.

REFERENCES

CIE, Compte Rendu, 12th Session, Stockholm, Vol. 1, Committee No.7, Colorimetry, pp. 11–52 (1951).

CIE Publication No. 15, *Colorimetry* (1971).

CIE Publication No. 15.2, *Colorimetry*, 2nd ed. (1986a).

CIE Publication No. 75, *Spectral luminous efficiency functions based upon brightness matching for monochromatic point sources*, 2°, and 10° fields (1988a).

CIE Publication No. 86, *CIE 1988 2° spectral luminous efficiency functions of photopic vision* (1988b).

CIE *Standard on Colorimetric Observers*, CIE S002 (1986b).

Grum, F. & Bartleson, C. J. *Color Measurement*, pp. 48–68, Academic Press, New York (1980).
Palmer, D. A. *J. Opt. Soc. Amer.* **71**, 966 (1981).
Stiles. W. S. *Phys. Soc. Year Book* p. 44 (1955).
Trezona, P. W. *Vision Research* **13**, 9 (1973).
Wyszecki, G. & Stiles, W. S. *Color Science*, 2nd ed., pp. 725-747, Wiley, New York (1982).

GENERAL REFERENCES

Billmeyer, F. W. & Saltzman, M. *Principles of Color Technology*, 2nd. ed., Wiley, New York (1981)
Chamberlin, G. J. & Chamberlin, D. E. *Colour, its Measurement, Computation, and Application*, Heyden, London (1980).
CIE Publication Nos. 15 and 15.2, *Colorimetry* (1971 and 1986a).
Grum, F. & Bartleson, C. J. *Color Measurement*, Academic Press, New York (1980).
Judd, D. B. & Wyszecki, G. *Color in Business, Science, and Industry* 2nd. ed., Wiley, New York (1975).
Wright, W. D. *The Measurement of Colour*, 4th. ed., Hilger, Bristol (1969).
Wyszecki, G. & Stiles, W.S. *Color Science*, 2nd. ed., Wiley, New York (1982).

3

Relations between colour stimuli

3.1 INTRODUCTION

We have seen in the previous chapter that, if two colours have the same tristimulus values, then they will look alike when seen under the same photopic viewing conditions by an observer whose colour vision is not significantly different from the CIE 1931 or 1964 Standard Colorimetric Observer, according to whether the field size is about 2° or about 10°, respectively. But, for colours in general, the tristimulus values differ over a wide range. In this chapter we see how such differences can be related to various colour attributes.

3.2 THE Y TRISTIMULUS VALUE

If the Y tristimulus value is evaluated on an absolute basis as Y_L, in candelas per square metre, for example, it represents the luminance of the colour. This provides a basis for a correlation with the perceptual attribute of brightness. As has already been explained, the correlation is complicated by the effect of the viewing conditions, by the non-linear relationship between brightness and luminance, and by the partial dependence of brightness on colourfulness. These factors will be discussed in some detail in Chapter 12. For the moment, however, we shall simply regard luminance as an approximate correlate of brightness.

When, as is customary, Y is evaluated such that, for the similarly illuminated and viewed perfect diffuser, $Y = 100$, then Y is equal to the reflectance factor, or transmittance factor, expressed as a percentage. When viewed by the human eye, the cone of light collected from samples is very small, and for small cones of collection these factors are approximately the same as the luminance factor. Hence it is customary to regard Y as representing the percentage luminance factor, and this is an approximate correlate of the perceptual attribute of lightness. The Y values then range from 100 for white, or transparent objects, that absorb no light, to zero for objects that absorb all the light. It is sometimes convenient to use the ratio Y/Y_n, where Y_n is the value of Y for a suitably chosen reference white or reference transparent specimen; in this case, if the reference object absorbs some light, as is usually the case, Y/Y_n will be slightly more than 1 for the perfect diffuser (this will be discussed further in section 5.7).

3.3 CHROMATICITY

If the Y tristimulus value correlates approximately with brightness or, more usually, with lightness, with what do the X and Z tristimulus values correlate? The answer is that they do not correlate, even approximately, with any perceptual attributes, and it is necessary to derive from them other measures that can provide such correlates.

Important colour attributes are related to the *relative* magnitudes of the tristimulus values. It is therefore helpful to calculate a type of relative tristimulus values called *chromaticity co-ordinates*, as follows:

$$x = X/(X + Y + Z)$$

$$y = Y/(X + Y + Z)$$

$$z = Z/(X + Y + Z) \ .$$

The chromaticity co-ordinates, for which lower-case letters are always used, thus represent the relative amounts of the tristimulus values. For example, if $X = 8$, $Y = 48$, $Z = 24$, then $X + Y + Z = 80$, and $x = 8/80 = 0.1$; $y = 48/80 = 0.6$; $z = 24/80 = 0.3$. This indicates that, for this particular colour, there is in its specification 10% of X, 60% of Y, and 30% of Z. If, for the moment, we regard the equi-energy stimulus, S_E, as a white, we can deduce from these figures that the colour considered has a higher y value, and a lower x value than the white, for which x, y and z are all 1/3 because X, Y, and Z are equal to one another for S_E. Since the $\bar{x}(\lambda)$, $\bar{y}(\lambda)$, and $\bar{z}(\lambda)$ curves correspond very approximately to the $\bar{r}(\lambda)$, $\bar{g}(\lambda)$, and $\bar{b}(\lambda)$ curves, respectively, we can deduce that the colour considered is probably greener, and less red (that is, more blue–green), than the white. This, together with the fact that the Y value is 48, suggests that we have a bluish green of lightness somewhere about midway between white and black. These deductions are only partly correct, and need to be refined, but they show the way in which chromaticity co-ordinates are useful in understanding the implications of tristimulus values.

It is clear from the way in which x, y, and z are calculated that

$$x + y + z = 1$$

and hence, if x and y are known, z can always be deduced from $1 - x - y$. With only two variables, such as x and y, it becomes possible to construct two-dimensional diagrams, or *chromaticity diagrams* as they are usually called, an example of which is shown in Fig. 3.1. In this figure, y is plotted as ordinate against x as abscissa, and this important diagram is usually referred to as the *x,y chromaticity diagram*. This diagram provides a sort of colour map on which the chromaticities of all colours can

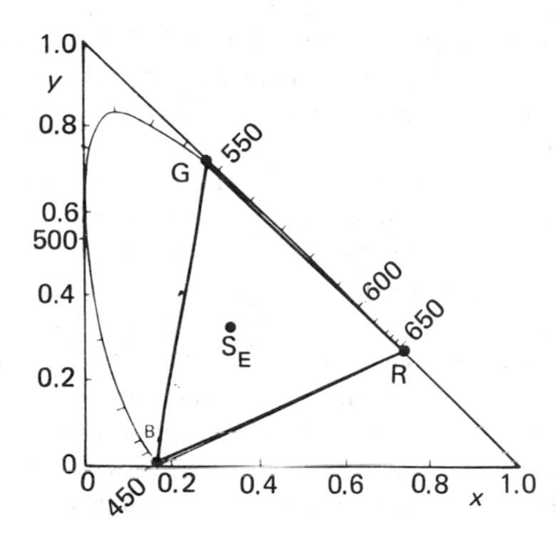

Fig. 3.1 — The CIE x,y chromaticity diagram, showing the spectral locus and equi-energy stimulus, S_E, and the CIE red, green, and blue matching stimuli, R, G, B.

be plotted. As already mentioned, the equi-energy stimulus, S_E, has tristimulus values that are equal to one another, so that its chromaticity co-ordinates are $x = 1/3$, $y = 1/3$, and $z = 1/3$, and it is marked at this position in Fig. 3.1.

The curved line in the diagram shows where the colours of the spectrum lie and is called the *spectral locus*; the wavelengths are indicated in nanometres along the curve, and the corresponding values of x and y are listed at 5 nm intervals in Appendix 4.

If two colour stimuli are additively mixed together, then the point representing the mixture is located in the diagram by a point that always lies on the line joining the two points representing the two original colours. This means that, if the two ends of the spectrum are joined by a straight line, that line represents mixtures of light from the two ends of the spectrum; as these colours are mixtures of red and blue, this line is known as the *purple boundary*. The area enclosed by the spectral locus and the purple boundary encloses the domain of all colours; this is because the spectral locus consists of a continuously convex boundary so that all mixtures of wavelengths must lie inside it. The positions of the three matching stimuli, R, G, and B, having wavelengths of 700 nm, 546.1 nm, and 435.8 nm, respectively, are also shown in the diagram. The line joining the points R and G represents all the colours formed by the additive mixtures of various amounts of the R and G stimuli; and the lines joining G and B, and B and R, have similar relationships to the appropriate mixtures of the other pairs of stimuli. The area within the triangle formed by the three lines represents all the colours that can be matched by additive mixtures of these three stimuli. The location of the spectral locus outside this triangle, especially in the blue–green part of the spectrum, is a consequence of negative amounts of one of the three matching stimuli being necessary in order to match the colours of the spectrum.

If, now, points are considered between the point representing S_E and a given point on the spectral locus, these must represent colours formed by additive mixtures of S_E and light of the particular wavelength considered. If the mixture consists mainly of the spectral colour, then the corresponding point will lie near the spectral locus and will tend to represent a saturated colour of that particular hue. If, on the other hand, the mixture consists mainly of S_E, then the corresponding point will lie near S_E and will tend to represent a pale colour of that hue. The chromaticity diagram thus represents a continuous gradation of stimuli from those like S_E to those like spectral colours, in the space between S_E and the spectral locus. It is very important to remember, however, that chromaticity diagrams show only the *proportions* of the tristimulus values; hence, bright and dim lights, or brightly lit and dimly lit surface colours, or light and dark surface colours, having tristimulus values in the same ratios to one another, all plot at the same point. For this reason, the point S_E, for instance, represents all levels of the equi-energy stimulus; and, if this stimulus is regarded as an illuminant, the same point also represents whites, greys, and blacks illuminated by it at any level. Similarly, a single point can represent orange and brown surface colours illuminated at any level.

It is also very important to remember that chromaticity diagrams are maps of relationships between colour stimuli, not between colour perceptions. Although it is often helpful to consider the approximate nature of the colour perceptions that are likely to be represented by various chromaticities, it must be remembered that the exact colour perceptions will depend on the viewing conditions, and on the adaptation and other characteristics of the observer.

A similar chromaticity diagram is derived from the X_{10}, Y_{10}, and Z_{10} tristimulus values, by evaluating:

$$x_{10} = X_{10}/(X_{10} + Y_{10} + Z_{10})$$

$$y_{10} = Y_{10}/(X_{10} + Y_{10} + Z_{10})$$

$$z_{10} = Z_{10}/(X_{10} + Y_{10} + Z_{10}).$$

The values of x_{10}, and y_{10} at 5 nm intervals are listed in Appendix 4. In Fig. 3.2, the spectral loci for the x,y and the x_{10},y_{10} chromaticity diagrams are superimposed; it is clear that they are broadly similar in shape, but the locations of the points representing some of the individual wavelengths are appreciably different (see, for instance, those for 480 nm). The equi-energy stimulus, S_E, is located at the same point in both diagrams, because for this stimulus $X = Y = Z$, and $X_{10} = Y_{10} = Z_{10}$.

The following sections describe various colorimetric measures in terms of tristimulus values for the 1931 Observer; they can all also be evaluated using tristimulus values for the 1964 Observer, in which case they would all be distinguished by having a subscript 10.

3.4 DOMINANT WAVELENGTH AND EXCITATION PURITY

Chromaticity diagrams can be used to provide measures that correlate approximately with the perceptual colour attributes, hue, and saturation. In Fig. 3.3, the x,y chromaticity diagram is shown again. The point, C, represents a colour being

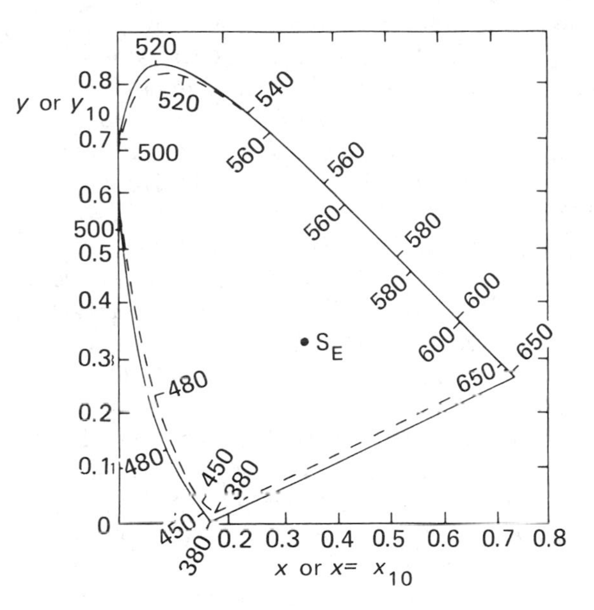

Fig. 3.2 — The CIE x_{10},y_{10} chromaticity diagram (broken lines), compared with the CIE x,y chromaticity diagram (full lines).

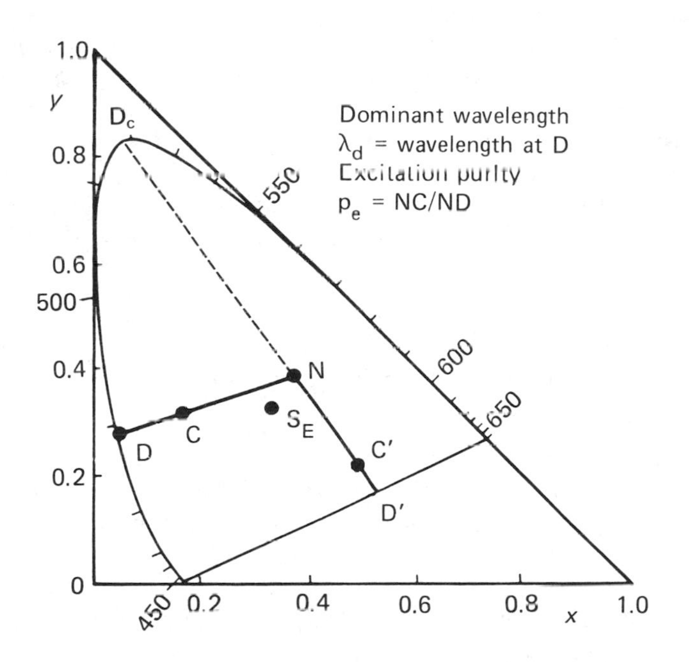

Fig. 3.3 — The derivation on the x,y diagram of dominant wavelength, λ_d, complementary wavelength, λ_c, at D_c and excitation purity, p_e, NC/ND or NC′/ND′.

considered. The point, N, represents a suitably chosen reference white (or grey), the *achromatic stimulus*, which is almost always different from S_E, the equi-energy stimulus. For reflecting surface colours, it is common practice to consider the perfect diffuser, illuminated by the light source in use, to be the reference white. A line is then drawn from N through C, to meet the spectral locus at a point, D. The wavelength on the spectral locus corresponding to D, is then called the *dominant wavelength*, λ_d, of the colour, C, relative to the white point, N. It is clear from the diagram that the colour, C, can be thought of as an additive mixture of the colour, D, and the white, N; in this sense, dominant wavelength is a very descriptive name for this measure. If, for a colour, C′, the point, D′ is on the purple boundary, then NC′ is extended in the opposite direction, as shown in Fig. 3.3, to meet the spectral locus at a point, D_c, which defines the *complementary wavelength*, λ_c. Dominant (and complementary) wavelength may be considered to be approximately correlated with the hue of colours, but, in addition to the usual factors that make this inexact, loci of constant hue (for a given set of viewing conditions) are not straight lines, like ND, but are slightly curved (this will be discussed in more detail in Chapter 12).

The ratio of the lengths of the lines NC and ND (or NC′ and ND′), provides a measure called *excitation purity*, p_e:

$$p_e = \text{NC/ND or NC}'/\text{ND}' \ .$$

If p_e is nearly unity, the point is near the spectral locus or purple boundary, and will tend to represent a colour that is highly saturated; but, if p_e is nearly zero, the point is near the reference white, and will tend to represent a colour that is very pale. Excitation purity may be considered to be approximately correlated with the saturation of colours, but, in addition to the usual factors that make this inexact, loci of constant saturation are not the same shape as loci of constant excitation purity (this will also be discussed in more detail in Chapter 12). The term *purity* is quite helpfully descriptive, but the adjective *excitation* serves only to distinguish this measure from other types of purity in colorimetry (as will be described in section 11.3).

3.5 COLOUR MIXTURES ON CHROMATICITY DIAGRAMS

If two colours C_1 and C_2 are represented by points, C_1 and C_2, as shown in Fig. 3.4, then, as has already been mentioned, the additive mixture of the two colours is represented by a point, such as C_3, lying on the line joining C_1 and C_2. The exact position of C_3 on the line depends on the relative luminances of C_1 and C_2, and is calculated as follows.

If the chromaticity co-ordinates of C_1 and C_2 are x_1, y_1, z_1, and x_2, y_2, z_2, respectively, and if, in the mixture, there are m_1 luminance units of C_1 and m_2 luminance units of C_2, we proceed as follows.

For some amount of C_1, it will be the case that $X = x_1$, $Y = y_1$, and $Z = z_1$. The luminance of this amount will be proportional to Y, and may therefore be expressed

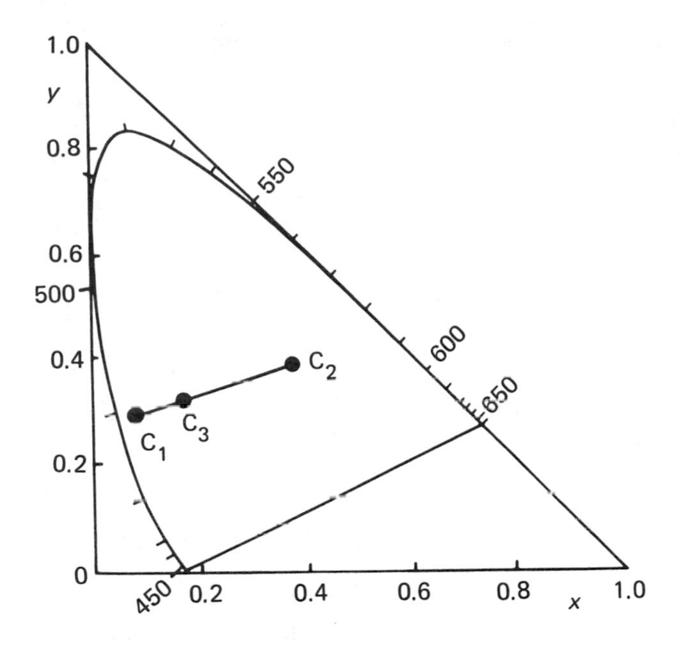

Fig. 3.4—The colour, C_3, that can be matched by an additive mixture of the colours, C_1 and C_2.

as $L_Y Y$, which is the same as $L_Y y_1$, where L_Y is a constant. Hence, for 1 luminance unit of C_1

$$X=\frac{x_1}{L_Y y_1} \qquad Y=\frac{y_1}{L_Y y_1} \qquad Z=\frac{z_1}{L_Y y_1}$$

and for m_1 luminance units of C_1

$$X=\frac{m_1 x_1}{L_Y y_1} \qquad Y=\frac{m_1 y_1}{L_Y y_1} \qquad Z=\frac{m_1 z_1}{L_Y y_1} \; .$$

Similarly for m_2 luminance units of C_2

$$X=\frac{m_2 x_2}{L_Y y_2} \qquad Y=\frac{m_2 y_2}{L_Y y_2} \qquad Z=\frac{m_2 z_2}{L_Y y_2} \; .$$

Hence for these two amounts of C_1 and C_2 additively mixed together:

$$X=\frac{m_1 x_1}{L_Y y_1}+\frac{m_2 x_2}{L_Y y_2} \qquad Y=\frac{m_1 y_1}{L_Y y_1}+\frac{m_2 y_2}{L_Y y_2} \qquad Z=\frac{m_1 z_1}{L_Y y_1}+\frac{m_2 z_2}{L_Y y_2} \; .$$

Remembering that $x_1 + y_1 + z_1 = 1$ and $x_2 + y_2 + z_2 = 1$,

$$X + Y + Z = \frac{m_1}{L_Y y_1} + \frac{m_2}{L_Y y_2},$$

hence for the mixture:

$$x = \frac{\dfrac{m_1 x_1}{y_1} + \dfrac{m_2 x_2}{y_2}}{\dfrac{m_1}{y_1} + \dfrac{m_2}{y_2}}$$

$$y = \frac{\dfrac{m_1 y_1}{y_1} + \dfrac{m_2 y_2}{y_2}}{\dfrac{m_1}{y_1} + \dfrac{m_2}{y_2}}.$$

The geometrical interpretation of this result on the x, y chromaticity diagram is that the point, C_3, representing the mixture, is on the line joining the points C_1, at x_1, y_1, and C_2, at x_2, y_2, in the ratio

$$\frac{C_1C_3}{C_2C_3} = \frac{m_2/y_2}{m_1/y_1}.$$

This means that C_3 is at the centre of gravity of weights m_1/y_1 at C_1 and m_2/y_2 at C_2, and hence the result is referred to as the *Centre of Gravity Law of Colour Mixture*.

3.6 UNIFORM CHROMATICITY DIAGRAMS

Although the x,y chromaticity diagram has been widely used, it suffers from a serious disadvantage: the distribution of the colours on it is very non-uniform. This is illustrated in Fig. 3.5. Each of the short lines in this figure joins a pair of points representing two colours having a perceptual colour difference of the same magnitude, the luminances of all the colours being the same. Ideally these identical colour differences should be represented by lines of equal length. But it is clear that this is far from being the case, the lines being much longer towards the green part of the spectral locus, and much shorter towards the violet part, than the average length. This is rather similar to some maps of the world. Because it is not possible to represent a curved surface accurately on a flat piece of paper, distortions occur. For example, in some maps of the world, countries near one of the poles, such as Greenland, are represented as far too large compared with those near the equator, such as India. On such maps, pairs of locations equally distant from one another on

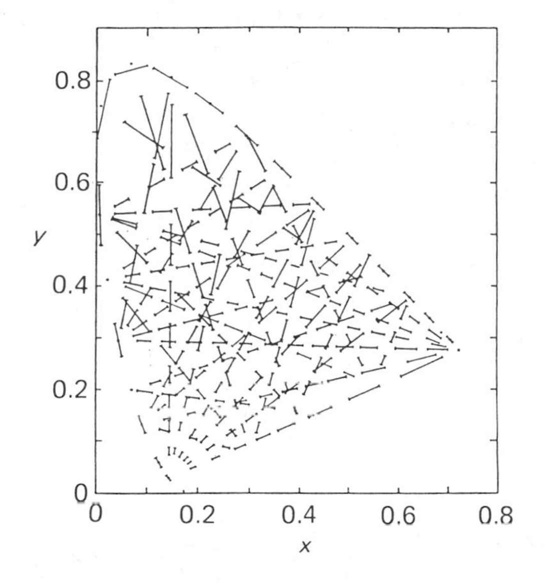

Fig. 3.5 — The x,y diagram with lines representing small colour differences (after the work of Wright 1941). Each line is three times the length of a distance representing a difference that is just noticeable in a 2° field.

the earth's surface are represented by points that are much closer together in India than in Greenland. No map on a flat piece of paper can avoid this problem entirely, but some types of map are better than others in minimizing the effect.

There is, in fact, no chromaticity diagram that can entirely avoid the problem illustrated in Fig. 3.5; but, just as with maps, some chromaticity diagrams are better than others in this respect. In Fig. 3.6, a different chromaticity diagram is used, in which a selection of the lines of Fig. 3.5 are shown again. It is immediately clear that the variation in the length of the lines, while not eliminated, has been much reduced; in fact, the ratio of the longest to the shortest line in Fig. 3.6 is only about four to one, instead of about twenty to one in Fig. 3.5. The chromaticity diagram shown in Fig. 3.6 is known as the *CIE* 1976 *uniform chromaticity scale diagram* or the *CIE* 1976 *UCS diagram*, commonly referred to as the *u', v' diagram*. It is obtained by plotting v' against u', where:

$$u' = 4X/(X + 15Y + 3Z) = 4x/(-2x + 12y + 3)$$

$$v' = 9Y/(X + 15Y + 3Z) = 9y/(-2x + 12y + 3) \ .$$

To obtain x, y from u', v' the following equations can be used:

$$x = 9u'/(6u' - 16v' + 12)$$

$$y = 4v'/(6u' - 16v' + 12) \ .$$

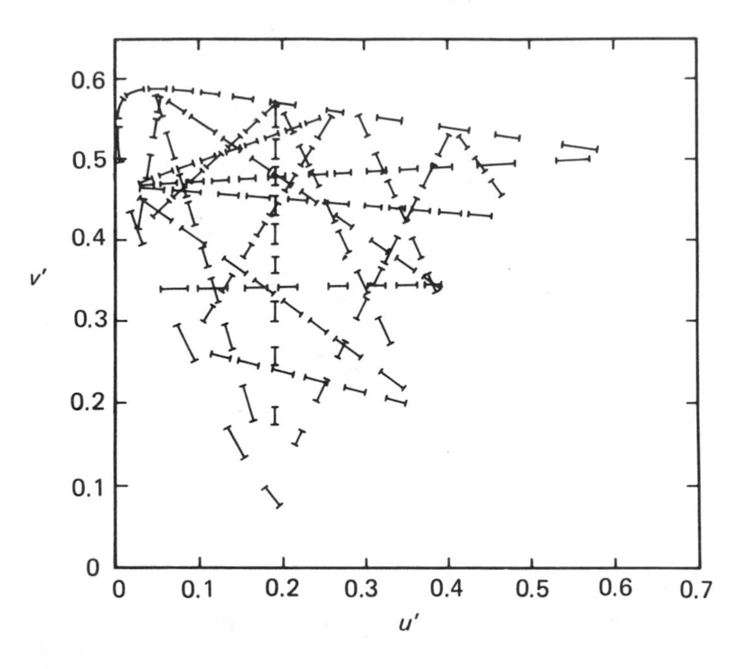

Fig. 3.6 — The approximately uniform CIE u',v' chromaticity diagram with a selection of the lines of Fig. 3.5.

The values of u' and v' for the spectral locus are given in Appendix 4. The u', v' diagram was recommended by the CIE in 1976; prior to that a similar diagram, the *u, v diagram* was used in which $u = u'$, and $v = (2/3)v'$. All chromaticity diagrams, whether x, y, or u', v', or u, v, have the property that additive mixtures of colours are represented by points lying on the straight line joining the points representing the constituent colours. The position of the mixture point has to be calculated by the method given in section 3.5, and in the u', v' diagram the weights used are m_1/v_1' and m_2/v_2'; and in the u, v diagram the weights are m_1/v_1, and m_2/v_2.

The u', v' diagram is useful for showing the relationships between colours whenever the interest lies in their discriminability. For example, in Fig. 3.7 are shown the loci of colours (of equal luminance) likely to be confused by colour defective observers. The loci for protan observers (those with protanopia or protanomaly) emanate from the point R, at $u' = 1.020$, $v' = 0.447$; those for deutan observers (those with deuteranopia or deuteranomaly) emanate from the point G, at $u' = -0.534$, $v' = 0.680$; and those for tritan observers (those with tritanopia or tritanomaly) from the point B, at $u' = 0.251$, $v' = 0$. Of course, the u', v' diagram represents the discriminability for normal observers; for colour deficient observers, diagrams compressed along the directions of the loci would be required. Thus, for a protanope, for example, all colours lying on any one line emanating from the point R would be confused; and, for a protanomalous observer, pairs of colours lying on any one of these lines would be seen to be less distinguishable than in the case of a normal observer, the difference depending on the degree of severity of the deficiency.

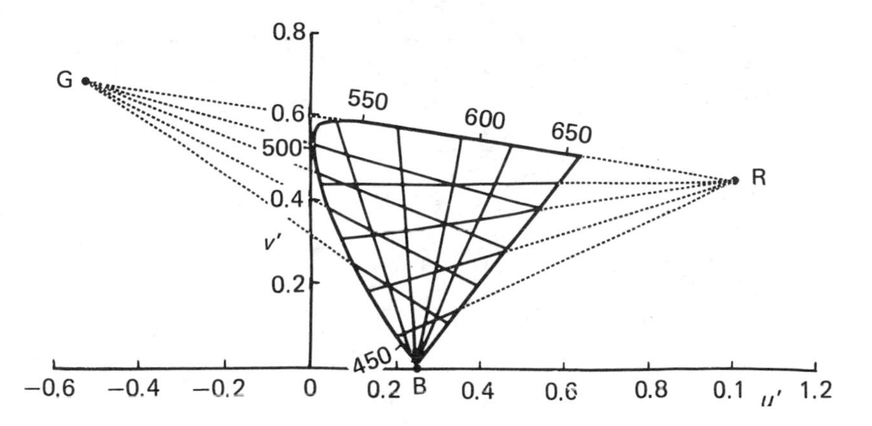

Fig. 3.7 — Loci of colours confused by protan, deutan, and tritan observers.

3.7 CIE 1976 HUE-ANGLE AND SATURATION

Because the x, y diagram is so non-uniform in its distribution of colours, excitation purity, p_e, does not correlate uniformly with the perception of saturation; and dominant wavelength correlates very non-uniformly with the perception of hue, because equal differences of hue correspond to very unequal differences of wavelength at different parts of the spectrum. Two new measures have therefore been provided, based on the u', v' diagram, that correlate with hue and saturation more uniformly. They are:

CIE 1976 u, v hue-angle, h_{uv}

$$h_{uv} = \arctan[(v' - v'_n)/(u' - u'_n)]$$

CIE 1976 u, v saturation, s_{uv}

$$s_{uv} = 13[(u' - u'_n)^2 + (v' - v'_n)^2]^{1/2} \ ,$$

where u'_n, v'_n are the values of u', v' for a suitably chosen reference white; arctan means 'the angle whose tangent is', and 13 is a scale factor introduced to harmonize s_{uv} with some other measures that will be considered later. h_{uv} lies between 0° and 90° if $v' - v'_n$ and $u' - u'_n$ are both positive; between 90° and 180° if $v' - v'_n$ is positive and $u' - u'_n$ is negative; between 180° and 270° if $v' - v'_n$ and $u' - u'_n$ are both negative; and between 270° and 360° if $v' - v'_n$ is negative and $u' - u'_n$ is positive.

In Fig. 3.8, the geometrical meanings of h_{uv} and s_{uv} can be seen. C is the point representing the colour considered, and N that for the reference white. Then h_{uv} is the angle between the line NC and a horizontal line drawn from N to the right, the angle being measured in an anticlockwise direction; and s_{uv} is equal to 13 times the distance NC. The uniformity of correlation between h_{uv} and hue, and between s_{uv} and saturation, will still not be perfect, because of the remaining non-uniformities of the u', v' diagram. In the case of hue, further discrepancies are caused by the fact,

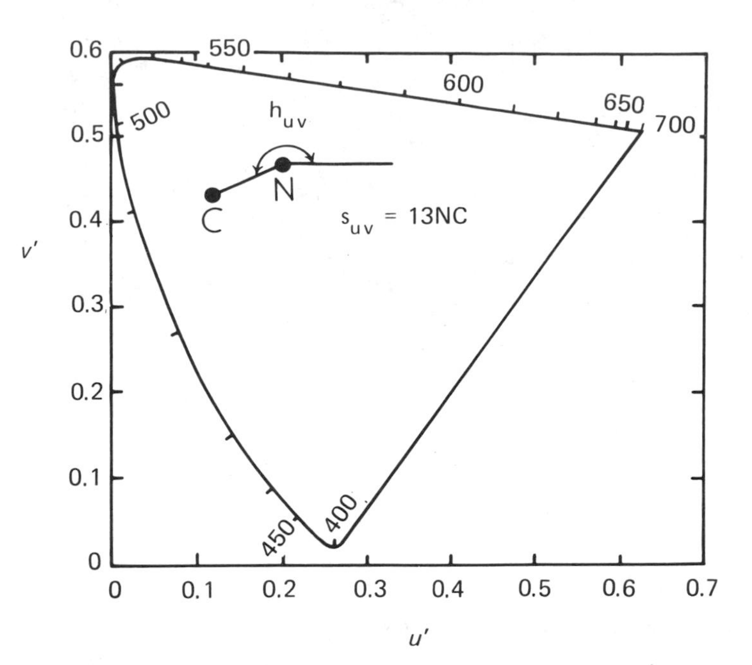

Fig. 3.8 — The derivation on the u',v' diagram of h_{uv} and s_{uv}.

already mentioned, that loci of constant hue are not straight lines, but are slightly curved in chromaticity diagrams; an approach that attempts to allow for this will be described in Chapter 12. But h_{uv} and s_{uv} are approximations to uniform correlates of hue and saturation that are of practical use.

3.8 CIE 1976 LIGHTNESS, L*

A scale of greys can be constructed in which the values of the ratios 100 L/L_n are in uniform increments, such as 10, 20, 30, 40, 50, 60, 70, 80, 90, and 100, where L denotes the luminances of the greys and L_n those of the appropriate reference white; if this is done, the apparent differences in lightness between adjacent pairs are very much less at the light end of the scale than at the dark end. In 1976 the CIE therefore recommended a non-linear function to provide a measure that correlated with lightness more uniformly. It is defined in terms of the ratio of the Y tristimulus value of the colour considered to that of the reference white, that is, Y/Y_n (which is equal to L/L_n), as follows:

*CIE 1976 lightness, L**

$$L^* = 116(Y/Y_n)^{1/3} - 16 \qquad \text{for } Y/Y_n > 0.008856$$

$$L^* = 903.3(Y/Y_n) \qquad \text{for } Y/Y_n \leqslant 0.008856 \ .$$

The reference white, for which $Y = Y_n$, then has a value of $L^* = 100$, and a perfect black for which $Y/Y_n = 0$, has a value of $L^* = 0$, a medium grey having a value of L^* of about 50. (The scale is similar in its distribution of lightnesses to the Munsell Value scale, to be considered in section 7.4, but the numbers obtained are about 10 times as great as those for Munsell Value.)

3.9 UNIFORM COLOUR SPACES

Chromaticity diagrams have many uses, but, as they show only *proportions* of tristimulus values, and not their actual magnitudes, they are only strictly applicable to colours all having the same luminance. In general, colours differ in both chromaticity and luminance, and some method of combining these variables is therefore required. To meet this need, the CIE has recommended the use of one of two alternative *colour spaces*. The first of these that we shall consider is the *CIE* 1976 *($L^*u^*v^*$) colour space* or the *CIELUV colour space*. It is produced by plotting, along three axes at right angles to one another, the quantities:

$$L^* = 116(Y/Y_n)^{1/3} - 16 \qquad \text{for } Y/Y_n > 0.008856$$

$$L^* = 903.3(Y/Y_n) \qquad \text{for } Y/Y_n \leqslant 0.008856$$

$$u^* = 13L^*(u' - u'_n)$$

$$v^* = 13L^*(v' - v'_n) \ ,$$

where u'_n, v'_n are the values of u', v' for the appropriately chosen reference white. It is clear from the expressions for u^* and v^* that a given difference in chromaticity will be reduced in magnitude by the factor L^* as the colour becomes darker. This allows for the perceptual fact that a given difference in chromaticity represents a smaller and smaller colour difference as its value of Y/Y_n is reduced. The constant 13 is included to equalize the importance of the L^*, u^*, and v^* scales.

Since $(v' - v'_n)/(u' - u'_n) = v^*/u^*$, hue-angle, h_{uv}, can also be defined as follows:

CIE 1976 *u,v hue-angle*, h_{uv}

$$h_{uv} = \arctan(v^*/u^*) \ .$$

These variables u^* and v^* also enable a correlate of chroma to be evaluated:

CIE 1976 *u,v chroma*, C^*_{uv}

$$C^*_{uv} = (u^{*2} + v^{*2})^{1/2} = L^* s_{uv}$$

In Fig. 3.9 the L*u*v* space is illustrated with the L^* axis being considered vertical, and the u^* and v^* axes lying in a horizontal plane. Examples of surfaces of constant CIE 1976 hue-angle, saturation, and chroma, are shown. Surfaces of constant hue-angle, h_{uv}, are planes having the L^* axis as one edge. Surfaces of constant chroma, C^*_{uv}, are cylinders having the L^* axis as their axes. Surfaces of constant saturation, s_{uv}, are cones having the L^* axis as their axes, and their apices at the origin of the space. If colours of a single hue-angle are considered, they will lie on a plane, with the lightest at the top and the darkest at the bottom, and with those of low chroma near the L^* axis and those of high chroma displaced farthest from it. (This arrangement is very similar to the way in which the samples in the Munsell System are arranged, to be considered in section 7.4.) A series of colours of constant chromaticity, but gradually decreasing value of Y/Y_n will fall on a line that slopes down towards the origin, on the hue-angle plane, and these will be a series of colours of constant saturation. (See Plates 2 and 3, pages 112, 113.)

In an earlier colour space, based on the u,v chromaticity diagram, the variables used were:

$$W^* = 25Y^{1/3} - 17$$

$$U^* = 13W^*(u - u_n)$$

$$V^* = 13W^*(v - v_n) \ ,$$

where Y was L/L_n expressed as a percentage.

The second space recommended by the CIE is the *CIE 1976* ($L^*a^*b^*$) *colour space* or the *CIELAB colour space*. (See Plate 1, page 112.) It is produced by plotting, along three axes at right angles to one another, the quantities:

$$L^* = 116(Y/Y_n)^{1/3} - 16 \qquad \text{for } Y/Y_n > 0.008856$$

$$L^* = 903.3(Y/Y_n) \qquad \text{for } Y/Y_n \leqslant 0.008856$$

$$a^* = 500[(X/X_n)^{1/3} - (Y/Y_n)^{1/3}]$$

$$b^* = 200[(Y/Y_n)^{1/3} - (Z/Z_n)^{1/3}] \ ,$$

where X_n, Y_n, Z_n are the values of X, Y, Z, for the appropriately chosen reference white; and where, if any of the ratios X/X_n, Y/Y_n, Z/Z_n is equal to or less than 0.008856, it is replaced in the above formula by

$$7.787F + 16/116 \ ,$$

where F is X/X_n, Y/Y_n, or Z/Z_n, as the case may be. In these formulae, the reduced perceptual significance of a given difference in chromaticity caused by a reduction in luminance factor is incorporated by using the tristimulus ratios X/X_n, Y/Y_n, and Z/Z_n instead of chromaticity co-ordinates. Because these ratios of the tristimulus values are incorporated as cube-roots, there is no chromaticity diagram associated with the CIELAB space, and therefore no correlate of saturation. Correlates of hue and chroma are, however, available, and are formulated in exactly analogous ways to those used in connection with the CIELUV space:

CIE 1976 *a,b hue-angle*, h_{ab}

$$h_{ab} = \arctan(b^*/a^*)$$

CIE 1976 *a,b chroma*, C^*_{ab}

$$C^*_{ab} = (a^{*2} + b^{*2})^{1/2} \quad .$$

The CIELUV and CIELAB spaces are intended to apply to object colours of the same size and shape, viewed in identical white to mid-grey surroundings, by an observer photopically adapted to a field of chromaticity not too different from that of average daylight. If the samples considered have an angular subtense greater than 4°, then X_{10}, Y_{10}, and Z_{10} should be used instead of X, Y, and Z, and the resulting measures are then all distinguished by an additional subscript 10, such as L^*_{10}, $h_{uv,10}$, etc.

3.10 CIE COLOUR DIFFERENCE FORMULAE

If the differences between two colours in L^*, u^*, and v^* are denoted by ΔL^*, Δu^*, and Δv^*, respectively, then the total colour difference may be evaluated as:

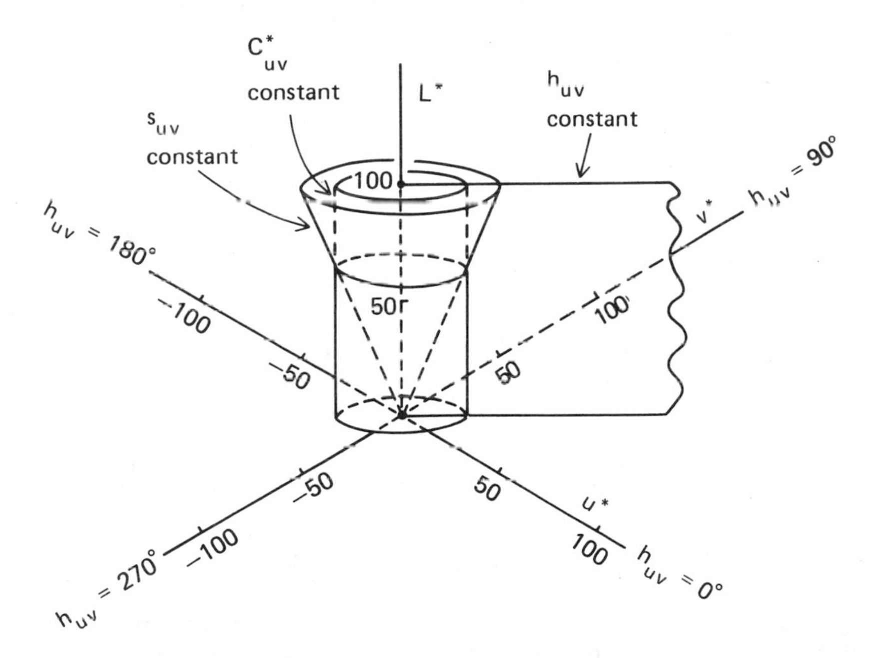

Fig. 3.9 — A three dimensional representaton of the CIELUV space. The CIELAB space is similar, except that there is no representation of saturation.

CIE 1976 ($L^*u^*v^*$) colour difference or *CIELUV colour difference*

$$\Delta E^*_{uv} = [(\Delta L^*)^2 + (\Delta u^*)^2 + (\Delta v^*)^2]^{1/2}$$

Thus ΔE^*_{uv} is equal to the distance between the two points representing the colours in the CIELUV space.

It is convenient to be able to express the same colour difference in terms of differences in L^*, C^*, and a measure that correlates with hue. To do this, h_{uv} is not used because, being an angular measure, it cannot be combined with L^* and C^* easily. What is used instead is:

CIE 1976 u,v hue-difference, ΔH^*_{uv}

$$\Delta H^*_{uv} = [(\Delta E^*_{uv})^2 - (\Delta L^*)^2 - (\Delta C^*_{uv})^2]^{1/2} \quad ,$$

where ΔC^* is the difference in C^* between the two colours. It follows that:

$$\Delta E^*_{uv} = [(\Delta L^*)^2 + (\Delta H^*_{uv})^2 + (\Delta C^*_{uv})^2]^{1/2} \quad .$$

For small colour differences away from the L^* axis, $\Delta H^*_{uv} = C^*_{uv}\Delta h_{uv}(\pi/180)$. ΔH^*_{uv} is to be regarded as positive if indicating an increase in h_{uv} and negative if indicating a decrease in h_{uv}. In different practical applications it may be necessary to use different weightings for ΔL^*, ΔC^*_{uv}, and ΔH^*_{uv}.

The following similar measures can also be evaluated in connection with the CIELAB space:

CIE 1976 ($L^*a^*b^*$) colour difference or *CIELAB colour difference*,

$$\Delta E^*_{ab} = [(\Delta L^*)^2 + (\Delta a^*)^2 + (\Delta b^*)^2]^{1/2}$$

$$\Delta E^*_{ab} = [(\Delta L^*)^2 + (\Delta H^*_{ab})^2 + (\Delta C^*_{ab})^2]^{1/2}$$

CIE 1976 *a,b hue-difference*, ΔH^*_{ab}

$$\Delta H^*_{ab} = [(\Delta E^*_{ab})^2 - (\Delta L^*)^2 - (\Delta C^*_{ab})^2]^{1/2} \quad .$$

It was unfortunate that in 1976 the CIE could not recommend a single colour space and associated colour difference formula to meet all needs. At that time, some representatives from the colorant industries argued strongly in favour of a difference formula that was similar to one known as the Adams-Nickerson (AN40) formula. The CIELAB formula fulfills this requirement, its colour difference evaluations being on average (but not uniformly throughout its space) about 1.1 times those produced by the AN40 formula. On the other hand, representatives from the television industry preferred a space having an associated chromaticity diagram because of the very simple way in which additive colour mixtures (such as occur in television display devices) are represented on it. Since 1976, evaluations of the

CIELUV and CIELAB colour spaces and difference formulae have shown that they are about equally good (or bad) in representing the perceptual sizes of colour differences, the ratios of the maximum to the minimum values of ΔE^* corresponding to a given perceptual difference being about six to one.

The CIELUV space has been used in television and the CIELAB space in the colorant industries. However, in the colorant industries, there has been a continuous search for better colour difference formulae, and, in some applications, these have now superseded the CIELAB formula. These more advanced formulae are usually non-Euclidean, in the sense that colour differences cannot be represented as simple distances in an ordinary three-dimensional space. One of these advanced spaces will be described in the next section. The CIELUV space probably represents the best that can be done with a space having an associated uniform chromaticity diagram, and it seems unlikely that it will be replaced by another space of this type.

3.11 CMC COLOUR DIFFERENCE FORMULA

A more complicated colour difference formula which has been shown to give better correlation than the CIELAB formula for small colour differences in the colorant industries is known as the CMC(l:c) formula (McLaren 1986). In this formula the colour difference is evaluated as:

$$[(\Delta L^*/lS_L)^2 + (\Delta C^*_{ab}/cS_C)^2 + (\Delta H^*_{ab}/S_H)^2]^{1/2} ,$$

where

$$S_L = 0.040975L^*/(1 + 0.01765L^*) \quad \text{unless } L^* < 16 \text{ when } S_L = 0.511$$

$$S_C = 0.0638C^*_{ab}/(1 + 0.0131C^*_{ab}) + 0.638$$

$$S_H = (fT + 1 - f)S_C ,$$

where

$$f = \{(C^*_{ab})^4/[(C^*_{ab})^4 + 1900)]\}^{1/2}$$

and

$$T = 0.36 + |0.4\cos(h_{ab} + 35)|$$

unless h_{ab} is between 164° and 345° when

$$T = 0.56 + |0.2\cos(h_{ab} + 168)| .$$

The vertical lines (| and |) enclosing some of the expressions indicate that, for these, the value is always to be taken as positive whatever the numerical result obtained by the initial calculation.

Table 3.1 — Colorimetric measures and their perceptual correlates

Non-uniform measures	Approximately uniform measures	Approximate perceptual correlates
Luminance L	None	Brightness
Chromaticity x,y	CIE 1976 chromaticity u',v'	Hue and saturation
Dominant wavelength λ_d	CIE 1976 hue-angle h_{uv} or h_{ab}	Hue
Excitation purity p_e	CIE 1976 saturation s_{uv}	Saturation
Luminance factor L/L_n	CIE 1976 lightness L^*	Lightness
None	CIE 1976 chroma C^*_{uv} or C^*_{ab}	Chroma
None	None	Colourfulness

It is clear that what this formula does is to vary the relative weightings of the contributions of the differences in L^*, C^*_{ab}, and H^*_{ab} according to the position of the colour in the CIELAB space. l and c are chosen to give the most appropriate weighting of differences in lightness and chroma, respectively, relative to differences in hue. For predicting the *perceptibility* of colour differences, l and c are both set equal to unity, and this is referred to as the CMC(1:1) formula. For predicting the *acceptability* of colour differences, it is sometimes preferable to set l and c at values greater than unity.

3.12 SUMMARY OF MEASURES AND THEIR CORRELATES

A summary of the colorimetric measures described in this Chapter, together with their perceptual correlates, is given in Table 3.1. The CIE has not yet recommended a measure that correlates approximately uniformly with brightness, or any measures that correlate with colourfulness; some unofficial measures in these categories will be discussed in Chapter 12.

The attributes, lightness and chroma, apply only to related colours, such as surface colours seen in normal surroundings. Hence, for unrelated colours, such as light sources, luminance factor, and CIE 1976 lightness and chroma, do not apply.

Sets of measures that are particularly useful in practice include:

for unrelated colours

u', v', L

for related colours

X, Y, Z, L
$u', v', Y/Y_n, L$
h_{uv}, C^*_{uv}, L^*, L
h_{ab}, C^*_{ab}, L^*, L ,

where L is the luminance.

3.13 ALLOWING FOR CHROMATIC ADAPTATION

At the end of section 3.9, it was pointed out that the CIELUV and CIELAB systems were intended for application to conditions of adaptation to fields of chromaticity not too different from daylight. For illuminants of other chromaticities, one of three alternative procedures is usually adopted to allow for chromatic adaptation to them.

First, the CIELUV and CIELAB systems are applied as they stand. This is the simplest approach, and, because the formulae in the systems are normalized for the reference white, they include an allowance for chromatic adaptation that results in the reference white (and greys of the same chromaticity) always having C^* equal to zero, related adjustments being made to the values of C^* and h for other colours. However, this procedure is only very approximately valid.

The second procedure is often referred to as a *Von Kries transformation*, after the name of its originator (Von Kries 1911). In this procedure, it is assumed that chromatic adaptation can be represented by the cone responses being multiplied (or divided) by factors that result in reference whites giving rise to the same signals in all states of adaptation. This may be expressed by saying that, if a stimulus gives rise to cone responses, ρ, γ, β, the visual signals will depend on

$$\rho/\rho_W \qquad \gamma/\gamma_W \qquad \beta/\beta_W ,$$

where ρ_W, γ_W, β_W are the cone responses for the reference white. It is convenient to apply this procedure relative to a reference state of adaptation; in this reference state, the cone responses are distinguished by a prime, thus ρ', γ', β', and ρ'_W, γ'_W, β'_W for the reference white. It then follows that, for a stimulus in the reference state of adaptation to have the same colour appearance as another stimulus in the state of adaptation considered, it is necessary that

$$\rho/\rho_W = \rho'/\rho'_W \qquad \gamma/\gamma_W = \gamma'/\gamma'_W \qquad \beta/\beta_W = \beta'/\beta'_W .$$

Such pairs of stimuli are often referred to as *corresponding colour stimuli*. It follows that:

$$\rho' = (\rho'_W/\rho_W)\rho \qquad \gamma' = (\gamma'_W/\gamma_W)\gamma \qquad \beta' = (\beta'_W/\beta_W)\beta .$$

If, then, a stimulus has tristimulus values X, Y, Z, in a state of adaptation such that the reference white has tristimulus values X_W, Y_W, Z_W, it is possible to calculate the corresponding colour stimulus, X', Y', Z', that has the same appearance in the reference state. If this reference state is a daylight, then the CIELUV and CIELAB formulae can be applied to X', Y', Z' in the usual way. To calculate X', Y', Z' it is necessary to know the tristimulus values X'_W, Y'_W, Z'_W of the reference white in the reference state, and to be able to transform X,Y,Z tristimulus values to ρ, γ, β cone responses, and vice versa. These transformations are achieved by using a set of transformation equations of which the following are an example:

$$\rho = 0.40024X + 0.70760Y - 0.08081Z$$

$$\gamma = -0.22630X + 1.16532Y + 0.04570Z$$

$$\beta = 0.91822Z \ ,$$

the reverse equations being:

$$X = 1.85995\rho - 1.12939\gamma + 0.21990\beta$$

$$Y = 0.36119\rho + 0.63881\gamma$$

$$Z = 1.08906\beta \ .$$

The steps involved in this procedure are then as follows:

Step 1. From X'_W, Y'_W, Z'_W calculate ρ'_W, γ'_W, β'_W.

Step 2. From X_W, Y_W, Z_W calculate ρ_W, γ_W, β_W.

Step 3. Calculate ρ'_W/ρ_W, γ'_W/γ_W, β'_W/β_W.

Step 4. For the test colour, from X, Y, Z, calculate ρ, γ, β.

Step 5. Calculate $\rho' = (\rho'_W/\rho_W)\rho$, $\gamma' = (\gamma'_W/\gamma_W)\gamma$, $\beta' = (\beta'_W/\beta_W)\beta$.

Step 6. From ρ', γ', β' calculate X', Y', Z'.

Step 7. Use X', Y', Z' in the chosen colour difference formula.

Although Von Kries transformations are often very useful, they are not always accurate; the third procedure, although similar to the one just described, and using the same or similar transformation equations, involves more complicated formulae for Step 5. One example of this type of procedure is given in Chapter 12; another, based on the work of Takahama, Sobagaki, and Nayatani, and which has been published by the CIE, is given in Table 3.2 (Takahama, Sobagaki & Nayatani 1984; CIE, *CIE Journal*, 1986a). These formulae depend on the illuminances, as well as on the chromaticities, of the adapting fields.

Table 3.2 — A procedure for allowing for chromatic adaptation

The following formulae may be used instead of those given in Step 5 of section 3.13.

$$\rho' = [100R'a' + 1][(\rho + 1)/(100Ra + 1)]^d - 1$$
$$\gamma' = [100R'b' + 1][(\gamma + 1)/(100Rb + 1)]^e - 1$$
$$\beta' = [100R'c' + 1][(\beta + 1)/(100Rc + 1)]^f - 1$$

R and R' are the luminous reflectances of spectrally non-selective surrounds in the test and reference fields respectively (R and R' should be equal and between 0.2 and 1.0).

$$a = (\ \ 0.48105x + 0.78841y - 0.08081)/y$$
$$b = (-0.27200x + 1.11962y - 0.04570)/y$$
$$c = \ \ 0.91822(1 - x - y)/y$$

where x, y are the chromaticities of the test field illuminant. a', b', c' are similarly calculated from x', y', the chromaticities of the reference field illuminant.

$$d = \frac{(6.469 + 6.362\rho_0^{0.4495})/(6.469 + 1.000\rho_0^{0.4495})}{(6.469 + 6.362(\rho'_0)^{0.4495})/(6.469 + 1.000(\rho'_0)^{0.4495})}$$

$$e = \frac{(6.469 + 6.362\gamma_0^{0.4495})/(6.469 + 1.000\gamma_0^{0.4495})}{(6.469 + 6.362(\gamma'_0)^{0.4495})/(6.469 + 1.000(\gamma'_0)^{0.4495})}$$

$$f = \frac{(8.414 + 8.091\beta_0^{0.5128})/(8.414 + 1.000\beta_0^{0.5128})}{(8.414 + 8.091(\beta'_0)^{0.5128})/(8.414 + 1.000(\beta'_0)^{0.5128})}$$

where

$$\rho_0 = aRE/\pi \qquad \rho'_0 = a'R'E'/\pi$$
$$\gamma_0 = bRE/\pi \qquad \gamma'_0 = b'R'E'/\pi$$
$$\beta_0 = cRE/\pi \qquad \beta'_0 = c'R'E'/\pi$$

and E and E' are the illuminances in lux of the test and reference fields, respectively. The procedure may be used for any combination of E and E' when R and R' are equal and between 0.2 and 0.3. However, when R and R' are equal and are greater than 0.3, then the procedure should only be used when E is equal to E'.

REFERENCES

CIE *CIE Journal*, **5**, 16 (1986a).

McLaren, K. *The Colour Science of Dyes and Pigments*, 2nd ed., p.143, Hilger, Bristol (1986).

Takahama, K., Sobagaki, H. & Nayatani, Y. *Color Res. Appl.*, **9**, 106 (1984).

Von Kries, J. A., in *Handbuch der Physiologisches Optik*, Vol. II (W.Nagel, ed), pp. 366–369, Leopold Voss, Hamburg (1911).

Wright, W.D. *Proc. Phys. Soc.*, **53**, 99 (1941).

GENERAL REFERENCES

CIE Publication No. 15.2, *Colorimetry*, 2nd ed. (1986).

CIE Supplement No. 2 to Publication No. 15, *Recommendations on uniform colour spaces, colour difference equations, and psychometric terms* (1978).

4

Light sources

4.1 INTRODUCTION

Without light there is no colour. Light sources therefore play a very important part in colorimetry. If the colour is self-luminous, such as in the case of fireworks, for example, then the light source itself is the colour. But, more often, colours are associated with objects that, instead of being self-emitting, reflect or transmit light emitted by light sources. That the nature of these light sources can have a profound effect on the appearance of coloured objects is a well-known experience. For instance, to take an extreme example, objects that normally appear red, become dark brown or black when seen under the yellow low-pressure sodium lamps that are widely used in street lighting. It is also well known that, compared to their appearance in daylight, red and yellow objects seen under tungsten filament light, or candle light, look lighter and more colourful, while blue objects look darker and less colourful. If the visual system did not adapt to the level and colour of the prevailing illumination, these changes would be even greater (see section 1.8; a quantitative treatment of adaptation will be given in Chapter 12).

We have already seen that light sources are involved in colorimetry in various ways. First, for reflecting and transmitting objects, the Y tristimulus value is usually evaluated so that $Y = 100$ for the perfect reflecting diffuser, or the perfect transmitter, similarly illuminated and viewed (section 2.6); the similar illumination implies the use of a particular light source. Second, the evaluation of dominant wavelength and excitation purity requires the adoption of an achromatic stimulus, and this is usually taken as the perfect diffuser or transmitter illuminated by a particular light source (section 3.4). Third, the variables in the CIELUV and CIELAB systems require the use of a reference white, and this is normally defined in terms of a specified white or a highly transmitting object illuminated by a particular light source (section 3.9). Fourth, pairs of colours that are a match under one light source may not match under another having a different spectral composition (section 2.7). Fifth, samples that fluoresce will usually vary in colour as light sources of different spectral compositions are used, particularly when variations in the ultraviolet content of the light sources occur (to be discussed in Chapter 9). Thus, in all these cases, the colour specifications relate to a particular light source. In fact, for reflecting and transmitting objects, X, Y, and Z tristimulus values, and any colorimetric measures derived from them, are meaningless unless the light source used is also specified.

It is therefore clear that the proper specification of light sources is an essential part of colorimetry. But there are many different light sources, and their choice and specification therefore warrant this chapter devoted to them.

4.2 METHODS OF PRODUCING LIGHT

Light can be produced by a variety of methods. These include:

Incandescence: solids and liquids emit light when their temperatures are above about 1000 K (where K indicates the temperature in Kelvin, which is equal to the Celsius temperature plus 273).

Gas discharges: gases can emit light when an electric current passes through them; the spectral distribution of the light is characteristic of the elements present in the gas.

Electroluminescence: certain solids, such as semiconductors or phosphors, can emit light when an electric current is passed through them.

Photoluminescence: when radiation is absorbed by some substances, light is emitted with a change of wavelength; if the emission is immediate it is termed *fluorescence*; if the emission continues appreciably after the absorbing radiation is removed, it is termed *phosphorescence*.

Cathodoluminescence: phosphors emit light when they are bombarded with electrons.

Chemiluminescence: certain chemical reactions result in light being emitted without necessarily generating heat.

Of the above methods, the most widely occurring in light sources are incandescence (in the Sun, and in tungsten filament lamps), gas discharges (in sodium and mercury street lamps, in fluorescent lamps, in mercury-based lamps for stadia and studios, and to some extent on the surface of the Sun), photoluminescence (in fluorescent lamps), and cathodoluminescence (in cathode ray tubes, as used in oscilloscopes and in many television and Video Display Unit, VDU, displays).

4.3 GAS DISCHARGES

Gases are normally insulators, so that electric currents cannot pass through them. But, in some circumstances, their insulating properties break down, a current passes, and, when it does, light may be emitted. One example of this is when a switch in an electrical circuit is opened: frequently a spark can be seen, and this is caused by an electric current being able to flow through the air across the very small gap between the contacts of the switch. Another example is lightning; in this case, a current flows because of very high voltages produced by friction in clouds, with moisture in the air facilitating the conduction. But to produce light in gases continually in a controlled manner, it is necessary to increase their conductivity, and one way of doing this is by having them at low pressures. When the pressure of a gas is low, its atoms are less concentrated, and the passage of electric current may then be facilitated.

When electrodes are used to produce an electric field across a gas, a current will still not flow, even with the gas at low pressure, unless some of its molecules are broken into electrons and ions; this ionization can be achieved in various ways, such as by having glowing filaments at each end of the electric field. With some electrons and ions present, a high voltage can then start a current to flow, and the consequent movement of the electrons and ions causes collisions which produce more ionization, so that the current continues to increase. This means that suitable circuits have to provide a high starting voltage and a current limiting device, the latter often being in the form of a choke (a coil having a high impedance for alternating current).

Light is produced in a gas discharge as a result of electrons colliding with atoms of the gas and exciting them to energy states that are higher than their ground states; the excited atoms then return to their ground states and, in doing so, emit light. The light emitted by this process has the very distinctive property of being confined to narrow lines in the spectrum.

Lamps producing light by the process of gas discharge contain the gas in a transparent envelope of glass, fused silica, or other ceramic, usually in the form of a tube, with the electrodes at the two ends.

4.4 SODIUM LAMPS

When low-pressure sodium is used for the gas in a gas-discharge lamp, the visible spectrum consists almost entirely of two spectral lines at wavelengths 589.0 and 589.6 nm. This spectral power distribution is shown in Fig. 4.1; the lines at about 770 nm are too far towards the infrared to have any appreciable visual effect. The theoretical maximum efficacy for any radiation is 683 lumens per watt (lm/W), and this occurs only for monochromatic radiations of wavelength 555 nm, this wavelength being at the maximum of the $V(\lambda)$ function (section 2.3). The maximum possible efficacy for light of wavelength 589 nm is 525 (lm/W), but in practice low-pressure sodium lamps have efficacies of only about 150 (lm/W), because of wastage

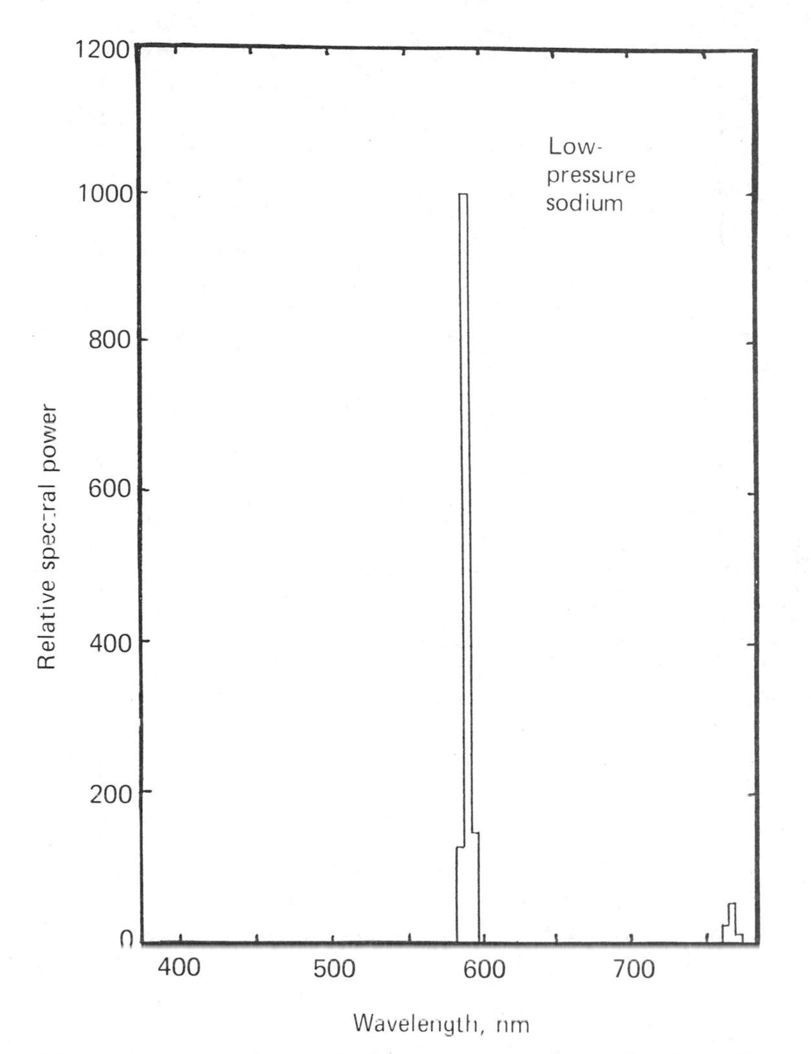

Fig. 4.1 — Relative spectral power distribution of a low pressure sodium lamp. The power is represented as a histogram in blocks of wavelengths 5 nm wide.

of power in operating the lamps. However, this efficacy is better than for all other types of lamp, and low-pressure sodium lamps are therefore widely used in street lighting, where high efficacy is very important; however, the colour rendering of these lamps is extremely poor. The relative spectral power distribution of a typical low pressure sodium lamp is given in Appendix 5.6

If the pressure of the sodium gas in the envelope is increased, the proximity of neighbouring atoms affects their energy levels and the lines of emission become broader; in addition to this effect, the sodium gas in the tube is sufficiently concentrated to absorb the primary radiation at 589.0 and 589.6 nm, with the result that the spectral power distribution is as shown in Fig. 4.2. Although this type of

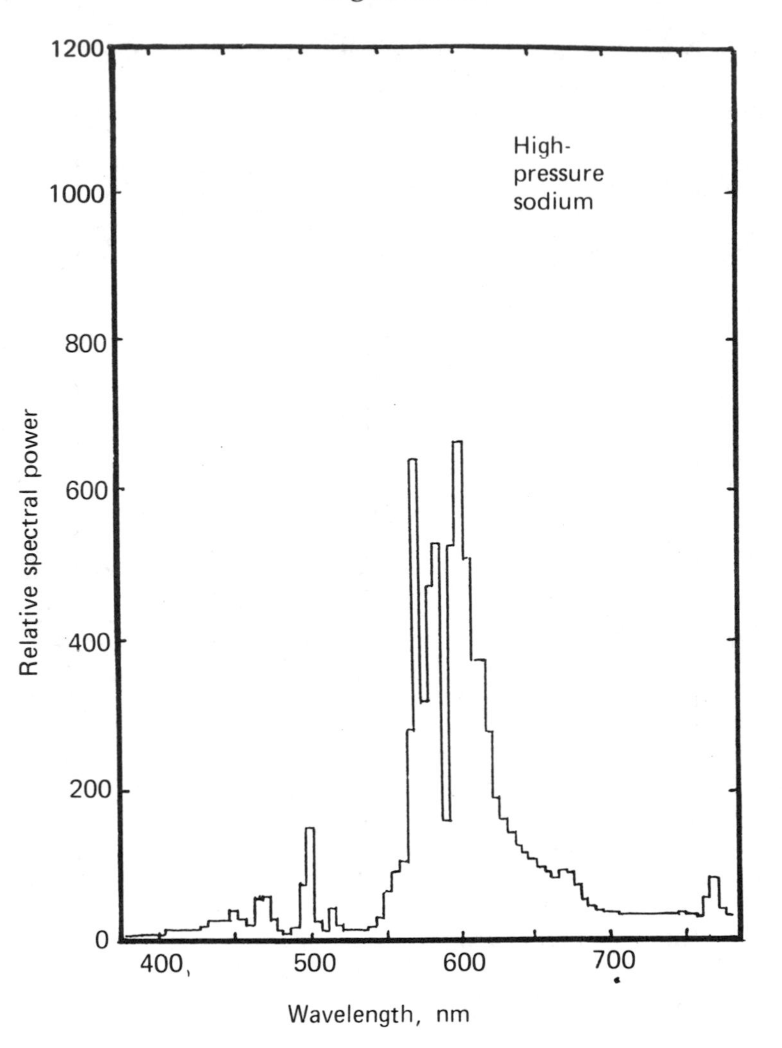

Fig. 4.2 — Same as Fig. 4.1, but for a high-pressure sodium lamp.

spectral power distribution is still far from giving good colour rendering, it is much improved over that of the low-pressure lamp, and is very useful for floodlighting, and for street lighting in shopping areas, for instance; the colour of the light appears a pleasant pale pink-orange, instead of the deep yellow of the low pressure lamps. These high-pressure lamps have an efficacy of about 100 lm/W. The relative spectral power distribution of a typical high-pressure sodium lamp is given in Appendix 5.6.

4.5 MERCURY LAMPS

When low-pressure mercury is used for the gas in a gas-discharge lamp, the spectrum consists mainly of a series of lines, the more prominent of which are at wavelengths of

253.7, 365.4, 404.7, 435.8, 546.1, and 578.0 nm, together with a very-low-level continuous radiation throughout the spectrum. If the pressure is increased, the lines tend to broaden and the continuum tends to represent a greater proportion of the light output. The absence of lines at longer wavelengths than 578.0 nm results in the typical blue-green colour appearance of mercury lamps; and the absence of light at the long wavelength end of the spectrum results in the colour rendering being very poor, especially of Caucasian skin tones. The radiation at 253.7 nm is in the ultraviolet and cannot be seen, but is used in fluorescent lamps to excite phosphors that emit in the visible part of the spectrum (this will be discussed in section 4.6; the line at 365.4 can also be used to excite some phosphors). In Fig. 4.3, the spectral

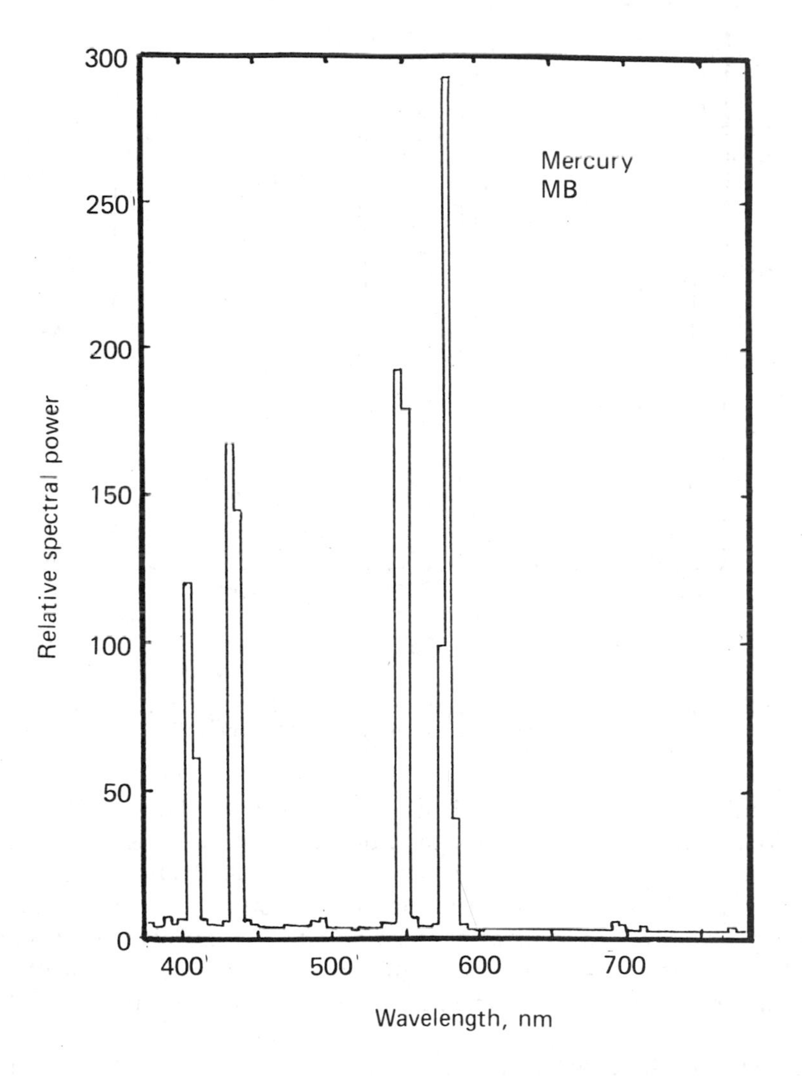

Fig. 4.3 — Same as Fig. 4.1, but for a high-pressure mercury lamp type MB.

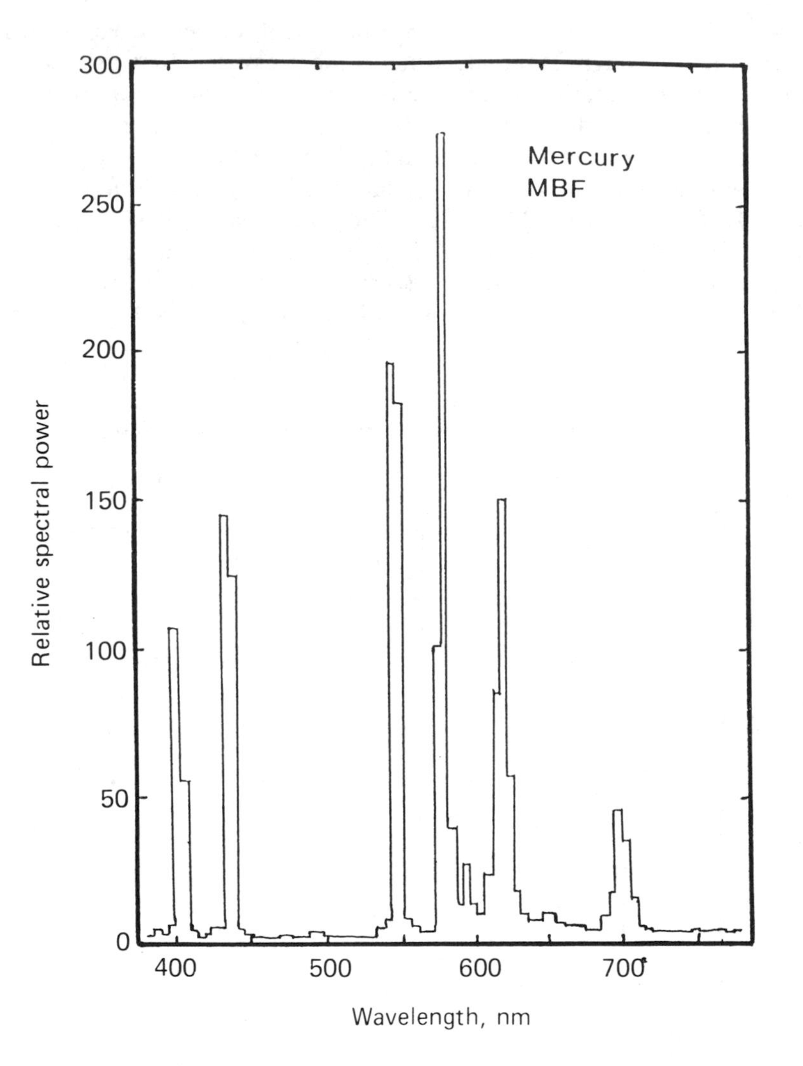

Fig. 4.4 — Same as Fig. 4.3, but for a high-pressure mercury lamp type MBF, which has a red-emitting phosphor coated on the inside of the envelope.

power distribution is shown for a typical high pressure mercury lamp (MB); Fig. 4.4 shows that for a high pressure mercury lamp with a red-emitting phosphor coated on the inside of the envelope (MBF); and Fig. 4.5 that for a similar lamp using a tungsten filament ballast (MBTF). The red-emitting phosphor improves the colour rendering of the lamp appreciably. The relative spectral power distributions of typical lamps of the types shown in Figs 4.3, 4.4, and 4.5 are given in Appendix 5.6. Typical efficacies range from about 50 lm/W (for MB and MBF lamps) to about 20 lm/W (for MBTF lamps).

By adding metal halides to the mercury vapour, extra lines can be provided in the spectrum, and the colour rendering can then be improved until it is comparable with

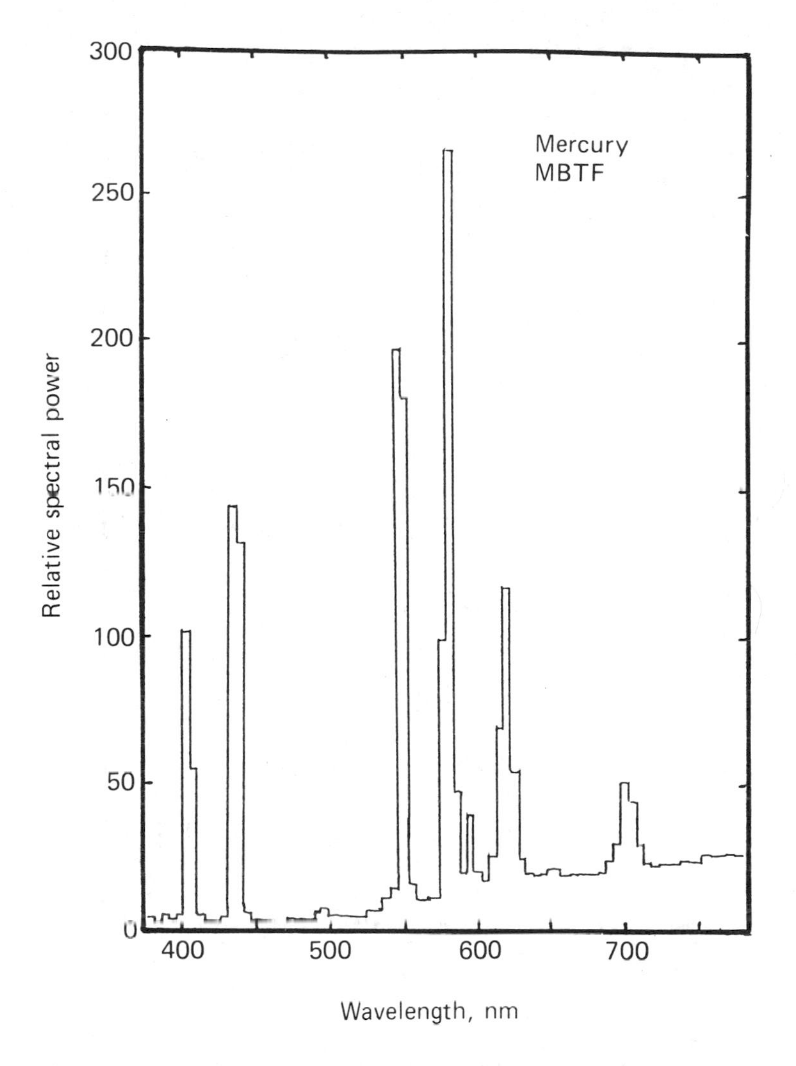

Fig. 4.5 — Same as Fig. 4.4, but for a high-pressure mercury lamp type MBTF, which is an MBF lamp with a tungsten filament ballast.

that from large-area, lower-luminance, types, such as fluorescent lamps (to be discussed in section 4.6). These metal-halide discharge lamps are useful for flood-lighting sports stadia, because, being compact, they can be used in reflectors so that the light is beamed towards the scene of the action. Lamps can be made that give light of various colours, including the range from daylight to tungsten filament light. One important member of this series of lamps is the HMI lamp, which was developed for use in television to supplement daylight. In Fig. 4.6 the spectral power distribution of the HMI lamp is shown, and corresponding relative numerical data are given in Appendix 5.6. Typical efficacies for metal halide mercury lamps are around 80 lm/W.

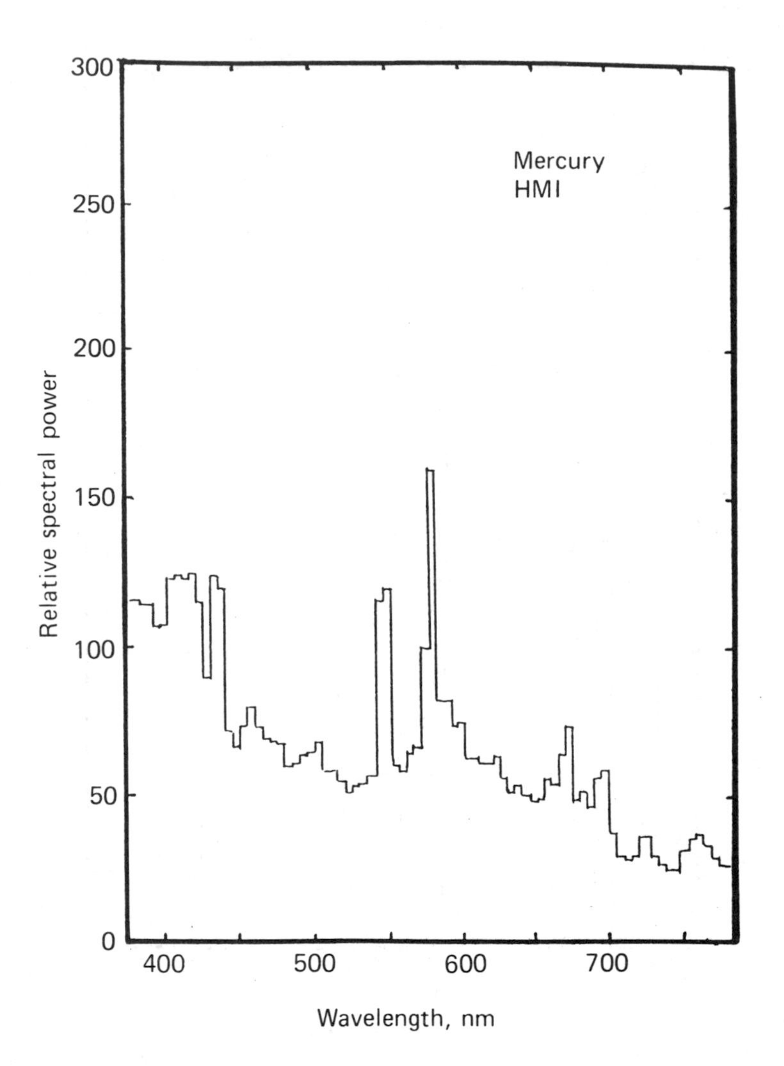

Fig. 4.6 — Same as Fig. 4.3, but for an HMI (mercury, medium arc, iodides) lamp, which has iodides added to the high-pressure mercury gas, and is used for supplementing daylight for television camera shooting.

4.6 FLUORESCENT LAMPS

Fluorescent lamps are very widely used for general lighting, especially in industrial and commercial environments where high levels of illumination, high efficacies, and good colour rendering are required. These lamps consist of a glass tube containing low-pressure mercury gas, in which a gas-discharge is produced; and the inside of the tube is coated with phosphors that are excited by the ultra-violet lines of the mercury spectrum, particularly that at 253.7 nm, to produce additional light. The phosphors are carefully chosen to supplement the light from the gas-discharge, special attention

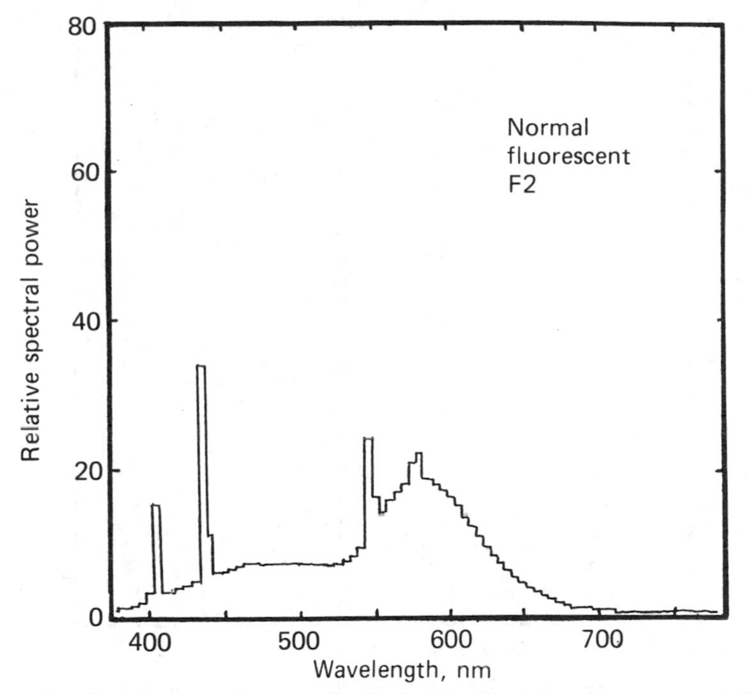

Fig. 4.7 — Relative spectral power distribution of a fluorescent lamp representative of the *normal* type (F2). The power is represented as a histogram in blocks of wavelengths 5 nm wide.

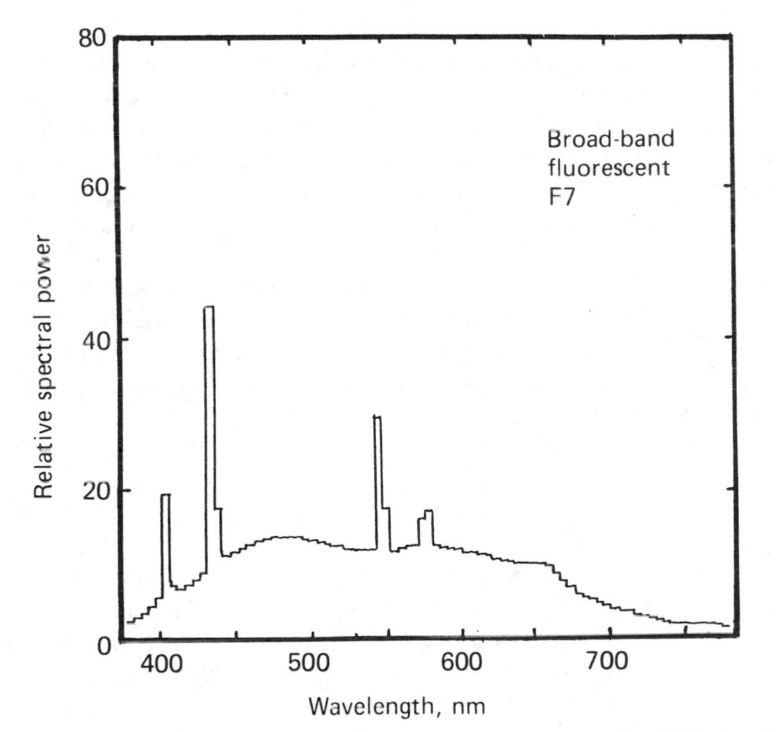

Fig. 4.8 — Same as Fig. 4.7, but for a fluorescent lamp representative of the *broad-band* type (F7).

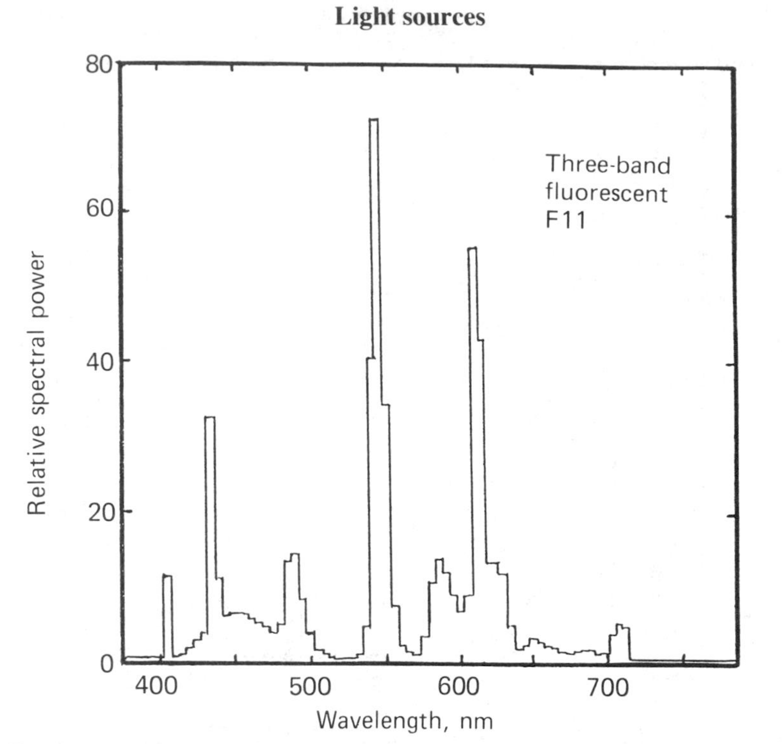

Fig. 4.9 — Same as Fig. 4.7, but for a fluorescent lamp representative of the *three-band* type (F11).

being paid to the need to provide additional light in the long wavelength parts of the spectrum where there are no lines. Thus the light from these lamps comes partly from the gas-discharge, but mainly from the phosphors. There is a wide range of different types of fluorescent lamp, according to the types of phosphors used; these vary from those having high efficacy but poor colour rendering, to those having lower efficacy but good colour rendering. Typical efficacies range from about 45 to about 95 lm/W.

In Appendix 5.3, spectral power distributions are given for twelve different types of fluorescent lamp, and these have been chosen to illustrate some of the main groups. The first six are designated *normal*; they have been designed to have a reasonably high efficacy, and their colour rendering is adequate for many purposes; but it is not very good for reddish colours, because of some deficiency in emission at the long wavelength end of the spectrum.

The second group is designated *broad-band*; they have been designed to have very good colour rendering, and their emission at the long wavelength end of the spectrum is appreciably greater than in the case of the normal group; however, their efficacy is lower.

The third group is designated *three-band*. As their name implies, their emission tends to be concentrated in three bands of the spectrum, and these bands are quite narrow, and are designed to occur around wavelengths of approximately 610, 545, and 435 nm. Lamps in this group tend to have relatively high efficacies and fairly good colour rendering. They tend to increase the saturation of most colours, and this makes them attractive for some purposes, such as lighting goods in stores; but the

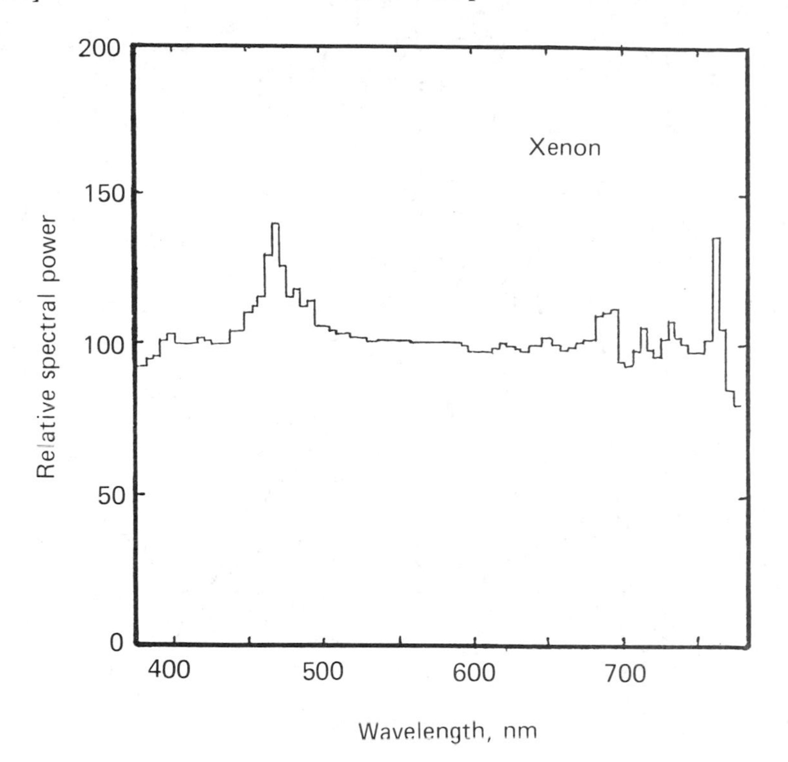

Fig. 4.10 Relative spectral power distribution typical of a xenon lamp. The power is represented as a histogram in blocks of wavelengths 5 nm wide.

appearance of some colours can be somewhat distorted, so that they are less suitable for critical evaluation of colours in general.

There are no CIE standard illuminants representing fluorescent lamps, but three of the distributions given in Appendix 5.3, F2, F7, and F11, one from each group, are chosen as deserving priority over the others when a few typical illuminants are to be selected. These three distributions are illustrated in Figs 4.7, 4.8, and 4.9.

Compact fluorescent lamps with tubes about 10 cm long, often in a U-shape configuration, are available as alternatives to tungsten filament lamps; their higher cost is offset by their greater efficacy and long life.

4.7 XENON LAMPS

An electric current can be made to pass through xenon gas by using a high-voltage pulse to cause ionization. The electrodes in the gas can either be separated by only a few millimetres (compact source xenon), or by about 10 centimetres or more (linear xenon). The operation can be either continuous, or in the form of a series of short pulses at each half-cycle of mains frequency (pulsed xenon), or as isolated flashes

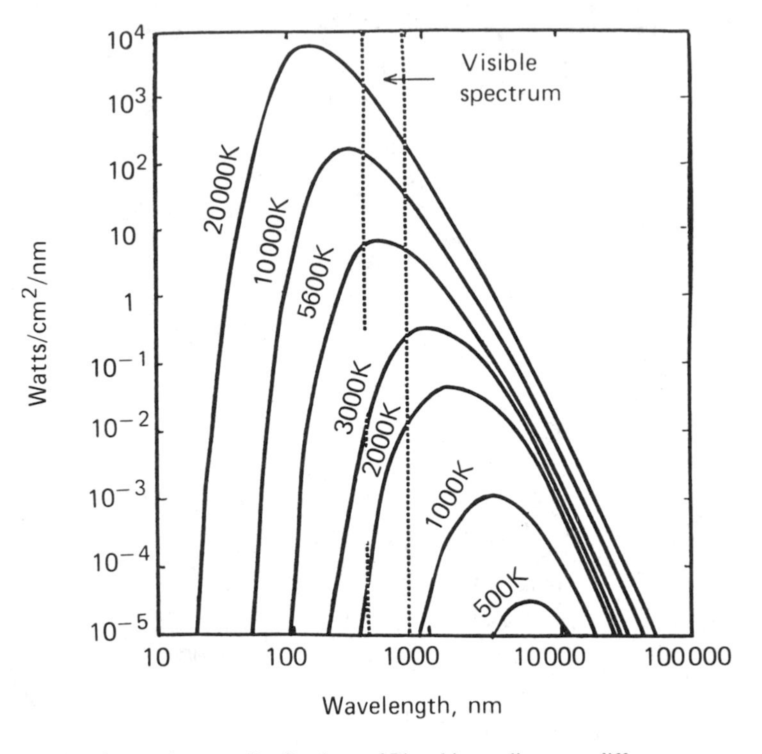

Fig. 4.11 — Spectral power distributions of Planckian radiators at different temperatures.

(xenon flash tubes). The cold-filling pressure of the gas varies from about ten atmospheres for some compact source lamps, to about a tenth of an atmosphere for linear lamps.

At low pressures, the xenon emission spectrum is comprised of many discrete lines, but, at the pressures used in xenon lamps, the lines broaden to provide the type of continuum shown in Fig. 4.10. The exact spectral power distribution depends on the pressure of the gas and on other operating characteristics of the lamp. The colour of the light emitted is quite similar to that of average daylight, but slightly more purple because of relatively greater emissions at the two ends of the spectrum. The relative spectral power distribution of a typical xenon lamp is given in Appendix 5.6. Efficacies for Xenon lamps range from about 20 to about 50 lm/W.

Compact-source xenon lamps are used in film projectors in cinemas, and in lighthouses; continuously run and pulsed xenon lamps are used for floodlighting, for accelerated fading tests, and for copy-board lighting in the graphic arts industry; xenon flash tubes are used extremely widely for flash photography. The near-daylight quality of the spectral power distribution makes xenon lamps a useful source for artificial daylight in colorimetry, and this is especially so for fluorescent materials because of their adequate emission in the ultraviolet (to be discussed in Chapter 9).

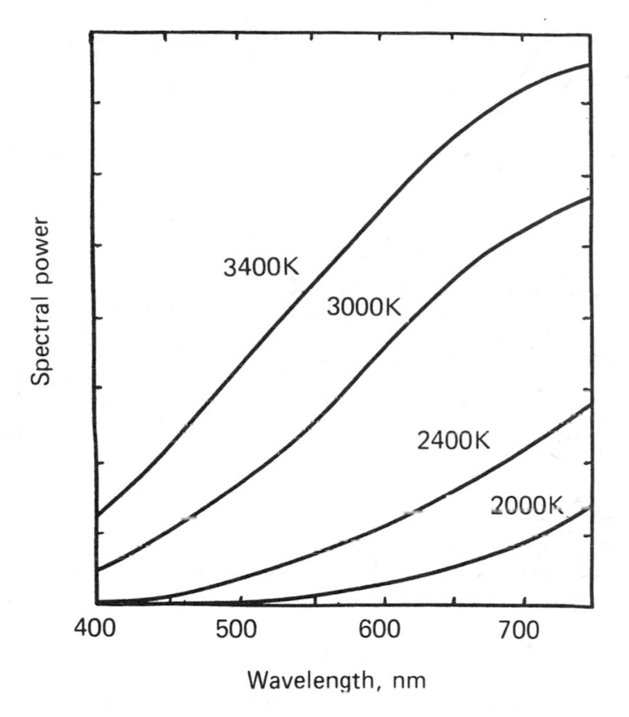

Fig. 4.12 — Spectral power distributions of Planckian radiators at four temperatures.

4.8 INCANDESCENT LIGHT SOURCES

We have already seen that, in gas discharges, when the gas pressure is raised, the spectral lines in the emission are broadened because of the effects of the neighbouring atoms in the gas. In solids and liquids, the atoms are much more closely packed than in gases, and the line-broadening effects are usually so great that the emission is continuous throughout the spectrum. Solids and liquids can be excited to produce light in various ways, including the passage of an electric current (electroluminescence), exposure to radiation (photoluminescence), bombardment with electrons (cathodoluminescence), and chemical reaction (chemiluminescence); but by far the most common and important way is by heating them to a temperature above about 1000 K (incandescence). The most important natural incandescent light source is the Sun. The most important man-made incandescent source is the tungsten filament lamp; but sources depending, at least partly, on incandescence that have been important in the past include *limelight* (calcium oxide heated by coal gas or hydrogen), gas mantles (meshes of thorium heated by combustible gases), and carbon arcs (carbon electrodes heated by arcs in air across small gaps).

The amount of power radiated by incandescent sources depends on their temperature — the higher the temperature, the more the power radiated; this is illustrated in Fig. 4.11. It can also be seen from this figure that, at temperatures below about 5600 K, more power is radiated at the long, than at the short, wavelength end

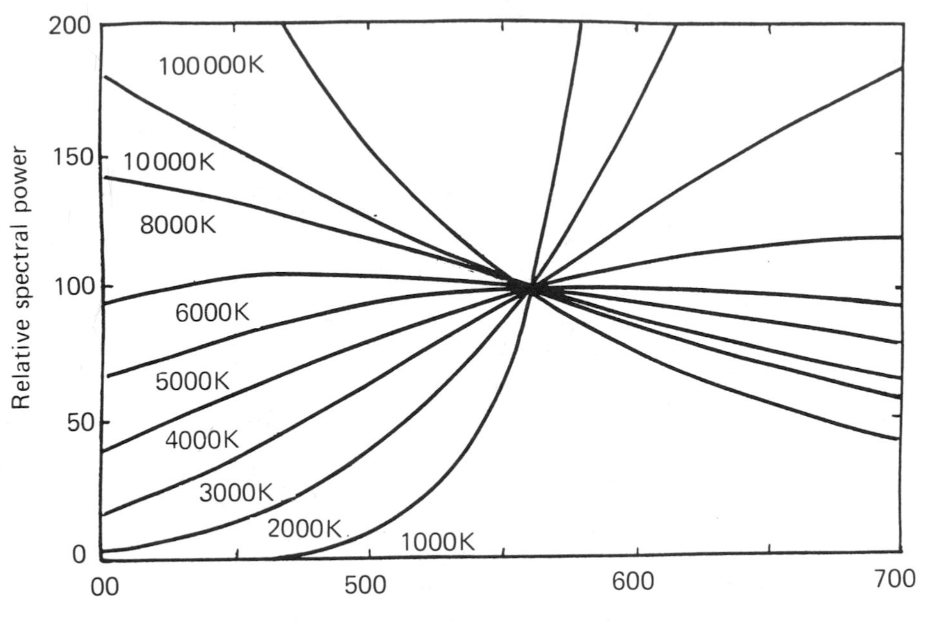

Fig. 4.13 — Relative spectral power distributions of Planckian radiators at various temperatures, normalized at 560 nm.

of the visible spectrum; and at temperatures above about 5600 K more power is radiated at the short, than at the long, wavelength end. In Fig. 4.12, data are shown for the visible spectrum only, for temperatures of 2000, 2400. 3000, and 3400 K (a range of values of particular importance for tungsten filament lamps); this figure shows that, as the temperature is raised from 2000 to 3400 K, more light is emitted, and the ratio of short to long wavelength light gradually increases. In Fig. 4.13, data for the visible spectrum are shown normalized for a wavelength of 560 nm; it can be seen that the amount of light from the two ends of the spectrum becomes similar when the temperature is between about 5000 K and 6000 K.

Although, for incandescent sources, their temperature is the most important factor determining the magnitude and spectral composition of the light they radiate, various materials do show some differences in their facility to radiate, both generally and at different wavelengths. These differences in *emissivity* modify the magnitude and the spectral composition of the radiation somewhat, but there is one class of light sources for which the nature of the radiation depends only on their temperature. They consist of heated enclosures, whose radiation escapes through an opening whose area is small compared to that of the interior surface of the enclosure; they are important theoretical sources, and in the past have been variously referred to as *black-body radiators*, or *full radiators*; but they are now generally referred to as *Planckian radiators*, after Max Planck, who originated a formula for deriving the power radiated at each wavelength from a knowledge of the temperature of the source. Planck's formula is as follows:

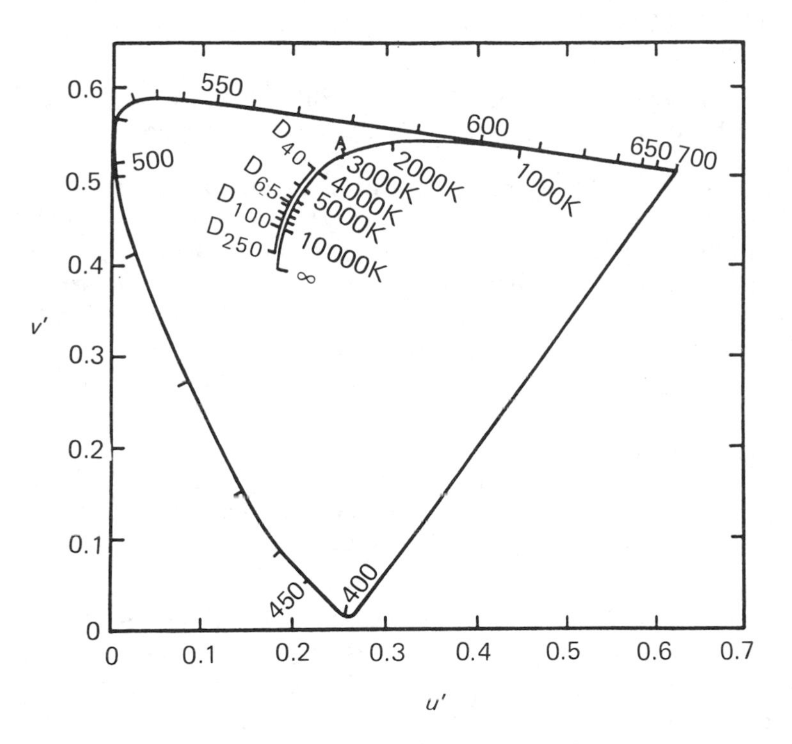

Fig. 4.14 — The loci of the chromaticities of Planckian radiators (the *Planckian locus*) and CIE D Illuminants (the *daylight locus*) in the u',v' diagram.

$$M_e = c_1\lambda^{-5}(e^{c_2/\lambda T} - 1)^{-1}$$

where M_e is the spectral concentration of radiant exitance, in watts per square metre per wavelength interval ($W.m^{-3}$), as a function of wavelength, λ, in metres, and temperature, T, in kelvins, and where $c_1 = 3.74183 \times 10^{-16}$ $W.m^2$ and $c_2 = 1.4388 \times 10^{-2}$ m.K; and $e = 2.718282$.

The unit used for the temperature of Planckian radiators is always the *kelvin*, K (which, as already mentioned, is equal to the Celsius temperature plus 273). In Fig. 4.14 the locus of chromaticities of Planckian radiators, the *Planckian locus*, is shown in the u',v' diagram. Relative spectral power distributions and chromaticity coordinates for a selection of temperatures are given in Appendix 5.5.

Planckian radiators are not usually available as practical sources, but many incandescent sources emit light whose spectral composition or colour bears a particular relationship to a Planckian radiator. A source whose emissivity is constant with wavelength, but less than that of a Planckian radiator, has the same spectral power distribution, but at a lower level. It is convenient, in this case, to characterize the spectral power distribution of the radiation by quoting the temperature of the Planckian radiator having the same relative spectral power distribution, and this is called the *distribution temperature* of the radiation.

Distribution temperature, T_D

> The temperature of a Planckian radiator whose relative spectral power distribution is the same as that of the radiation considered.

Many sources have relative spectral power distributions that are not quite the same as those of Planckian radiators, but are very similar. Tungsten filament lamps fall into this category. Such stimuli usually have chromaticities that are on the Planckian locus, and it is convenient, in this case, to characterize the colour of the stimulus by quoting the temperature of the Planckian radiator having the same chromaticity, and this is called the *colour temperature*.

Colour temperature, T_c

> The temperature of a Planckian radiator whose radiation has the same chromaticity as that of a given stimulus.

For stimuli whose chromaticities are near, but not exactly on, the Planckian locus, it is useful to characterize the colour of the stimulus by quoting the temperature of the Planckian radiator having the most similar perceived colour, and this is called the *correlated colour temperature*.

Correlated colour temperature, T_{cp}

> The temperature of the Planckian radiator whose perceived colour most closely resembles that of a given stimulus seen at the same brightness and under specified viewing conditions. The recommended method of calculating the correlated colour temperature of a stimulus is to determine on the u,v (not the u',v') chromaticity diagram the temperature corresponding to the point on the locus of Planckian radiators that is nearest to the point representing the stimulus.

The u,v diagram (as described in section 3.6) is obtained by plotting:

$$u = 4X/(X + 15Y + 3Z) = u'$$
$$v = 6Y/(X + 15Y + 3Z) = (2/3)v' \ .$$

The u,v diagram is used instead of the u',v' diagram only for historical reasons. Correlated colour temperatures are used in calculating CIE Colour Rendering Indices (to be described in section 4.17), and these were introduced before the u',v' diagram was available; changing to the u',v' diagram would involve changes to Colour Rendering Indices, and the consequent inconvenience has so far prevented this change being made. The concept of correlated colour temperature is valid only for stimuli whose chromaticities are reasonably close to the Planckian locus; no formal limits of closeness have been specified by the CIE, but accepted practice is to quote correlated colour temperatures for commonly used 'white' light sources, including tungsten lamps, tungsten halogen lamps, fluorescent lamps, and the various aspects of daylight. A program for computing correlated colour temperatures from chromaticities has been published (Krystek 1985).

4.9 TUNGSTEN LAMPS

The simplest way of providing an incandescent source is to heat a conductor by passing an electric current through it. If this is done in air, when the temperature is high enough to provide a reasonable amount of light, rapid oxidation usually occurs and the material soon burns away. But, if the conductor is enclosed in a glass envelope that contains a vacuum or only inert gases, then a stable and convenient light source can be obtained. The earliest lamps of this type used carbon filaments in vacuum; but, although carbon has a high emissivity, it tends to evaporate and produce a dark deposit on the inside of the glass envelope.

The most suitable material found so far for the filament is tungsten. The spectral emissivity of tungsten is nearly constant in the visible spectrum, but decreases slightly with increasing wavelength; consequently, the spectral power distributions of tungsten filaments are almost identical to those of Planckian radiators whose temperatures are about 50 K greater than those of the corresponding filaments. The melting point of tungsten is 3410°C, that is 3683 K, and the filament must obviously be run below this temperature. It is desirable to operate at as high a temperature as possible, because then, as can be seen from Fig. 4.11, the amount of light emitted increases, and the ratio of short, relative to long, wavelength light becomes less imbalanced; but higher temperatures tend to result in earlier failure of lamps because of greater evaporation of tungsten from the filaments leading to their breakage. This evaporation can be reduced by filling the envelope with inert gases, and these *gas-filled* lamps can be run at higher temperatures for a given life, as compared to vacuum lamps. One disadvantage of the gas filling is that it provides a means for heat being removed from the filament, thus reducing the efficacy. However, this can be minimized by having the filament as compact as possible by making it in the form of a tight coil, which is then often made into a larger *coiled coil*. A reasonable life can be obtained for lamps with thick filaments (about a fifth of a millimetre in diameter) operating at colour temperatures of about 3200 K; but with thin filaments (about a fiftieth of a millimetre in diameter) colour temperatures of only about 2500 K can be achieved because of the fragility of the filament and the serious weakening resulting from any evaporation from it. Thick filaments result in low resistance, and hence can be used only with low voltage or high wattage lamps. Typical efficacies for tungsten filament lamps are about 25 lm/W for lamps operating at colour temperatures of 3200 K, and about 10 lm/W at 2650 K.

4.10 TUNGSTEN-HALOGEN LAMPS

In an ordinary tungsten lamp, tungsten gradually evaporates from the filament (thus weakening it) and forms a dark deposit on the glass envelope. A tungsten-halogen lamp is made of pure fused silica (often referred to as 'quartz'), and is much smaller than in an ordinary lamp; it therefore attains a much higher temperature when the lamp is running. The envelope contains a halogen gas, and, when tungsten evaporates from the filament, it combines with the halogen gas at the hot envelope wall to form a tungsten halide (a gas). This gas then migrates back to the filament, where it decomposes to deposit the tungsten back on to the filament and the halide back into the mixture of gases in the envelope. The halide is thus available for further use, and can provide a continuous cycle of combination with, and dissociation from, tungsten from the filament. Thus dark deposits on the envelope are avoided. The compact shape of the lamp results in the gas pressure being higher than in a conventional lamp, and this reduces the evaporation of tungsten, and hence the filament is weakened less. This means that the lamp can be run at a higher colour temperature with better efficacy. Tungsten-halogen lamps can be run at colour temperatures in the range from about 2850 K to about 3300K and have efficacies from about 15 to about 35 lumens per watt.

The compact nature of the tungsten-halogen lamp makes it very suitable for use with mirrors and lenses in optical systems, and it is very widely used in slide projectors and overhead projectors; these projector lamps generally operate on a supply of less than 30 volts in order to keep them as compact as possible. Low voltage tungsten-halogen lamps are also sometimes used for automobile headlamps. Mains voltage tungsten-halogen lamps are used for floodlighting, and for studio lighting for television and filming.

4.11 DAYLIGHT

All daylight originates from the Sun, the temperature of which is millions of degrees at its centre; but, at its surface, it is only about 5800 K. Because the light from the Sun has to pass through both its own atmosphere and that of the Earth, its colour temperature as seen from the surface of the Earth is not 5800 K, but is about 5500 K when high in the sky, and much lower when near the horizon. A spectral power distribution that is typical for Sun and sky light received at the Earth's surface is shown in Fig. 4.15. Also shown in this figure is the spectral power distribution of the Planckian radiator having the same correlated colour temperature. It is clear that the Sun and sky light does not have a smooth distribution like that of the Planckian radiator, and this is the result of the selective absorptions of the light by the Sun's, and by the Earth's, atmospheres. The absorptions by the Sun's atmosphere are mostly in the form of very narrow absorption lines (Fraunhofer lines). The most significant absorptions by the Earth's atmosphere are of two types. First, light is lost at the short-wavelength end of the spectrum as a result of the scattering of blue light and ultraviolet back into space; this is indicated by the more rapid downward slope of the full curve at the left-hand end in Fig. 4.15. Second, light is absorbed by molecules of various gases, particularly water vapour and ozone; this is the cause of the considerable undulations of the full curve at the right-hand end in Fig. 4.15.

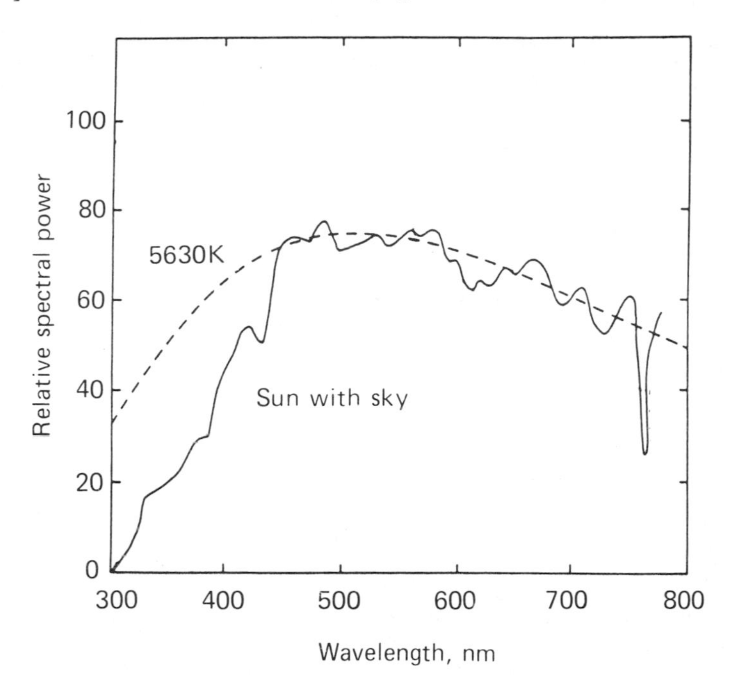

Fig. 4.15 — Relative spectral power distribution of a typical daylight (full curve), and of the Planckian radiator of the same correlated colour temperature, in this case 5630 K (broken line).

In sunlit areas in clear sunny weather, about 90% of the illumination comes directly from the Sun and only about 10% from the sky. Because the sky is normally blue in these conditions, the addition of the sky light makes the light slightly bluer, and raises the correlated colour temperature somewhat. The correlated colour temperature of clear sunlight plus sky light thus depends on the altitude of the Sun in the sky, and on the colour of the sky; a correlated colour temperature of 5500 K is usually taken as a representative figure for solar altitudes not lower than about 30°.

In fully overcast weather, the light from the Sun and sky above the clouds is diffused by them to produce a much more uniform light. If the clouds are low, the light usually has a correlated colour temperature similar to that of sunlight plus skylight on a clear day, that is, about 5500 K. But, if the clouds are very high, the blue light scattered by the atmosphere can be reflected back down to the Earth, and the lighting can become of higher correlated colour temperature.

There are many types of weather between clear Sun and fully overcast, and the resulting daylight can vary in spectral power distribution very considerably according to the solar altitude and the nature of the cloud formation and atmospheric conditions.

In Fig. 4.16 are shown relative spectral power distributions for three examples of daylight, having correlated colour temperatures of 5610, 7140, and 10350 K. It is

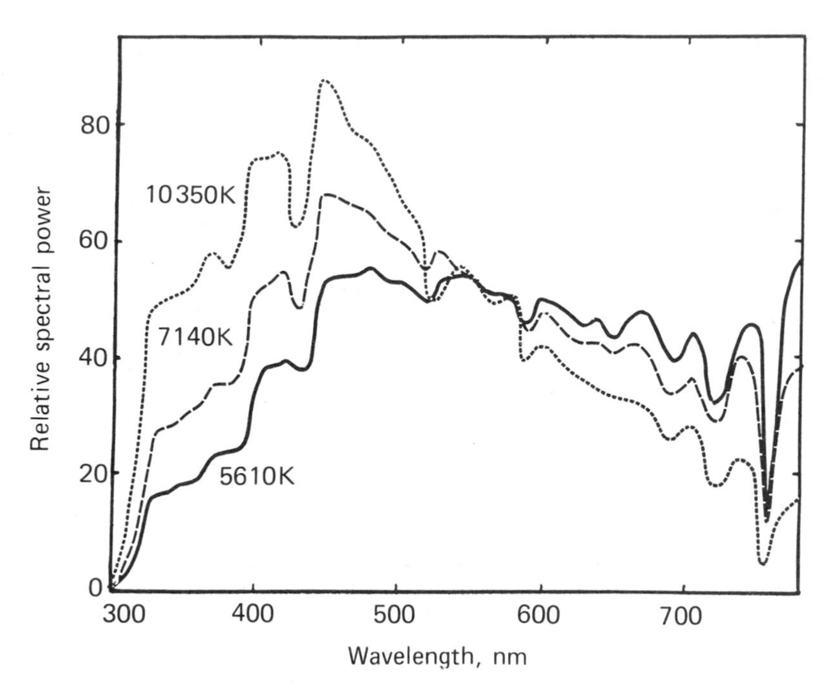

Fig. 4.16 — Relative spectral power distributions of three examples of daylight, having correlated colour temperatures of 5610, 7140, and 10350 K.

clear that the shape of these distributions remains much the same, except for a general tilting of the curve at different angles.

Indoor daylight varies according to the nature of the sky (and sunlight, if any), the geometry of the windows relative to the sky, the spectral transmission of the windows, and the nature of the interior of the room. Indoor daylight can therefore be even more variable than outdoor daylight, and, in particular, the decor of the room can have a large effect. For instance, if a room has a red carpet, and pink walls, the effective colour of the daylight in the room can be considerably redder than outside. A representative correlated colour temperature for indoor daylight is generally taken as 6500 K; this is higher than the value of 5500 K taken for Sun and sky light, because, when indoor situations are averaged, the sky plays a more important part than the Sun in providing the illumination.

From the point of view of colorimetry, daylight presents three problems. First, the spectral power distribution is complicated. Second, it is variable. Third, it has correlated colour temperatures much higher than can be achieved by artificial incandescent sources.

4.12 STANDARD ILLUMINANTS AND SOURCES

Enough has been said in this chapter to show that commonly used light sources provide a great variety of spectral power distributions. The result is that, in spite of the adaptation processes of the visual system, coloured objects undergo appreciable changes in colour appearance as the light source is changed. Furthermore, pairs of colours that match under one light source will not necessarily match under another. To reduce the complexity of dealing with this situation, and yet retain some simplified representation of it, the CIE has introduced some standardization. The CIE distinguishes between *illuminants*, which are defined in terms of spectral power distributions, and *sources*, which are defined as physically realizable producers of radiant power.

4.13 CIE STANDARD ILLUMINANT A

The most common artificial light source is the tungsten filament lamp. In spite of the availability of fluorescent lamps, and other discharge types of lamp, tungsten lamps are widely used because they are inexpensive, compact, and usually considered flattering to the human complexion. It is thus appropriate that one of the standard illuminants adopted by the CIE should represent the light from tungsten lamps. The tungsten lamp is also a very convenient standard illuminant, because the spectral power distribution of its light is almost entirely dependent on just one variable, the temperature of its filament. The CIE standard tungsten illuminant is defined as:

Standard Illuminant A

> An illuminant having the same relative spectral power distribution as a Planckian radiator at a temperature of about 2856 K.

The temperature is quoted as 'about 2856 K' because the temperature associated with a particular power distribution depends on the value adopted for one of the constants, c_2, in the formula for Planck's law. The value assigned to this constant has changed over the years since this Standard Illuminant was first defined in 1931, and, to keep the power distribution unchanged, the temperature is now defined as (1.4388/1.4350)2848, which is approximately equal to 2856, using the 'International Practical Temperature Scale, 1968'. The relative spectral power distribution of Standard Illuminant A, S_A, is illustrated in Fig. 4.17, and its values are given at every 5 nm throughout the visible spectrum in Appendix 5.2 (CIE 1986a). The values are also available at every 1 nm (CIE 1971; CIE 1986b). In the case of Standard Illuminant A, 2856 K is both its distribution temperature and its colour temperature.

4.14 CIE STANDARD ILLUMINANTS B AND C

Of even more practical importance than tungsten light is daylight, and standard illuminants representing daylight are therefore essential; they are, however, more complicated to define. In 1931 the CIE established two different standard illuminants to represent daylight. They were:

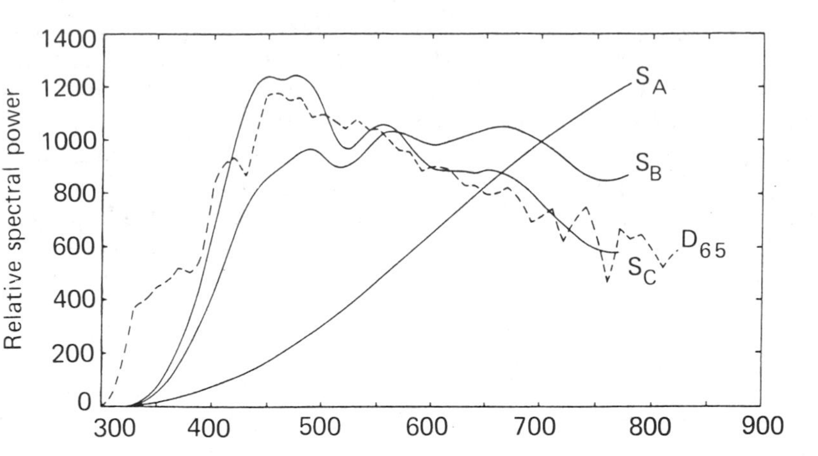

Fig. 4.17 — Relative spectral power distributions of Standard Illuminants A (S_A), B (S_B), C (S_C), and D_{65}.

Standard Illuminant B

An illuminant having the relative spectral power distribution given in Appendix 5.2.

Standard Illuminant C

An illuminant having the relative spectral power distribution given in Appendix 5.2.

These distributions are shown in Fig. 4.17. Standard Illuminant B, S_B, has a correlated colour temperature of about 4874 K and was intended to represent sunlight. Standard Illuminant C, S_C, has a correlated colour temperature of about 6774 K, and was intended to represent average daylight. Standard Illuminant B is now obsolete.

4.15 CIE STANDARD SOURCES

In the case of Standard Illuminant A, its radiation can be produced by Standard Source A, which is defined as:

Standard Source A

A gas-filled tungsten filament lamp operating at a correlated colour temperature of 2856 K ($c_2 = 1.4388 \times 10^{-2}$ m × K). A lamp with a fused-quartz envelope or window is recommended if the spectral power distribution of the ultraviolet radiation of Illuminant A is to be realized more accurately.

The radiation of Standard Illuminant C can be produced by Standard Source C, which is defined as:

Standard Source C

Standard Source A combined with a filter consisting of a layer, 1 cm thick, of each of two solutions C_1 and C_2, contained in a double cell made of colourless optical glass. The solutions are to be made up as follows:

Solution C_1:

Copper sulphate ($CuSO_4.5H_2O$)	3.412 g
Mannite [$C_6H_8(OH)_6$]	3.412 g
Pyridine (C_5H_5N)	30.0 ml
Distilled water to make	1000.0 ml

Solution C_2

Cobalt ammonium sulphate [$CoSO_4.(NH_4)_2SO_4.6H_2O$]	30.58 g
Copper sulphate ($CuSO_4.5H_2O$)	22.52 g
Sulphuric acid (density 1.835 $g.ml^{-1}$)	10.0 ml
Distilled water to make	1000.0 ml

The use of these liquid filters is inconvenient in practice, and considerable care is necessary in making them up to obtain the correct result (Billmeyer & Gerrity 1986). As a result, glass filters are often used instead, but they do not exactly reproduce the required spectral transmission characteristics.

4.16 CIE STANDARD ILLUMINANTS D

It has been mentioned that daylight has a lower ultraviolet content than Planckian radiators of the same correlated colour temperatures. But Standard Illuminants B and C have too little power in the ultraviolet region. This is not significant in the case of colours that do not fluoresce, but, for those that do, these illuminants result in less fluorescence than occurs in real daylight. With the extensive use of fluorescing agents in white colorants, often referred to as *optical brightening agents*, there was a need for a Standard Illuminant that was more representative of daylight in the ultraviolet region. Therefore, in 1963, the CIE recommended a new Standard Illuminant, D_{65}, to represent average daylight throughout the visible spectrum and into the ultraviolet region as far as 300 nm. The spectral power distribution of this illuminant is also shown in Fig. 4.17, and it can be seen that below about 380 nm it has considerably more power than S_C. Illuminant D_{65} has a correlated colour temperature of about 6504 K and is one of a series of D illuminants representing daylights of different correlated colour temperatures; these are designated D_{50}, D_{55}, etc., for example, for daylights having correlated colour temperatures of about 5000 K, 5500 K, etc., respectively. In Appendix 5.2 the spectral power distributions are given at 5 nm intervals for illuminants D_{50}, D_{55}, D_{65}, and D_{75} (CIE 1986a). The method of calculating the spectral power distributions of these D illuminants is given in Appendix 5.4. If values of the spectral power distributions are required at closer intervals than every 5 nm, they are interpolated linearly from the 5 nm values. For D_{65} these have been made available at every 1 nm (CIE 1971 and 1986b). The locus of the chromaticities of the D illuminants (the *daylight locus*) is given in Fig. 4.14; it can be seen that the D illuminants lie slightly on the green side of the Planckian locus.

There are no CIE sources that realize the D illuminants, and their main use is in computing tristimulus values and other data. However, the simulation of D illuminants remains a pressing need in colorimetry, and various attempts have been made to provide such sources (Terstiege 1989). The degree to which a practical source approximates a D illuminant in relative spectral power distribution for colorimetric purposes can be assessed by using carefully chosen samples and a colour difference formula (CIE 1981).

4.17 CIE COLOUR RENDERING INDICES

Different light sources have different *colour rendering* properties. For instance, low pressure sodium lamps, which emit light at virtually only one wavelength, result in the appearance of colours being very different from that in daylight. With the advent of fluorescent lamps, whose spectral power distributions could be varied at will over quite a wide range, it became desirable to have some means of expressing the degree to which any illuminant gave satisfactory colour rendering. For this purpose, the CIE recommended both a *Special Colour Rendering Index* and a *General Colour Rendering Index* (*CRI*) in 1965 (CIE 1965), revised versions of which were introduced in 1974 and 1988 (CIE, 1974 and 1988), and are as follows.

The *CIE Special Colour Rendering Index* is given by

$$R_i = 100 - 4.6d_i$$

where d_i is the distance in the CIE $U^*V^*W^*$ space between the points representing the colour concerned when illuminated by the test source and by the CIE D-illuminant closest in the CIE u,v (not u',v') chromaticity diagram (but for sources whose correlated colour temperatures are below 5000 K the closest Planckian radiator is used instead). U^*, V^*, W^* are given by:

$$U^* = 13W^*(u - u_n)$$
$$V^* = 13W^*(v - v_n)$$
$$W^* = 25Y^{1/3} - 17$$

The *CIE General Colour Rendering Index* (*CRI*) is given by

$$R_a = 100 - [4.6/8][d_1 + d_2 + d_3 + d_4 + d_5 + d_6 + d_7 + d_8]$$

where d_1 to d_8 are the values of R_i for the Munsell colours 7.5R6/4, 5Y6/4, 5GY6/8, 2.5G6/6, 10BG6/4, 5PB6/8, 2.5P6/8, 10P6/8 (the Munsell system will be described in section 7.4). Allowance is made for any differences in chromaticity between the source considered and the nearest D-illuminant (or Planckian source) by using a Von Kries type of adjustment (see section 3.13) for chromatic adaptation.

A Colour Rendering Index of 100 indicates that, for the colours considered in its evaluation, the source is equivalent to the nearest D illuminant (or Planckian source). Sources having General Indices of about 90 or more, are usually considered to have very good colour rendering properties in practical applications. Some fluorescent lamps have General Indices as low as about 50; this indicates that their colour rendering is appreciably deficient in some respects, but such lamps are used in some applications because they tend to have high efficacies, and hence are economical in power consumption.

4.18 COMPARISON OF COMMONLY USED SOURCES

Table 4.1 gives the correlated colour temperatures of a selection of commonly used light sources, together with their reciprocal temperatures multiplied by 10^6. These values of mega reciprocal kelvins (or *mireks*, previously denoted as *mireds*) are useful because, quite fortuitously, they represent an approximately uniform scale of colour differences.

Table 4.1 — Correlated colour temperatures of some typical light sources

Source	K	10^6/K
North-sky light	7500	133
Average daylight	6500	154
Fluorescent lamps (northlight, colour-matching)	6500	154
Xenon (electronic flash or continuous)	6000	167
Sunlight plus skylight	5500	182
Blue flash bulbs	5500	182
Carbon arc for projectors	5000	200
Sunlight at solar altitude 20°	4700	213
Fluorescent lamps (cool white)	4200	238
Sunlight at solar altitude 10°	4000	250
Clear flash-bulbs	3800	263
Fluorescent lamps (white or natural)	3500	286
Photoflood tungsten lamps	3400	294
Tungsten-halogen lamps (short life)	3300	303
Projection tungsten lamps	3200	312
Studio tungsten lamps	3200	312
Tungsten-halogen lamps (normal life)	3000	333
Fluorescent lamps (warm white)	3000	333
Tungsten lamps for flood-lighting	3000	333
Tungsten lamps (domestic, 100 watts)	2800	357
Tungsten lamps (domestic, 40 watts)	2700	370
Sunlight at sunset	2000	500
Candle flame	1900	526

Table 4.2, for a smaller selection of sources, gives, not only correlated colour temperatures, but also General Colour Rendering Indices (CRI) and efficacies.

Table 4.2 — Correlated colour temperatures, general colour rendering indices, and efficacies for some typical light sources

Source	Correlated colour temperature K	Colour rendering index CRI	Efficacy lm/W
Tungsten (40 W, 240 V)	2650	100	10
Tungsten (40 W, 110 V)	2700	100	12
Tungsten (100 W, 240 V)	2750	100	13
Tungsten (100 W, 110 V)	2850	100	15
Tungsten halogen	3000	100	21
Daylight (D_{65})	6500	100	—
Xenon	5290	93	25
Fluorescent			
Northlight, colour-matching	6500	93	48
Artificial daylight	6500	92	45
Cool white	4200	58	81
Kolor-rite	4000	89	51
Natural de luxe	3500	92	45
White	3450	54	83
Warm white de luxe	3000	80	48
Warm white	3000	51	83
Three-band	4000	85	93
High-pressure sodium	2000	25	100
Low-pressure sodium	—	—	150
Colour-corrected mercury (MBF)	3800	45	50
Colour-corrected mercury (MBTF)	3800	45	20
Metal halide mercury (HMI)	6430	88	85
Metal halide mercury	4000	80	70

REFERENCES

Billmeyer, F. W. & Gerrity, A. *Color Res. Appl.* **8**, 90 (1986).

CIE Publication No. 13 *Method of specifying and measuring colour rendering properties of light sources* (1965).

CIE Publication No. 13.2 *Method of specifying and measuring colour rendering properties of light sources*. 2nd ed. (1974).

CIE Publication No. 13.2 *Method of specifying and measuring colour rendering properties of light sources*. 2nd ed. revised (1988).

CIE Publication No. 15, *Colorimetry* (1971).

CIE Publication No. 15.2, *Colorimetry*. 2nd ed. (1986a).

CIE Publication No. 51, *A method for assessing the quality of daylight simulators for colorimetry* (1981).

CIE *Standard on Colorimetric Illuminants*, CIE S001 (1986b).

Krystek, M. *Color Res. Appl.* **10**, 38 (1985).

Terstiege, H. *Color Res. Appl.* **14**, **131** (1989).

GENERAL REFERENCES

Cayless, M. A. & Marsden, A. M. *Lamps and Lighting*. 3rd. ed., Arnold, London (1983).

Henderson, S. T. *Daylight and its Spectrum*. 2nd ed., Hilger, Bristol (1977).

5

Obtaining spectral data and tristimulus values

5.1 INTRODUCTION

CIE X, Y, Z tristimulus values can be obtained by one of three main methods: visual matching, calculation from spectral power data, or direct measurement with filtered photo-detectors. The visual matching may be done by using either additive mixtures of red, green, and blue light (as described in section 2.4, and as will be further considered in Chapter 10) or by comparing colours with those in colour order systems (as will be described in Chapter 7). However, visual matching tends to be a time-consuming and rather difficult task, and therefore the vast majority of tristimulus value determinations use one of the other two methods. In this chapter we shall consider these two methods, together with some related topics.

The second method, calculation from spectral power data, is the most widely used, and is therefore of the greatest importance. The X, Y, Z tristimulus values are obtained from the spectral power data by using CIE colour-matching functions as weighting functions, as described in section 2.6. Measuring the spectral power data involves spectroradiometry and spectrophotometry, and there are various factors that can affect the results. Different methods of calculation can also affect the results. Different considerations also apply to the measurement of self-luminous samples, and non-self-luminous samples.

5.2 RADIOMETRY AND PHOTOMETRY

In *radiometry*, the amount of *radiant power* in a stimulus is measured; hence the detectors that are used have to be equally sensitive to all wavelengths.

In *photometry*, the amount of *light* in a stimulus is compared with that of a standard stimulus, and the ratio of the two amounts then provides a photometric measure. The amounts of light may be compared by visual assessment; but it is much more common for them to be compared by using filtered photo-detectors whose spectral sensitivities approximate the $V(\lambda)$ or the $V'(\lambda)$ function, according to whether photopic or scotopic levels of illumination are involved (see sections 2.2 and 2.3). Alternatively the comparison can be made by digital calculations from spectral power data using either the $V(\lambda)$ or the $V'(\lambda)$ function as a weighting function (see Tables 2.1 and 2.2).

5.3 SPECTRORADIOMETRY

In *spectroradiometry*, the amounts of radiation are measured using narrow bands of wavelengths situated at regular intervals throughout the spectrum, and the measures obtained are then referred to as *spectral*; the corresponding measuring instruments are known as *spectroradiometers*. Spectroradiometry is often carried out as a relative measurement, the radiant power considered being compared to that from a standard source.

5.4 TELE-SPECTRORADIOMETRY

If we want to obtain tristimulus values that accurately represent a colour that was seen in a given set of viewing conditions, then ideally we should set up an instrument at the same position as was occupied by the observer's eye, and direct it at the colour while it is illuminated by the same illuminant in the same surroundings. The instrument for doing this is called a *tele-spectroradiometer*; *tele* because it incorporates a telescope to collect the light from the colour at the observing position, *spectro* because it analyses the data throughout the spectrum, and *radiometer* because the measurement it makes is of the radiant power. Such instruments are commercially available, and they should be used when it is important that the measurements relate to a particular set of viewing conditions that occurs in practice. If measurements of *absolute* radiant power are required, the instrument must be calibrated to give this result, such a calibration is usually most conveniently done by using a standard light source whose absolute spectral power distribution is known to the required accuracy. If, as is often the case, only *relative* spectral power data are required, then it is only necessary to know the *relative* spectral power distribution of the standard source.

If it is required to evaluate variables that depend on values for a reference white (as described in Chapter 3), it is also necessary to measure with the tele-spectroradiometer the spectral power distribution of a suitably chosen reference white under exactly the same conditions as were used for the colours.

The above procedures can be somewhat tedious, and, for general work, it is usually more convenient to adopt a standard geometry for the illuminating and viewing conditions. We shall now therefore consider these simpler procedures.

5.5 SPECTRORADIOMETRY OF SELF-LUMINOUS COLOURS

For self-luminous colours, such as light sources or television displays and Video Display Units (VDUs), the light from the sample has to be made to enter the spectroradiometer by some suitable means.

In the case of light sources, one convenient way is to let the source illuminate a stable white reflecting surface, whose spectral power distribution is then measured. Similar measurements are then repeated, using a standard source whose relative spectral power distribution is known, such as CIE Standard Source A. The ratio of the two sets of measurements at each wavelength can then be used to modify the spectral power distribution of the known source to give that of the unknown source.

In the case of self-luminous colours that are not light sources, such as television displays, for instance, the amount of light available is not usually sufficient to adopt the above procedure, and the light emitted has to be passed directly into the spectroradiometer. The standard reflecting surface (whose spectral reflectance factor data must now be known) is then used with the standard source only for calibration purposes.

It must be borne in mind that the geometry of the collection of the light from self-luminous colours may affect the results; two instruments having different collection angles will give the same results only if the spectral composition of the light emitted by the colour is independent of the angle at which it leaves its surface.

5.6 SPECTROPHOTOMETRY OF NON-SELF-LUMINOUS COLOURS

Non-self-luminous colours require to be illuminated by a source that emits light throughout the spectrum. A comparison is then made, throughout the spectrum, between the radiant power of a beam of light after it has been reflected from, or transmitted by, the sample, and the radiant power of a similar beam after it has been reflected from, or transmitted by, a calibrated working standard. Although the comparison is still of radiant power, the instruments used are then usually called *spectrophotometers*, and the procedure referred to as *spectrophotometry*, implying a comparison, not of radiant power, but of light (radiant power weighted according to the $V(\lambda)$ or $V'(\lambda)$ function); however, in the narrow bands of wavelengths usually concerned in spectral measurements, there is generally little or no difference between ratios of radiant power and ratios of light.

In the case of diffusing samples, the working standard is usually a white reflecting surface or a white transmitting diffuser. In the case of non-diffusing objects, for reflecting samples, a calibrated mirror can be used; for transmitting samples, either no sample (that is, air) can be used, or, in the case of liquids, an empty cell of the same type as used for containing the sample, but filled with a suitable non-absorbing liquid.

5.7 REFERENCE WHITES AND WORKING STANDARDS

In measuring spectral reflectance (or transmittance) factor, the CIE recommends, as the reference white, the adoption of the *perfect reflecting (or transmitting) diffuser*. As already mentioned in section 2.6, this is defined as:

PLATES

The plates provide illustrations of various colour order systems. Because of the limitations of the reproduction processes used, the colours shown are only approximately like those in the actual systems. The plates, therefore, while depicting the nature of the different systems, must not be used for colour evaluation.

The plates are best viewed in reasonably high levels of daylight. At low levels of daylight illumination, the brightnesses and colourfulnesses of the samples will be reduced, but their lightnesses, hues, chromas, and saturations will remain approximately constant. If illuminants of other colours are used, such as tungsten light or fluorescent light, the samples will tend to remain approximately constant in hue, lightness, chroma, and saturation (the phenomenon of *colour constancy*, see sections 1.8 and 12.5), but some changes usually occur.

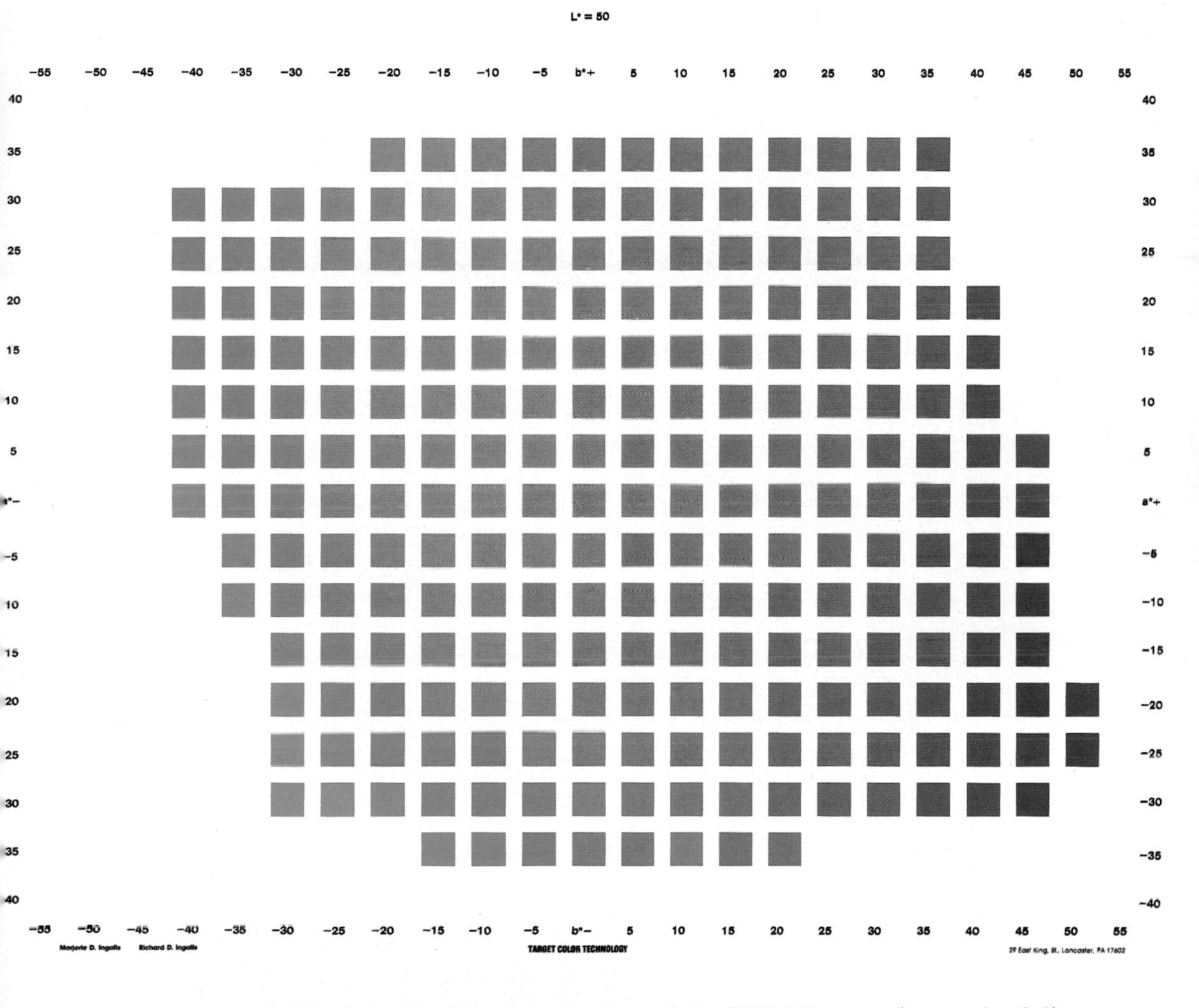

Plate 1. Reproduction from a horizontal section through the CIELAB system (see section 3.9) for a value of $L^*=50$. Original produced on Kodak *Ektacolor* paper by Richard Ingalls of Target Color Technology. Separations made on a Crosfield Magnascan scanner by courtesy of Crosfield Electronics Limited.

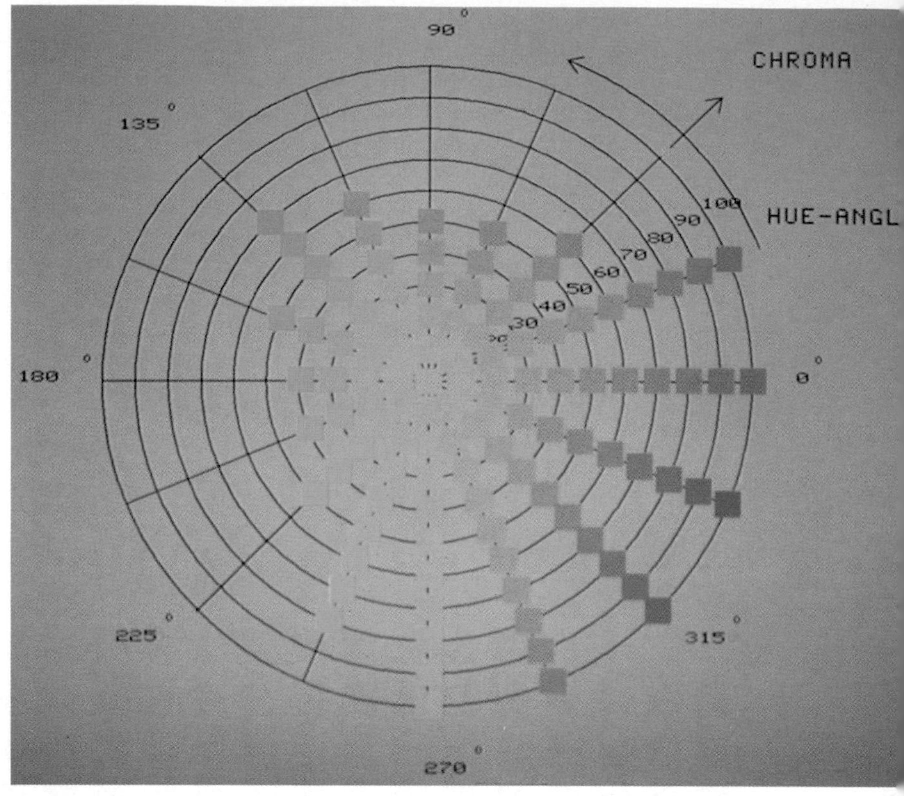

Plate 2 (upper). Reproduction from a representation of a horizontal section through the CIELUV system (see section 3.9) for a value of $L^*=50$. The original was displayed on a Sigmex Video Display Unit and photographed on Kodak *Ektachrome* film. Original and photographic transparency by courtesy of Dr A. A. Clarke and Dr M. R. Luo of Loughborough University of Technology. Separations made from the transparency on a Crosfield Magnascan scanner by courtesy of Crosfield Electronics Limited.

Plate 2 (lower). Same as Plate 2 (upper) but for a vertical section for red and cyan colours.

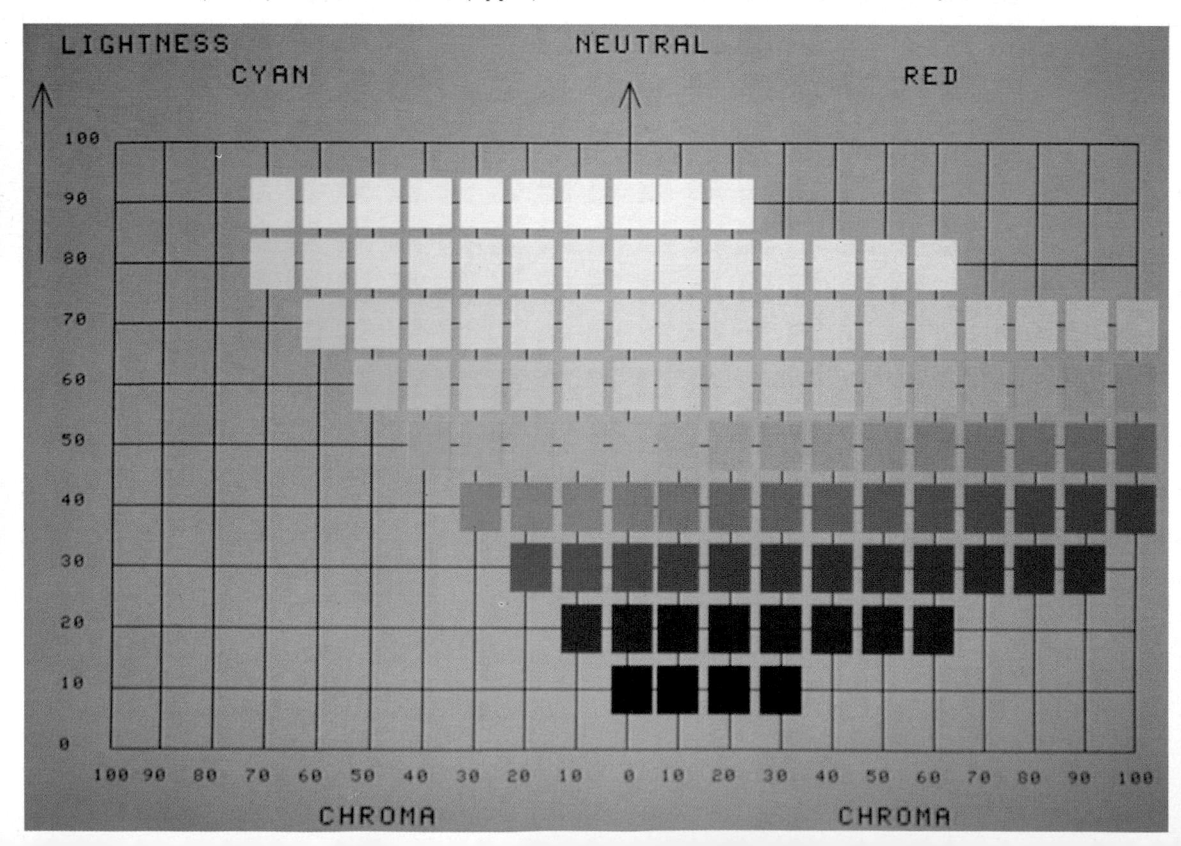

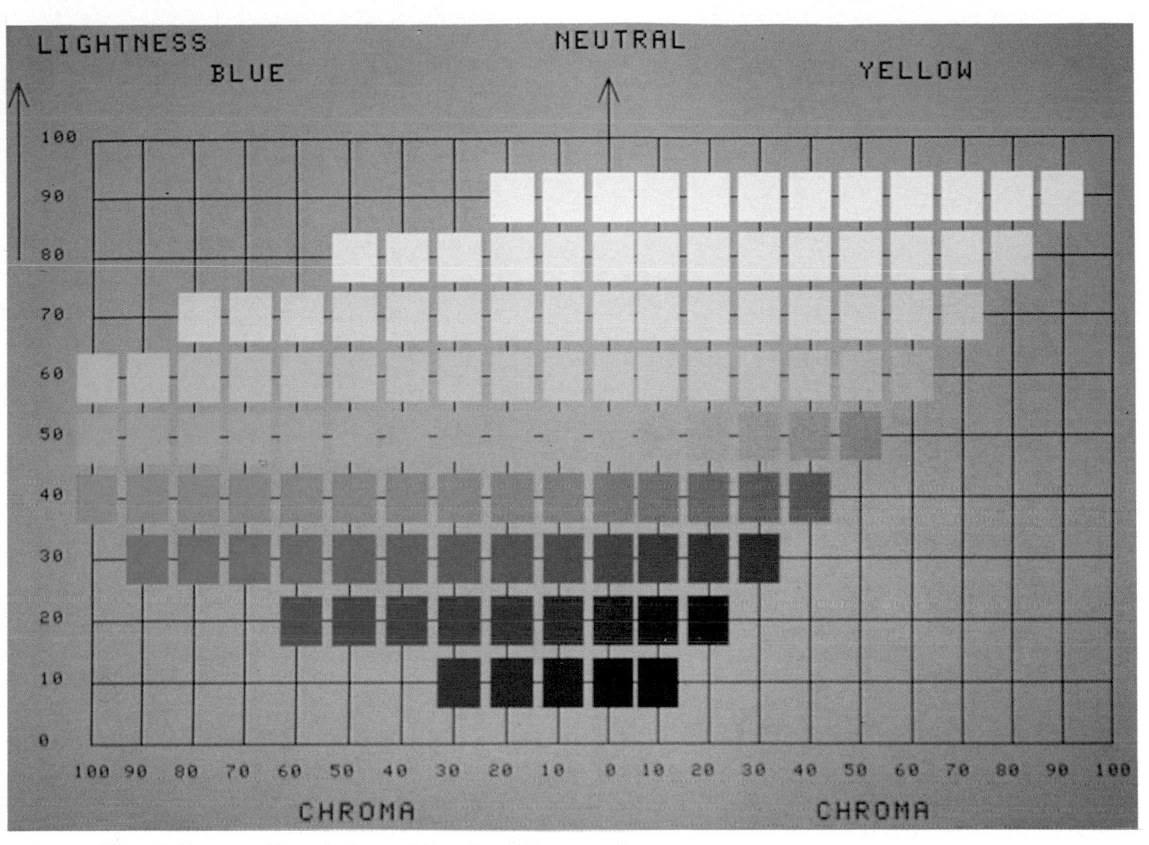

Plate 3. Same as Plate 2 (lower) but for different colours. *Upper*. For yellow and blue colours. *Lower*. For green and magenta colours.

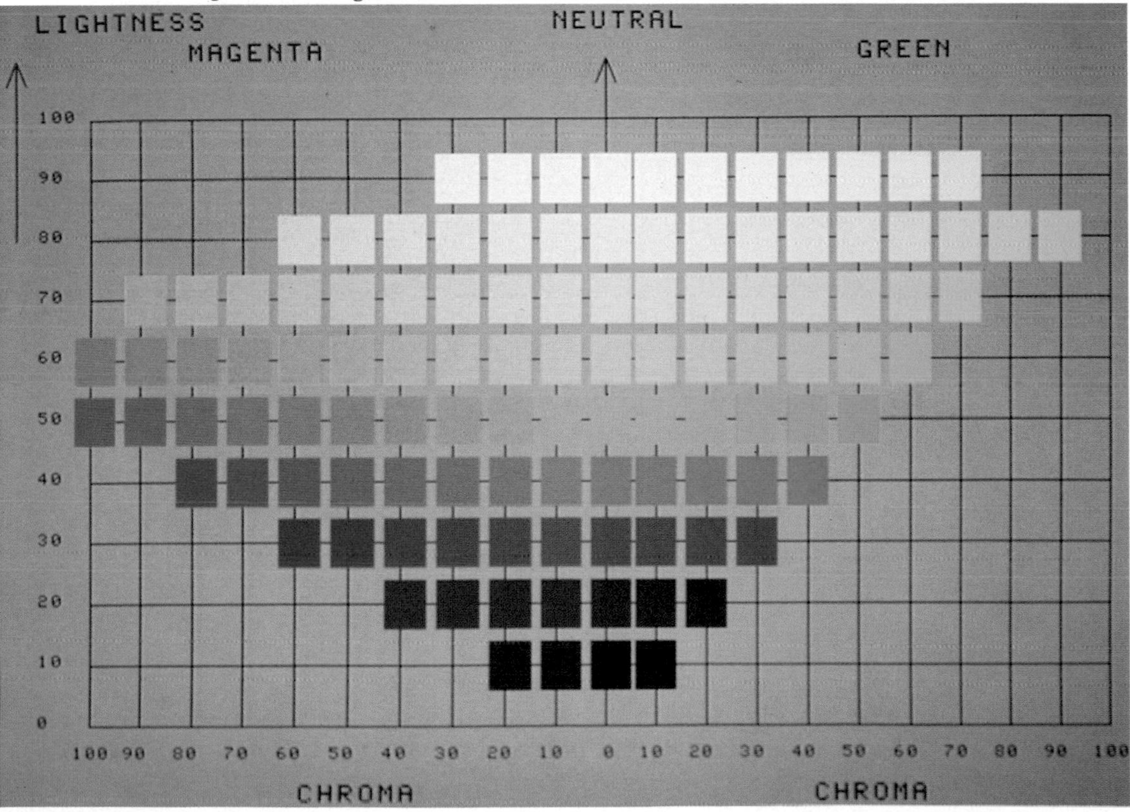

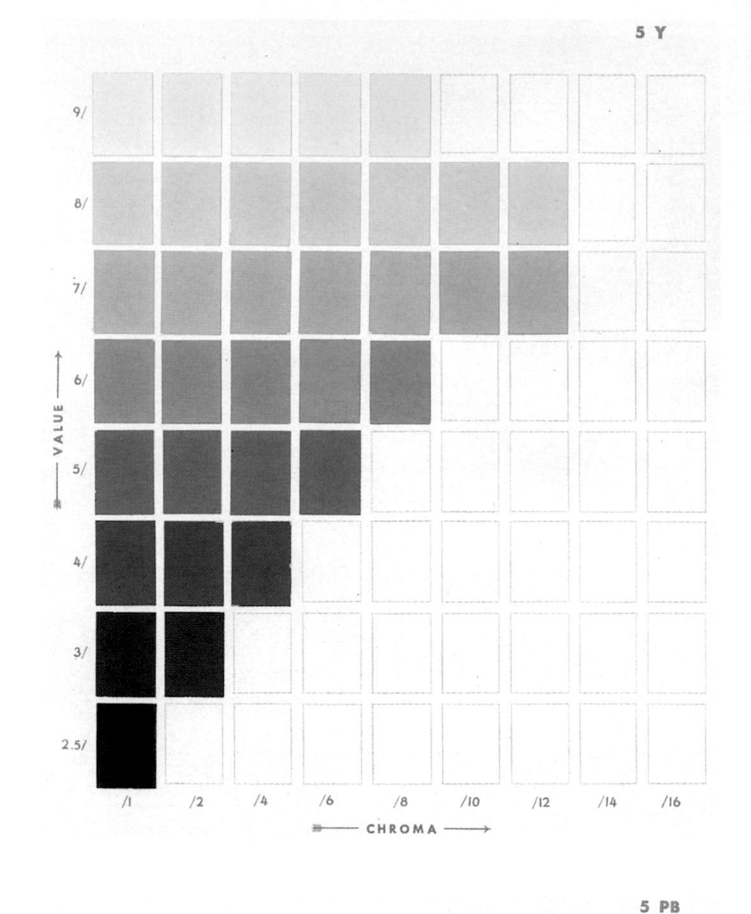

Plate 4. Reproductions from two constant-hue pages of the Munsell system (see sections 7.4 and 7.5). The horizontal rows of colours are approximately constant in perceived lightness (note that the bottom row is for Munsell Value 2.5; the others are for Munsell Values 3, 4, 5, 6, etc.). The vertical columns of colours are approximately constant in perceived chroma (note that the first column is for Munsell chroma 1; the others are for Munsell chromas 2, 4, 6, etc.). Separations made on a Crosfield Magnascan Scanner by courtesy of Crosfield Electronics Limited.

5 Y
9/
8/
7/
6/
5/
4/
3/
2.5/
VALUE
/1
/2
/4
/6
/8
/10
/12
/14
/16
CHROMA

5 PB
9/
8/
7/
6/
5/
4/
3/
2.5/
VALUE
/1
/2
/4
/6
/8
/10
/12
/14
/16
CHROMA

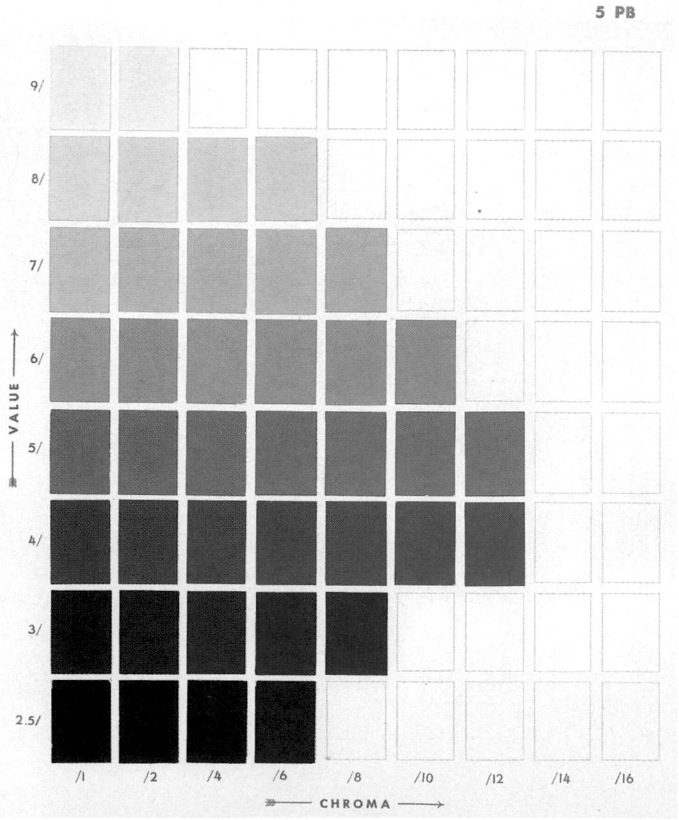

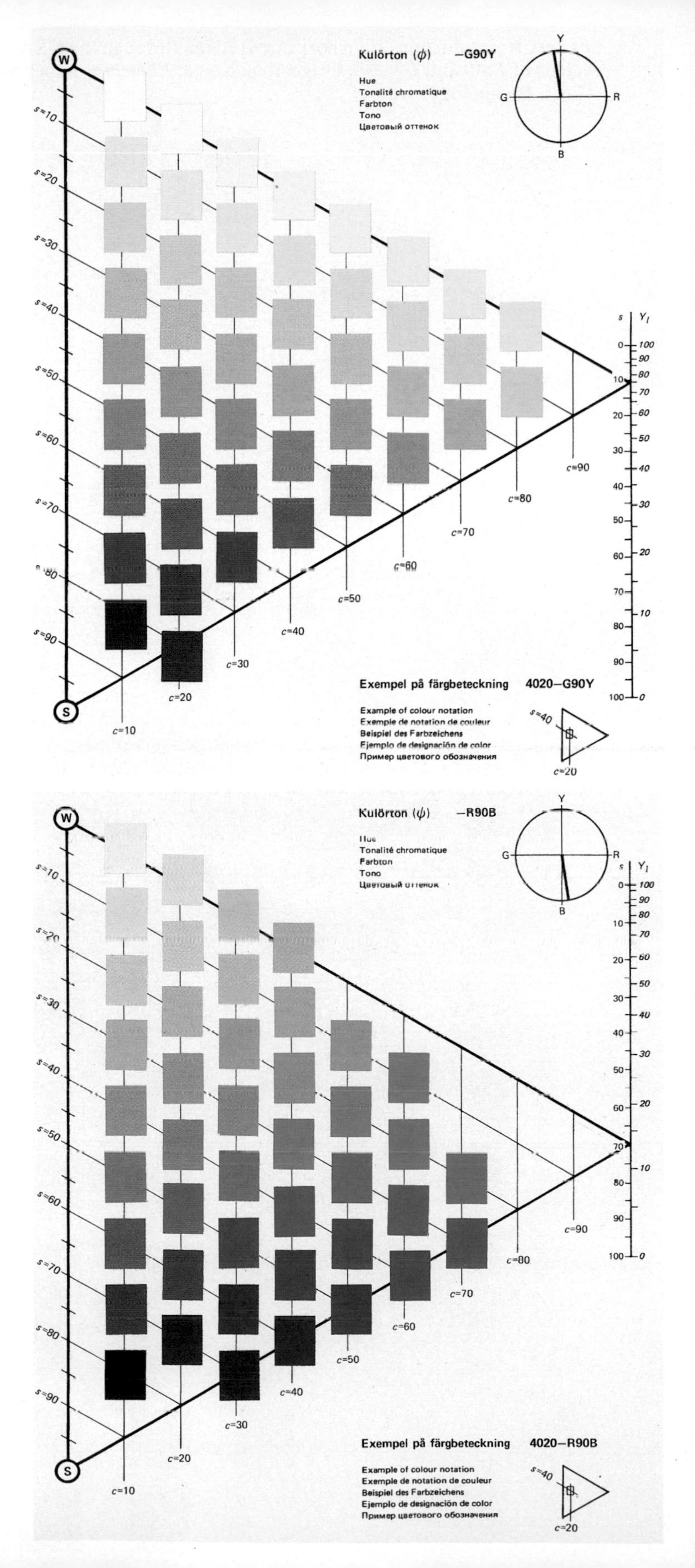

Plate 5. Reproductions from two constant-hue pages of the Natural Colour System (NCS) (see sections 7.7 and 7.8). Separations made on a Crosfield Magnascan scanner by courtesy of Crosfield Electronics Limited.

Plate 6. Reproductions from horizontal sections through the OSA system (see section 7.11) for values of L=0 and L= −3. Original on Kodak *Ektacolor* paper by Richard Ingalls of Target Color Technology.

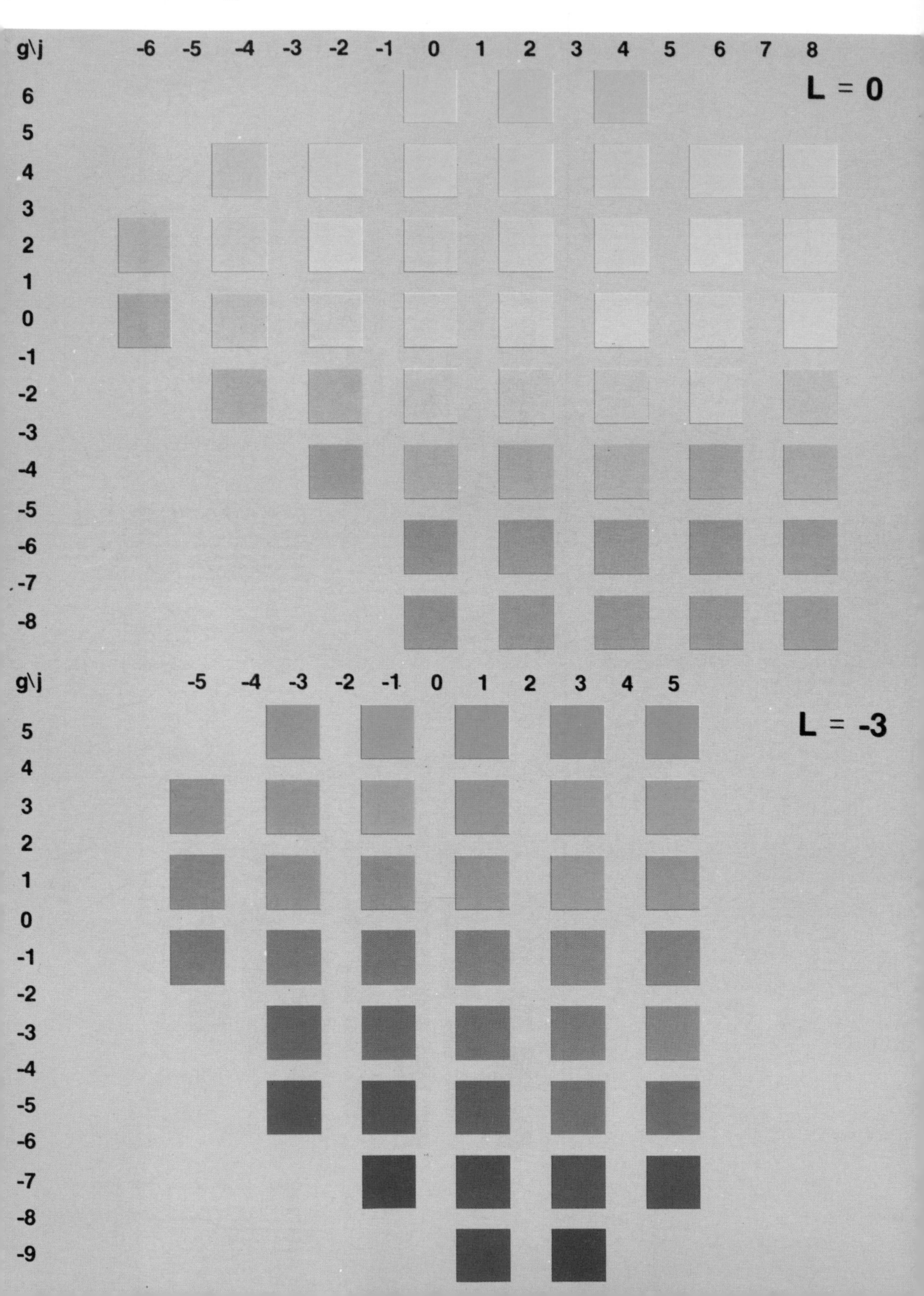

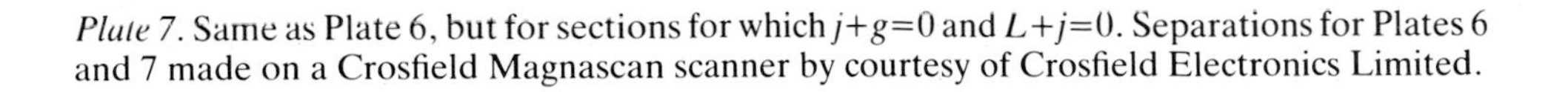

Plate 7. Same as Plate 6, but for sections for which $j+g=0$ and $L+j=0$. Separations for Plates 6 and 7 made on a Crosfield Magnascan scanner by courtesy of Crosfield Electronics Limited.

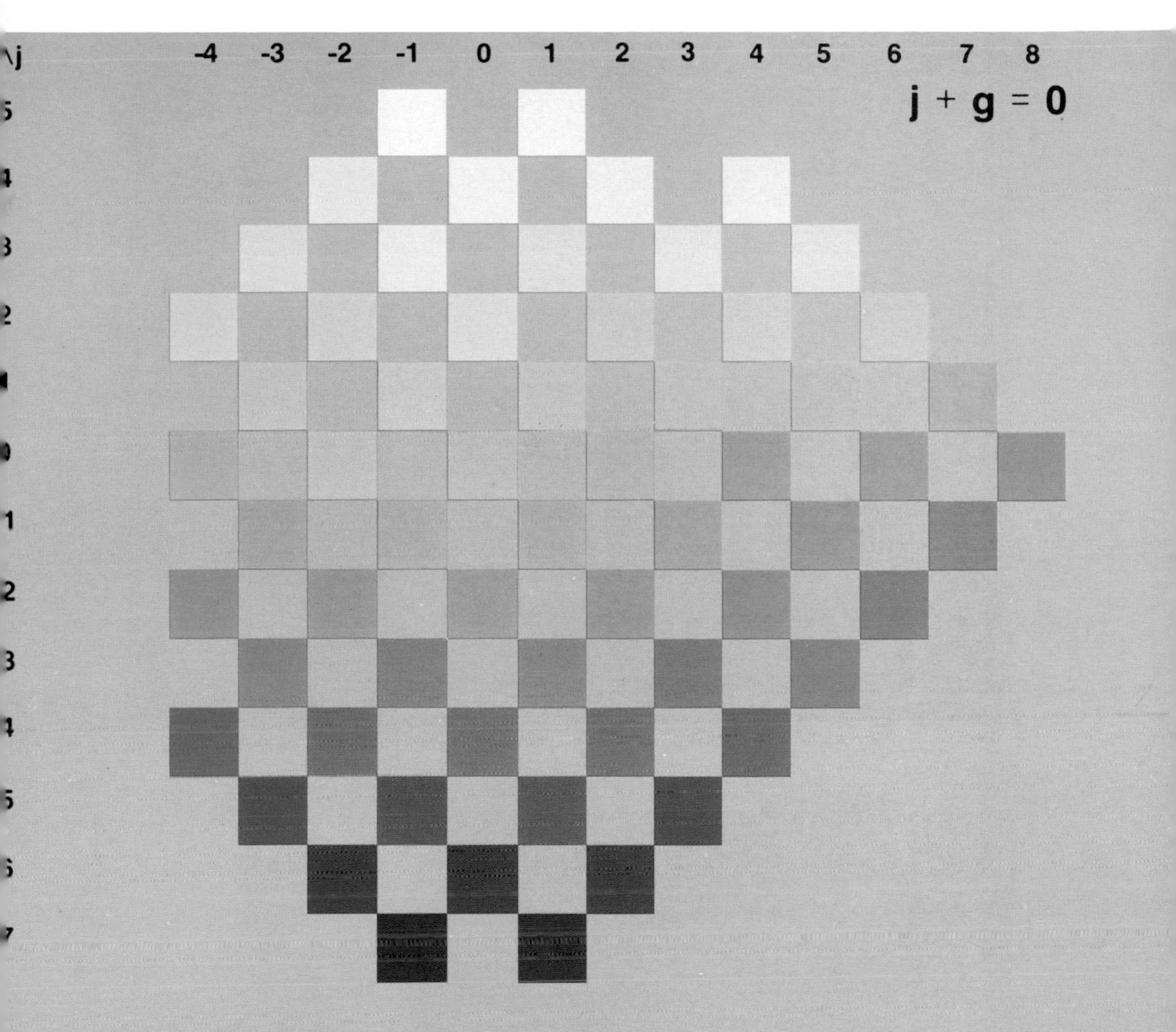

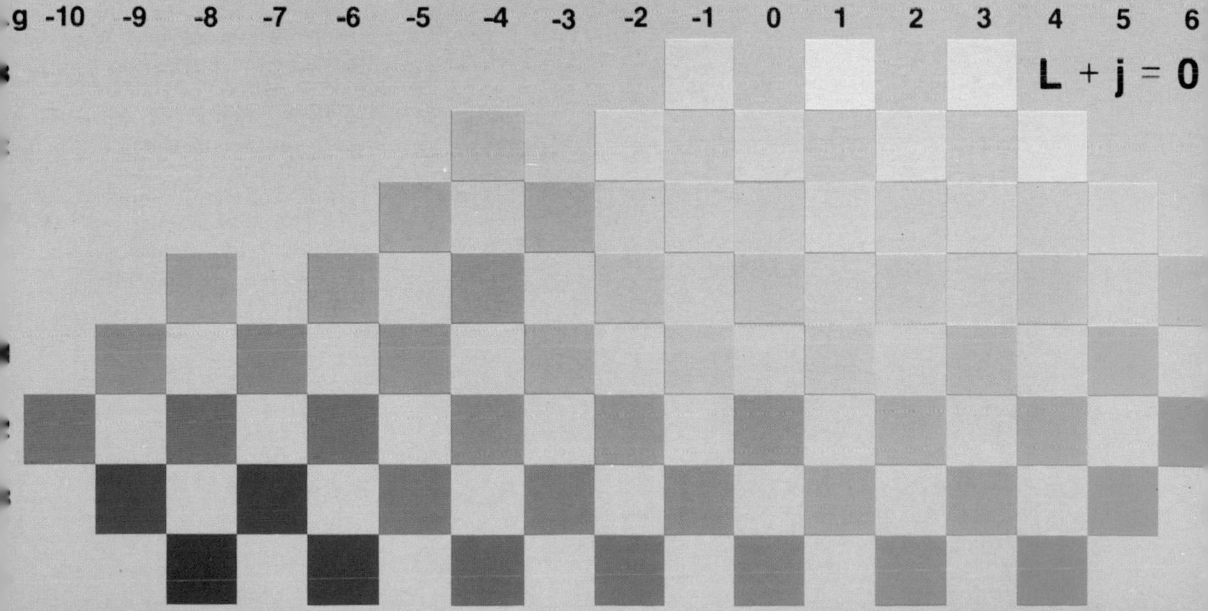

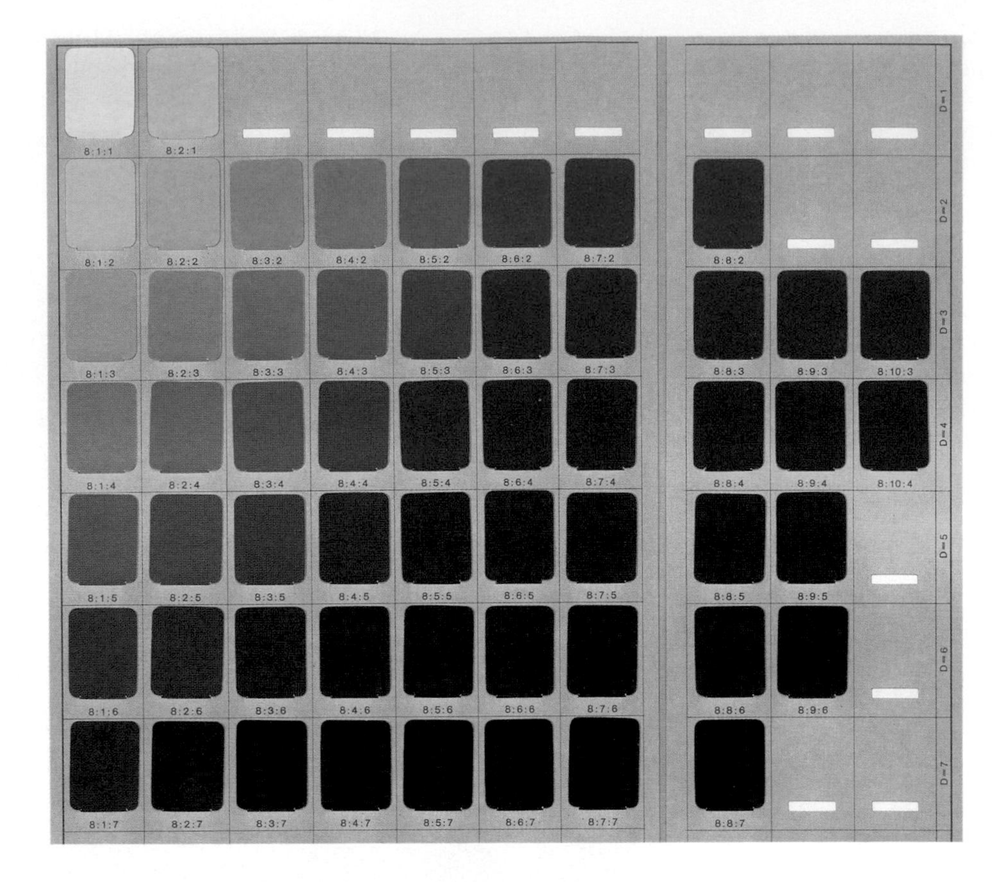

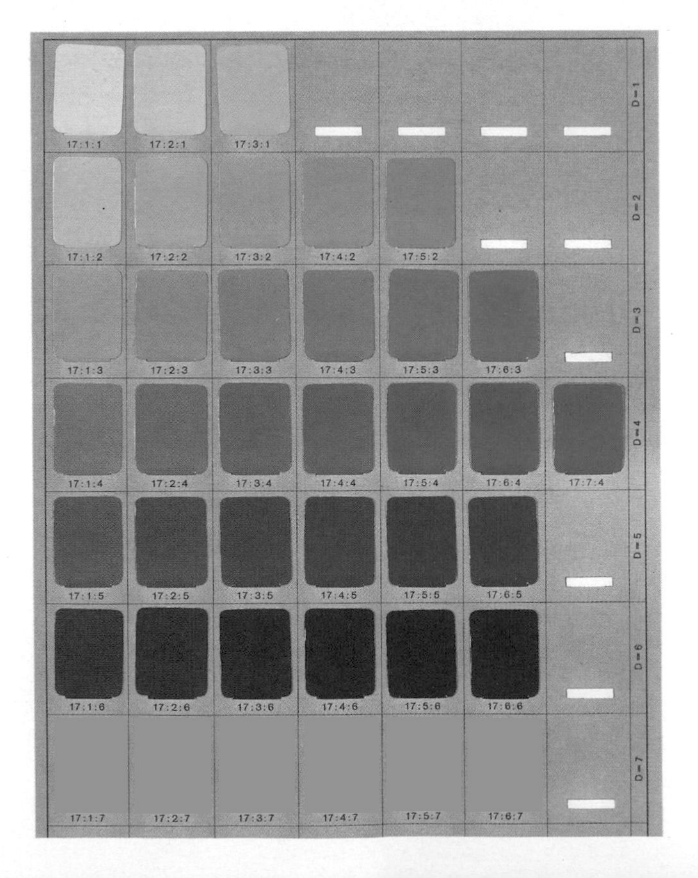

Plate 8. Reproductions from two constant-hue pages of the DIN system (see section 7.9). The vertical columns of colours are approximately constant in perceived saturation. Separations made on a Crosfield Magnascan scanner by courtesy of Crosfield Electronics Limited.

The perfect reflecting (or transmitting) diffuser

> An ideal isotropic diffuser with a reflectance (or transmittance) equal to unity.

'Isotropic' means that the spatial distribution of the reflected radiation is such that the radiance (or luminance) is the same in all directions in which the radiation is reflected (or transmitted); hence, the perfect reflecting diffuser is completely matt and is entirely free from any gloss or sheen. The reflectance (or transmittance) is equal to unity at all wavelengths, so that the diffusion is independent of wavelength, that is, it is *non-selective*.

The perfect diffuser is easy to use in making colorimetric calculations, but, for practical measurements, it is not available. It is therefore necessary to use a working standard that has been calibrated against the perfect diffuser. These calibrations are usually obtained from national standardizing laboratories, and are often supplied with the instruments. The results obtained with these working standards are then corrected to obtain those that would have been obtained if the perfect diffuser had been used instead. (This topic will be discussed further in section 8.5.)

As in tele-spectroradiometry, to evaluate variables that depend on values for a reference white (see Chapter 3), it is necessary to include the reference white in the samples considered. One choice for the reference white is the perfect diffuser, and this may be appropriate for general use in evaluating colorants such as textiles and paints. However, in other applications, the perfect diffuser may not be appropriate.

For instance, in evaluating printing inks, the paper being used is generally a better choice. This is because, if the paper were slightly yellowish, for example, a non-selective neutral grey ink would appear to be yellowish relative to the perfect diffuser; but the ink itself is not yellowish. The ink is therefore better evaluated by using the unprinted paper as the reference white; but the perfect diffuser is appropriate for evaluating the paper.

In the case of photographic reflection prints, the perfect diffuser is appropriate for evaluating the paper in areas where there is no image; but, for image areas, an area representing the reproduction of a typical white in the scene is appropriate. This may not only be of a different colour, but also appreciably darker than the perfect diffuser. By comparison, the perfect diffuser would then have a value of Y/Y_n (its Y tristimulus value divided by that for the reference white) that was appreciably higher than unity, indicating that its lightness was greater than that of a white, which indeed would be its appearance if it were incorporated into the picture. Exactly similar arguments apply in evaluating transparent areas in photographic film: Y_n should be the value of Y for a typical white in the picture, in which case Y/Y_n for an area without any image would again be in excess of unity, and this is appropriate.

When self-luminous areas are being evaluated, as in television displays, or on VDUs, the perfect diffuser is not relevant, because there is no illuminant (in the usual sense of the word). The use of the peak white of the television system is not usually appropriate, because whites used in the displays are usually reproduced at lower levels. Again, what has to be done is to select an area representing a typical white in the scene. In the case of pictorial images this can usually be done, although not always without some difficulty; but, in the case of non-pictorial images, such as data presentations, it may be very difficult to define a stimulus that corresponds to an area that appears perceptually white. Indeed, in some cases the display may appear to be entirely self-luminous, in which case only brightness, hue, colourfulness, and saturation will be perceived, and correlates of lightness and chroma are inappropriate. In this case, Y_n has no meaning, but values of u'_n, v'_n are required, and these can either be those of the nominal 'white' for the display (usually D_{65} or S_C), or those of the equi-energy stimulus (S_E).

When images are being evaluated, important practical problems are often caused by lack of uniformity of the illumination at various stages in the process of forming the image. For example, the lenses used in cameras, enlargers, and projectors, invariably give images that have lower luminances at the edges, than at the centre, of the picture. The human visual system, however, is extraordinarily adept at discounting the effects of this. For example, a white reproduced at the corner of a picture may have a lower luminance than a light grey at the centre of the picture; but the white and grey are usually correctly recognized without difficulty. Realistic evaluations of colours in images therefore often require that meticulous corrections be made for the effects of non-uniformities of illumination.

It must be stressed that Y tristimulus values are always to be evaluated such that $Y = 100$ for the perfect diffuser. It is Y_n, the value of Y for the appropriately chosen reference white, that must often be different from 100, when the ratio Y/Y_n is being used to calculate correlates of perceptual attributes.

5.8 GEOMETRIES OF ILLUMINATION AND VIEWING

If an object with a very glossy reflecting surface is viewed, its appearance is greatly affected by the angle of view relative to the angles at which the illuminating light falls on the surface. If the light comes from just one direction, then it is possible, by tilting the surface appropriately, to ensure that the mirror-like image of the light source is avoided, and the colour of the surface can then be seen properly. But, if the surface is illuminated from many different directions, as may occur, for instance, in a room lit by large windows or many artificial lights, then it may be impossible to find a direction in which mirror-like reflections of light sources are completely avoided. Furthermore, if the surface is viewed under a very large source of light, such as an overcast sky, or a uniform ceiling light, then the colour of the surface will always be seen in the presence of mirror-like reflections of part of the light source. These mirror-like (or *specular* or *regular*, as they are often referred to) reflections are produced by the top surface of objects, and, unless the objects are metals, the light reflected is not coloured by the surface and is therefore the same colour as the light source. In the usual case when the colour of the illuminant is white, the specular reflections add white light to the colour of the surface, and, unless it is itself white, the effect is to desaturate the colour. This is why glossy surfaces look more saturated in directional, than in diffuse, illumination.

In the case of completely matt surfaces, every beam of incident light, no matter what its angle of incidence on the surface, will result in some light entering the eye that has not penetrated the surface and which is therefore not affected by the colorant (except in the case of metals). Hence, when matt surfaces are viewed in white light, they are always desaturated by these top surface reflections. It is for this reason that matt surfaces do not usually appear as saturated as glossy surfaces, unless the glossy surfaces are lit very diffusely.

Most surfaces are neither completely matt nor highly glossy. The effects of the geometry of the illuminating and viewing conditions on these surfaces is therefore intermediate to those described above, and their colour saturations tend to be less than those of glossy, but more than those of matt, samples.

These differences in surface properties greatly affect the appearance of objects, and the particular degree of sheen is the characteristic that gives rise to surface finishes that may be described as satin, egg-shell, lustre, pearlescent, etc.

It is clear, therefore, that the illumination and viewing conditions play a very significant part in colour appearance. Hence, when it comes to colour measurement, if tele-spectroradiometry is not used, and simplified procedures are adopted, it is very important that these procedures represent the viewing conditions that are of major interest in the particular application. The CIE has recommended several alternative geometries of illumination and viewing for colorimetry, and these will now be described.

5.9 CIE GEOMETRIES OF ILLUMINATION AND VIEWING

The CIE recommends that colorimetric specifications of opaque specimens be given so as to correspond to one of the illuminating and viewing conditions given in Table 5.1. The diffuse illumination or viewing is normally provided in instruments by integrating spheres (hollow spheres that are painted white inside). These integrating spheres can often be used with *gloss traps*, which may consist of cavities designed to absorb all the light incident on the sphere wall at angles near those corresponding to specular reflections for the incident or viewing beams. Hence, if gloss traps are used, the specularly reflected light is excluded (referred to as SPEX); if gloss traps are not used, the specularly reflected light is included (referred to as SPINC).

Although there are six different conditions listed in Table 5.1, for practical purposes each pair, such as 45/0 and 0/45, can be regarded as equivalent, so that only three need to be considered as giving different results (but this is not true for polarizing samples, see section 8.6). The 45/0 and 0/45 represent typical viewing of surfaces in directional light. The d/0 (SPINC) and 0/d (SPINC) represent viewing surfaces in completely diffuse illumination. The d/0 (SPEX) and 0/d (SPEX) represent viewing in a diffuse illumination that has a black area in the direction corresponding to a mirror image of the direction of view. This may sound a rather artificial circumstance, but practical viewing situations often involve light sources having quite large areas, and viewers usually adjust their angle of viewing so as to exclude the specular reflections as much as possible. It is, therefore, a mode that is widely used for practical measurements. However, if it is required to compare measurements made on different integrating-sphere instruments, it is better to use the SPINC mode, because gloss traps vary considerably from one instrument to another. It has also been found that for computer match prediction the SPINC mode is to be preferred, because it gives more accurate results.

Table 5.1 —CIE geometries of illumination and viewing for reflection

Geometry	Illumination		Viewing		Specular reflection	Approximate CIE measure
	Axis	Spread	Axis	Spread		
45/0	45° ± 2°	± 8°	± 10°	± 8°	Excluded	Radiance factor, 45/0
0/45	± 10°	± 8°	45° ± 2°	± 8°	Excluded	Radiance factor, 0/45
d/0 SPINC	Diffuse	Ports ⩽10%	± 10°	± 5°	Included	Radiance factor, d/0
0/d SPINC	± 10°	± 5°	Diffuse	Ports ⩽10%	Included	Reflectance
d/0 SPEX	Diffuse	Gloss trap	± 10°	± 5°	Excluded	Radiance factor, d_s/0
0/d SPEX	± 10°	± 5°	Diffuse	Gloss trap	Excluded	Diffuse Reflectance

Note 1. The axis figures denote the angle between the central ray of the beam and the normal to the surface. The spreads denote the maximum permitted deviation of the rays of the beam from the central ray.
Note 2. Integrating spheres are usually used for the diffuse conditions. When they are used, the total area of their ports should not exceed 10% of the internal reflecting sphere area.
Note 3. For the SPEX condition, details of the gloss traps should be given.
Note 4. In the 0/d condition the specimen should not be measured with a strictly normal axis of illumination if it is required to include the regular component of reflection. Similarly, in the d/0 condition the specimen should not be measured with a strictly normal axis of view if it is required to include the regular component of reflection.
Note 5. It is important that the particular illuminating and viewing conditions used should be specified even if they are within the range of one of these recommended standard conditions. Measurements of some types of specimens (for example retro-reflective materials) may require different geometries or smaller tolerances.
Note 6. When integrating spheres are used, they should be fitted with white-coated baffles to prevent light passing directly between the specimen and the spot of the sphere wall illuminated or viewed. When the regular component of reflection is to be included, the sphere efficiency for the portion of the sphere wall corresponding to the regular component should be identical to that of the rest of the sphere wall.
Note 7. When fluorescent samples are measured, it is preferable not to use integrating spheres, because the samples then modify the spectral power distribution of the illuminant (unless small ports are used, see section 9.10).

Use of the different CIE geometries can give very different results for the same specimen, unless it is completely matt. For highly glossy samples, it is preferable to use 45/0, 0/45, d/0 (SPEX), or 0/d (SPEX); but the 45/0 geometry may give rise to problems of polarization (see section 8.6). For partly glossy, or textured, samples, the preferred geometry is the one that minimizes the surface effects, and this usually corresponds to that which gives the lowest luminance factor and highest excitation purity. (ASTM 1987, 1988, 1990.)

For opaque samples, all the measures are of *reflectance factor*, the ratio of the radiant flux reflected by the sample into a defined cone to the radiant flux similarly reflected by the perfect diffuser. But they also correspond approximately to more particular CIE measures, and these are listed in Table 5.1. All except the diffuse viewing cases give approximations to the measurement of *radiance factor* because the

Table 5.2 — CIE geometries of illumination and viewing for transmission

Geometry	Illumination Axis	Illumination Spread	Viewing Axis	Viewing Spread	Specular transmission	Approximate CIE measure
0/0	±5°	±5°	±5°	±5°	Included	Regular transmittance
0/d SPINC	±5°	±5°	Diffuse	Ports ⩽10%	Included	Transmittance
d/0 SPINC	Diffuse	Ports ⩽10%	±5°	±5°	Included	Radiance factor d/0
0/d SPEX	±5°	±5°	Diffuse	Ports ⩽10%	Excluded	Diffuse transmittance
d/0 SPEX	Diffuse	Ports ⩽10%	±5°	±5°	Excluded	Diffuse transmittance
d/d	Diffuse	Ports ⩽10%	Diffuse	Ports ⩽10%	Included	Doubly diffuse transmittance

Note 1. The axis figures denote the angle between the central ray of the beam and the normal to the specimen. The spreads denote the maximum permitted deviation of the rays of the beam from the central ray.
Note 2. Integrating spheres are usually used for the diffuse conditions. When they are used, the total area of their ports should not exceed 10% of the internal reflecting sphere area.
Note 3. For SPEX conditions, details of the light traps should be given.
Note 4. Errors may be caused by multiple reflection between the specimen and the incident beam optics if the incident beam is normal to the specimen surface; such errors can be eliminated by slightly tilting the specimen.
Note 5. The construction of an instrument for 0/0 measurements shall be such that the flux incident on the specimen and the flux reaching the detector when there is no specimen in place shall be equal.
Note 6. It is important that the particular illuminating and viewing conditions used should be specified even if they are within the range of one of these recommended standard conditions. Measurements of some types of specimens may require different geometries or smaller tolerances.
Note 7. When integrating spheres are used, they should be fitted with white-coated baffles to prevent light passing directly from source to specimen in the case of diffuse illumination or directly from specimen to detector in the case of diffuse collection.
Note 8. When fluorescent samples are measured, it is preferable not to use integrating spheres, because the samples then modify the spectral power distribution of the illuminant (unless small ports are used, see section 9.10).

size of the cone of rays collected is fairly small (it would have to be negligibly small to give radiance factor exactly). In the two cases of diffuse viewing, 0/d (SPINC) approximates *reflectance*, and 0/d (SPEX) approximates *diffuse reflectance* (the reflectance attributable to diffusion alone). (Definitions for the CIE measures referred to in Table 5.1 are given in Appendix A1.6.)

A similar set of conditions for transmitting samples has also been recommended by the CIE, and these are listed in Table 5.2. As with opaque samples, the similar pairs, 0/d (SPINC), d/0 (SPINC), and 0/d (SPEX), d/0 (SPEX), give similar results in practice. (Definitions for the CIE measures referred to in Table 5.2 are given in Appendix A1.6.)

5.10 SPECTRORADIOMETERS AND SPECTROPHOTOMETERS

Spectroradiometers and spectrophotometers of various designs have been produced.

They all require a means of dispersing the light into a spectrum, and a means of measuring the radiation from the samples and from the calibration beams. The light can be dispersed by prisms, but gratings are widely used because, unlike prisms, their spectra have a uniform scale of wavelength; some instruments use a series of interference filters, each transmitting only a narrow band of wavelengths. For detection and comparison of the light, photoelectric cells are used; either a single cell can be used sequentially for the different wavelength bands throughout the spectrum; or an array of photosensitive elements (such as a photo-diode array) can be used to make all the measurements throughout the spectrum simultaneously.

In the case of spectrophotometers, a light source must also be provided. Earlier instruments tended to use tungsten light sources, but, more recently, sources such as xenon flash lamps have found increasing application. The use of xenon flash lamps, together with simultaneous detection throughout the spectrum by an array of detectors, can produce the results extremely rapidly, and this is advantageous when many samples have to be measured. The results, in the form of reflectance factor or transmittance factor, are often produced automatically as plots against wavelength, or as lists of numerical data, often with computations of CIE tristimulus values and other useful measures calculated from them.

Some instruments scan through the spectrum continuously and produce a continuously varying graph of radiant power emitted, of reflectance factor, or of transmittance factor, against wavelength; but when tristimulus values are calculated these data have to be sampled at intervals throughout the spectrum. Instruments that use multiple photo-detectors to sample the whole spectrum simultaneously, can do so only at a given number of wavelengths. For most purposes, it is usually considered sufficient to sample the spectrum at 5 nm intervals; but in some cases 10 nm, or even 20 nm, intervals are used, particularly for simultaneous type instruments. The CIE colour-matching functions tabulated at 1 nm intervals cover a range of wavelengths from 360 to 830 nm; but, for most colorimetric purposes, it is sufficient to use a range from 380 to 780 nm, which is what is used for these functions tabulated at 5 nm intervals. Some instruments, particularly the simultaneous types, may use a smaller range of wavelengths, such as from 400 to 700 nm. (The effects of these differences in wavelength interval and range will be discussed in sections 5.12 and 8.13.)

The illuminating and viewing geometry of spectrophotometers is usually intended to comply with one of the six configurations described in Tables 5.1 and 5.2. However, the scope for variation within the limits given in each configuration is such that different instruments employing the same nominal configuration may give significantly different results. It is therefore always good practice to state, not only the configuration used, but also the particular realization adopted, and this can be done in terms of the particular make and model of the instrument, or of the precise geometry actually employed. (This will be discussed further in section 8.4.)

Other characteristics of spectroradiometers and spectrophotometers that can significantly affect the results include: the amount of stray light; the bandwidth of wavelengths used for the measurements; polarization of the light; and translucency, thermochromism, and fluorescence of the samples. (These topics will be discussed in Chapters 8 and 9.)

5.11 CHOICE OF ILLUMINANT

Spectral reflection or transmission data are independent of the illuminant, but tristimulus values can be obtained from such data for only a specified illuminant. In the case of fluorescent samples, a specified illuminant is necessary even to obtain the proper spectral data (as will be discussed in Chapter 9).

For the highest accuracy in colorimetry, when spectral data are used to obtain tristimulus values, the illuminant adopted should be the same as that used in the practical situation that the colorimetry is intended to represent. Thus, if samples are to be inspected in a viewing booth, then the illuminant should be the one actually used in the booth; hence its relative spectral power distribution is required. Either this can be measured with a spectroradiometer, or published data can be used.

But, for general work, the use of large numbers of different illuminants is inconvenient, and makes comparisons of results impossible. The CIE has therefore recommended that, wherever possible, either Standard Illuminant A (S_A) or Standard Illuminant D_{65} should be used. S_A is a very convenient representation of tungsten lamps because it is also available as a source. D_{65} is a very good representation of average indoor daylight, but has the disadvantage of not being available as a source (see section 4.16). Visual inspection of samples whose colorimetry has been evaluated by using D_{65} is therefore difficult; real daylight is variable, and artificial daylight sources usually differ significantly in their spectral power distributions from D_{65}. However, experience has shown that D_{65} colorimetry is useful in predicting colour matches for samples viewed under sensibly chosen real and artificial daylight sources, but the possibility of the source having a disturbing influence should always be borne in mind.

For some applications, S_A and D_{65} are not appropriate, and other sources have to be used. Sometimes a different D Illuminant (see section 4.16) can be used: for instance, in photography, the sunlight plus sky light widely used for amateur picture taking is well represented by D_{55}; and in the graphic arts industry D_{50} is used as a representative viewing illuminant.

5.12 CALCULATION OF TRISTIMULUS VALUES FROM SPECTRAL DATA

Tristimulus values are calculated from spectral data by the method given in section 2.6. The essence of this procedure is to weight the spectral power distribution of a sample by the CIE colour-matching functions $\bar{x}(\lambda)$, $\bar{y}(\lambda)$, and $\bar{z}(\lambda)$, or by the functions $\bar{x}_{10}(\lambda)$, $\bar{y}_{10}(\lambda)$, and $\bar{z}_{10}(\lambda)$ if the field size has an angular subtense greater than 4°.

Spectral data consist of quantities at a selection of wavelengths throughout the spectrum; for colorimetry, the number of wavelengths chosen is at least sixteen, and usually many more than sixteen. But tristimulus values consist of only three quantities. The reduction of spectral data to tristimulus values is therefore an irreversible process; it is not possible, from tristimulus values, to deduce the corresponding spectral data.

For light sources and self-luminous colours, such as television displays or VDUs, the spectral power distribution provides the natural starting data. But the data for reflecting and transmitting samples usually consist of measurements of spectral reflectance factor or spectral tranmittance factor; to convert these measurements to

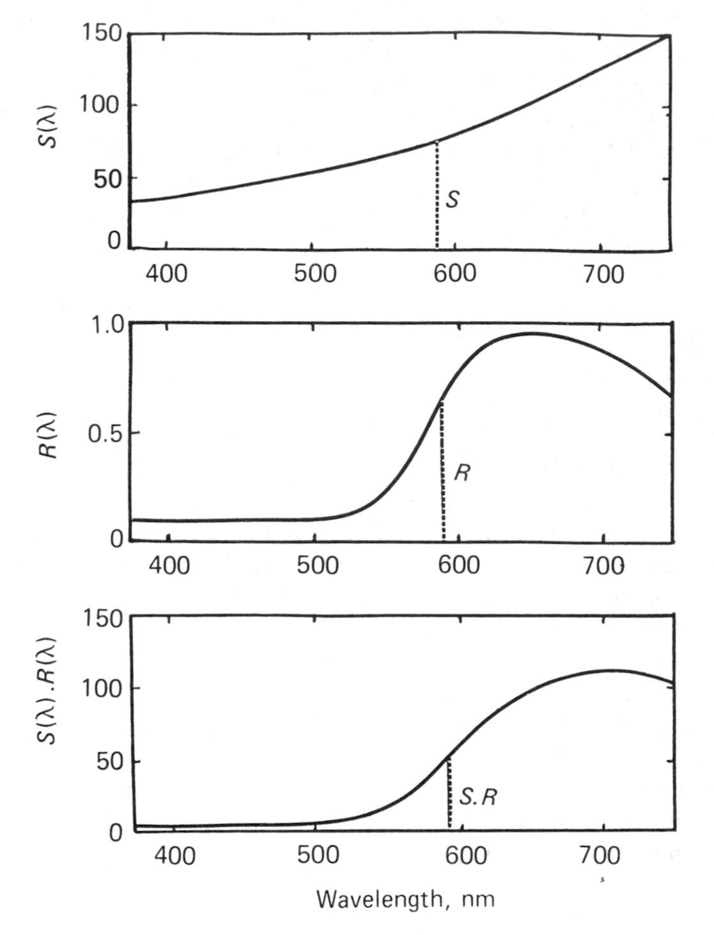

Fig. 5.1 — Top: spectral power distribution, $S(\lambda)$, of an illuminant. Centre: spectral reflectance factor, $R(\lambda)$, of a sample. Bottom: spectral power distribution, $S(\lambda).R(\lambda)$, of the sample in the illuminant.

spectral power data requires their values to be multiplied, wavelength by wavelength, by the spectral power distribution of a chosen light source. This procedure is illustrated in Fig. 5.1.

If the values of spectral power for a self-luminous sample, or of spectral power for the chosen illuminant and of spectral reflectance (or transmittance) factor for a non-self-luminous sample, are available at every 5 nm from 380 to 780 nm, then the calculation is quite straightforward. But it sometimes happens that an interval of greater than 5 nm has been used (abridged data), or that some of the values at the ends of the spectrum are not available (truncated data).

The most common form of abridged data is when a measuring interval of 10 nm or 20 nm has been used. If such data are weighted simply by those values in the tabulations of the CIE functions that correspond to the nominal wavelengths used at the 10 nm or 20 nm intervals, ignoring the rest, then significant errors can occur. If the size of these errors is too great to be tolerated, then either the measurements should be repeated, using a 5 nm interval, or corrections should be provided either by deconvoluting the data, or by using optimized weights (this will be discussed in section 8.13).

The most common form of truncated data is when a range of wavelengths of only 400 to 700 nm has been used. There are two different ways of dealing with this situation. The first way is to use only those values in the 5 nm tabulations of the CIE functions that occur within this range of wavelengths. The second is to estimate the missing values by extrapolation so as to compile data down to 380 and up to 780 nm (in the absence of any data that indicate otherwise, the extrapolation can be done by setting the missing values equal to the nearest measured value of the appropriate quantity in the truncated data). The second method is preferable, because it is more accurate (this will be discussed further in section 8.13).

It is very important that the same wavelength interval, and range of wavelengths, is used for any set of calculations in which data for different colours are to be compared precisely. This is illustrated in Table 5.3 where the tristimulus values and chromaticity coordinates are given for CIE Standard Illuminants A and D_{65} evaluated by using different wavelength intervals and ranges. The variations in the values for a given illuminant in Table 5.3 emphasize the importance of using self-consistent data when evaluating colour differences, or combining results for sample colours and for a reference white to obtain correlates of perceptual attributes of colour such as those included in the CIELUV and CIELAB systems.

5.13 COLORIMETERS USING FILTERED PHOTOCELLS

A simple, but usually rather less accurate, method of obtaining CIE tristimulus values is to use filtered photocells. The photocells are covered with carefully designed colour filters so that their resulting spectral sensitivities are a close match to the CIE colour-matching functions. The samples, and a suitably calibrated standard specimen, are then illuminated by the light source chosen, and the reflected or transmitted light made to fall on the filtered photocells. The ratios of the sample and standard readings of the three photocells are then used to derive the CIE tristimulus values. Instead of using three photocells, it is also possible to use a single photocell and to take three readings sequentially through the three filters. Because it is difficult to find filtration that exactly matches the CIE colour-matching functions, the method is usually somewhat limited in accuracy, but offers a very convenient way of providing simple and relatively inexpensive instruments. Instead of using the colour-matching functions themselves, a suitable linear combination of them can be used, and the readings obtained are then converted to CIE X, Y, Z tristimulus values by the use of a simple algebraic transformation; this procedure can be further elaborated by using more than three filters to increase the accuracy (Wharmby 1975). The geometry of illumination and viewing of such instruments should be in accordance with the CIE specifications given in Tables 5.1 and 5.2.

Table 5.3 — Chromaticity coordinates for Standard Illuminants D_{65} and A

		1 nm	5 nm	10 nm	10 nm	20 nm
		360 to 830	380 to 780	380 to 780	400 to 700	400 to 700
D_{65}	X	95.04696596	95.04	95.02	94.94	95.57
	Y	100.00000000	100.00	100.00	100.00	100.00
	Z	108.88268093	108.89	108.81	108.71	109.67
	x	0.3127268660	0.3127	0.3127	0.3127	0.3131
	y	0.3290235126	0.3290	0.3291	0.3293	0.3276
	u'	0.1978398560	0.1978	0.1976	0.1977	0.1986
	v'	0.4683365445	0.4683	0.4684	0.4684	0.4676
	ΔE_{uv}		0.025	0.078	0.245	1.385
	ΔE_{ab}		0.014	0.059	0.185	1.042
	X_{10}	94.81098989	94.81	94.80	94.76	95.10
	Y_{10}	100.00000000	100.00	100.00	100.00	100.00
	Z_{10}	107.30449228	107.33	107.35	107.32	108.31
	x_{10}	0.3138236717	0.3138	0.3138	0.3137	0.3134
	y_{10}	0.3309992566	0.3310	0.3310	0.3310	0.3296
	u'_{10}	0.1978604469	0.1979	0.1978	0.1978	0.1981
	v'_{10}	0.4695510574	0.4695	0.4695	0.4696	0.4687
	$\Delta E_{uv,10}$		0.022	0.022	0.114	1.091
	$\Delta E_{ab,10}$		0.014	0.014	0.070	0.788
A	X	109.85031527	109.85	109.84	109.69	109.78
	Y	100.00000000	100.00	100.00	100.00	100.00
	Z	35.58493013	35.58	35.54	35.54	35.64
	x	0.4475735141	0.4476	0.4476	0.4473	0.4473
	y	0.4074394443	0.4074	0.4075	0.4078	0.4075
	u'	0.2559710790	0.2560	0.2560	0.2556	0.2558
	v'	0.5242906461	0.5243	0.5243	0.5244	0.5253
	ΔE_{uv}		0.047	0.038	0.453	0.265
	ΔE_{ab}		0.029	0.052	0.268	0.154
	X_{10}	111.14394095	111.15	111.12	111.07	111.08
	Y_{10}	100.00000000	100.00	100.00	100.00	100.00
	Z_{10}	35.19994369	35.20	35.20	35.21	35.18
	x_{10}	0.4511739397	0.4512	0.4512	0.4510	0.4511
	y_{10}	0.4059366042	0.4059	0.4059	0.4060	0.4061
	u'_{10}	0.2589645415	0.2590	0.2589	0.2588	0.2588
	v'_{10}	0.5242482977	0.5242	0.5243	0.5243	0.5243
	$\Delta E_{uv,10}$		0.045	0.045	0.183	0.171
	$\Delta E_{ab,10}$		0.027	0.027	0.099	0.117

Although these instruments can be used with different illuminants, they cannot provide diagnostic data on metamerism (to be discussed in Chapter 6).

REFERENCES

ASTM Document E 1164. *Standard practice for obtaining spectrophotometric data for object-color evaluation*. American Society for Testing Materials, Philadelphia (1987, revised 1988 and 1990).

Wharmby, D. O. *J. Phys. E: Scientific Instrum.* **8**, 41 (1975).

GENERAL REFERENCES

Bartleson, C. J. & Grum, F. *Color Measurement*, Academic Press, New York (1980).

Billmeyer, F. W. & Saltzman, M. *Principles of Color Technology*, 2nd. ed., Wiley, New York (1981).

Chamberlin, G. J. & Chamberlin, D. E. *Colour, its Measurement, Computation, and Application*, Heyden, London (1980).

CIE Publication No. 15 *Colorimetry* (1971).

CIE Publication No. 15.2 *Colorimetry*, 2nd ed. (1986).

Judd, D. B. & Wyszecki, G. *Color in Business, Science, and Industry*, 2nd. ed., Wiley, New York (1975).

MacAdam, D. L. *Color Measurement*, Springer-Verlag, Berlin (1981).

McLaren, K. *The Colour Science of Dyes and Pigments*, 2nd ed., Hilger, Bristol (1986).

Wright, W. D. *The Measurement of Colour*, 4th ed., Hilger, Bristol (1969).

Wyszecki, G. & Stiles, W. S. *Color Science*, 2nd ed., Wiley, NewYork (1982).

6

Metamerism

6.1 INTRODUCTION

As mentioned in section 2.7, when two colours match one another, but are different in spectral composition, they are said to be *metameric*, and the phenomenon is referred to as *metamerism*. The phenomenon of metamerism is of great practical importance, because the greater the degree of metamerism, that is, the greater the degree of difference in spectral composition, the greater will be the likelihood that the colours will no longer match one another if one of the conditions of the match is altered, such as the spectral composition of the illumination, or the spectral sensitivity of the observer, or the size of the matching field. In the colorant industries, it is frequently necessary to work with samples that are metameric to one another, and it is important to know to what extent such colours will cease to match under the range of illuminants, observers, and field sizes, that are commonly met with in practice. It is therefore very desirable to have meaningful methods of evaluating any difference in spectral composition between two samples that are a metameric match. Such methods will be described later on in this chapter. But we shall first consider some of the more general aspects of metamerism.

6.2 THE CAUSE OF METAMERISM

As described in section 1.5, the eye responds to light, not on a wavelength by wavelength basis, but as a result of the integrated stimulation of each of the three different cone types, ρ, γ, and β. If two stimuli result in identical ρ, γ, and β cone stimulations, then, when seen under the same conditions, they will look identical whatever their spectral compositions might be.

This is illustrated in Fig. 6.1. The smooth curve is the spectral power distribution of a non-selective neutral grey stimulus, N, illuminated by a tungsten filament light; and the undulating curve is that of a selective grey stimulus, S, that is a metameric match to N under the same illuminant for the CIE 1931 Standard Colorimetric

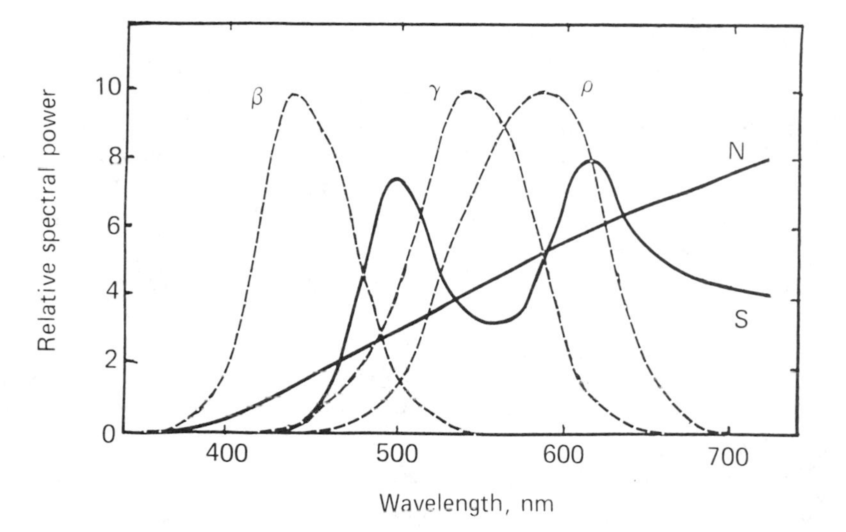

Fig. 6.1 — Full lines: the relative spectral power distributions of two stimuli, N and S, that are metameric for the CIE 1931 Standard Colorimetric Observer. Broken lines: spectral sensitivities representative of those believed to be typical of the three different types of cones, ρ, γ, and β, of the retina of the eye.

Observer. Also shown in the figure, by the broken lines, are the spectral sensitivity curves of the ρ, γ, and β type cones. It is clear by inspection of the figure that, although the spectral power distribution of S is greater in the long wavelength part of the ρ band, it is smaller in the short wavelength part of that band, and the total effect of the S stimulus on the ρ cones could therefore be the same as that of the N stimulus. Similarly, in the γ band the smaller power of the S stimulus in the long wavelength part of the band could be exactly compensated by its greater power in the short wavelength part. Finally, in the case of the β band, the greater power at the long wavelength part could be balanced by the smaller power at the shorter wavelengths. Hence it is clear that the stimuli, N and S, could match one another in spite of their large differences in spectral composition.

For equal ρ, γ, and β responses to result from two different spectral power distributions, it is necessary for their curves to exhibit a crossover point within each of the bands of the spectrum to which the ρ, γ, and β cones are sensitive. There can be more than one crossover point in each band, so that there may be more than three crossovers in total, but there can never be fewer than three. It is therefore a characteristic of metameric pairs of stimuli that their spectral power distributions exhibit three or more crossover points in the visible spectrum; the positions of the crossovers depend on the spectral compositions of the stimuli that are involved (Berns & Kuehni 1990).

6.3 THE DEFINITION OF METAMERISM

Because the exact shapes of the ρ, γ, and β cone sensitivity curves are not known, metameric pairs of colours are defined, not in terms of equality of ρ, γ, and β responses, but as stimuli that have the same tristimulus values. This means that their spectral power distributions, when weighted by the CIE $\bar{x}(\lambda)$, $\bar{y}(\lambda)$, and $\bar{z}(\lambda)$ colour-matching functions, must produce equal results. These colour-matching functions can be regarded as linear combinations of the ρ, γ, and β spectral sensitivities, and hence the arguments about crossovers given above also apply to the colour-matching functions. It is necessary to choose colour-matching functions either for the CIE 1931 Standard Colorimetric Observer, or for the CIE 1964 Supplementary Standard Colorimetric Observer, according to whether the field size is less than or equal to 4°, or greater than 4°, respectively; when the latter is the case all the measures involved are distinguished by a subscript 10.

Equality of tristimulus values is the colorimetric way of defining metamerism. But the term metamerism is also often used when two spectrally different samples are a visual match for an individual real observer, even if the two sets of tristimulus values are not equal; this may be referred to as *perceived metamerism* when it is desirable to distinguish it from equality of tristimulus values, which may be regarded as *psychophysical metamerism*. (For a summary of terms used in connection with metamerism, see section 6.11.)

6.4 EXAMPLES OF METAMERISM IN PRACTICE

In the colour reproduction industries, the multitude of spectral power distributions of colours in original scenes is generally reproduced by just three colorants (Hunt 1987). Thus, in photography, all the displayed colours are mixtures of cyan, magenta, and yellow dyes. In Fig. 6.2 are shown the spectral reflectance factors of a non-selective neutral grey (broken line), and of a mixture of cyan, magenta, and yellow dyes (full line) that is metameric to it for D_{65}; it is clear that there are considerable differences between the two curves, and that there are seven crossover points. In Fig. 6.3, the relative spectral power distributions of three phosphors are shown that are typical of those used in colour television. Mixtures of these three distributions will obviously result in composite spectral power distributions that undulate considerably throughout the spectrum, and especially so at the long wavelength end of the spectrum where the very spiky nature of the red rare-earth phosphor will be much in evidence; hence, the reproductions will be very metameric to the smooth spectral power distributions typical of most colours in original scenes. In the case of graphic arts printing, the cyan, magenta, and yellow inks absorb in similar parts of the spectrum as the dyes used in photography. But a black ink is usually used in addition, and this can reduce the degree of metamerism in the case of dark colours; and sometimes extra inks of special colours are used, and this can reduce metamerism further in some cases.

In the colorant industries, metamerism can arise when colour matches are required between different types of material. For instance, in fashion design, the same colour may be required in a dress material, in plastic buttons, and in leather shoes; or, in the automobile industry, the interior colours of the seat material, the carpet, the paint work, and the plastic trimmings may be required to match. In these applications, the use of dyes and pigments in different media precludes achieving the

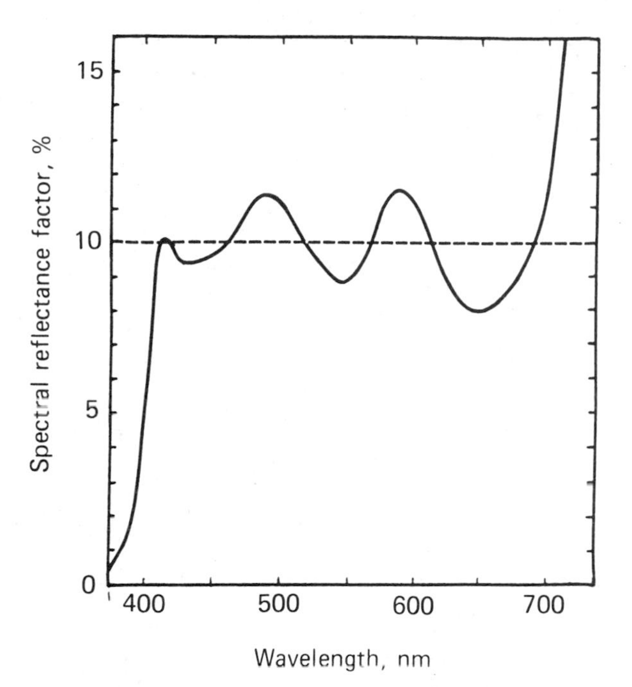

Fig. 6.2 — Spectral reflectance factor of a colour photographic grey composed of cyan, magenta, and yellow dyes (full line) that is a metameric match to a non-selective grey (broken line) in Standard Illuminant D_{65} for the CIE 1931 Standard Colorimetric Observer.

same spectral reflectance in all cases, but the colorists will seek to minimize the unavoidable differences by careful choice of dye or pigment mixtures.

Keeping metamerism to a minimum is thus an important consideration in the choice of colorants. Generally, of course, the smaller the differences in spectral power distribution the better. But where differences are unavoidable their effects should be evaluated by using the appropriate indices of metamerism as discussed in sections 6.6, 6.7, 6.8, and 6.9. Experience results in those skilled in the art knowing to some extent what differences in spectral power distribution are least harmful (Pinney & DeMarsh 1963).

6.5 DEGREE OF METAMERISM

Fig. 6.4 shows the spectral reflectance curves of three pairs of stimuli that are metameric in Standard Illuminant C for the CIE 1931 Standard Colorimetric Observer. It is intuitive that the greater the difference between the spectral power distributions of a metameric pair of stimuli, or between the spectral reflectance or transmittance curves of a pair of samples that are metameric for a given illuminant, then the greater will be the colour difference between the pair when there is a change in observer, field size, or illuminant. Although this is generally the case, the concept

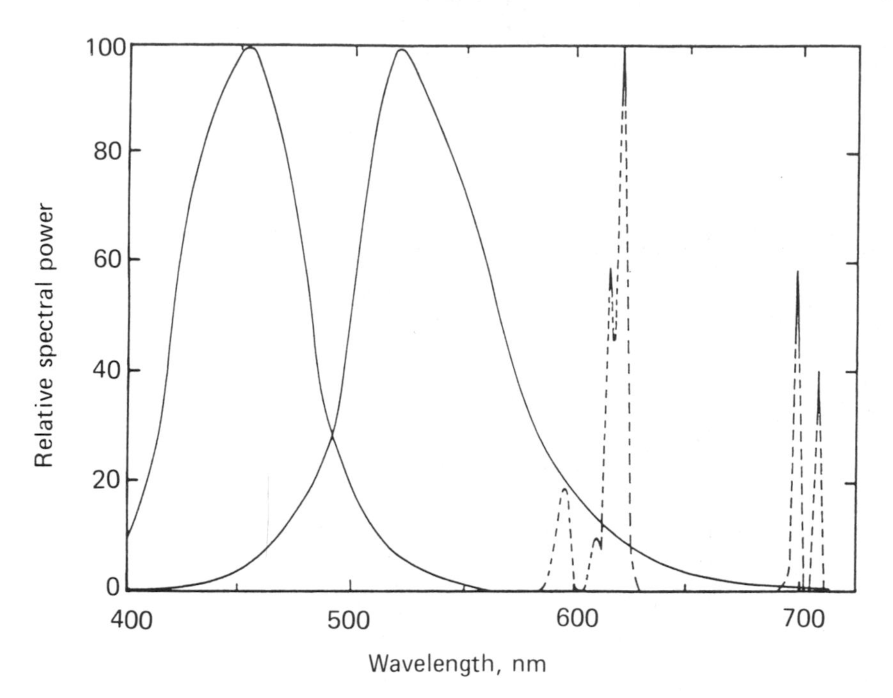

Fig. 6.3 — Relative spectral power distributions typical of those produced by the red, green, and blue emitting phosphors used in colour television displays and Video Display Units (VDUs).

of a general index of metamerism that would represent this phenomenon quantitatively is not soundly based. If the differences in spectral composition are averaged throughout the spectrum (Nimeroff & Yurow 1965), then the decreasing importance of differences as the two ends of the spectrum are approached is ignored. The differences therefore need to be weighted appropriately (Nimeroff 1968); but the correct weightings must depend on the relative importance of different parts of the spectrum, and this in turn is governed by whether the change is in observer, in field size, or in illuminant, and, in the latter case, which particular illuminants are used. Instead of seeking a general index of metamerism, it is therefore more justifiable to use different indices according to whether the change is in observer, in field size, or in illuminant. These *special indices of metamerism* will now be considered.

6.6 INDEX OF METAMERISM FOR CHANGE OF ILLUMINANT

The CIE has recommended that the degree of metamerism for changes of illuminant be evaluated by calculating an *Illuminant Metamerism Index*, *M*, consisting of the size of the colour difference between a metameric pair caused by substituting, in the place of a reference illuminant, a test illuminant having a different spectral composition. If the tristimulus values of the two samples, 1 and 2, under the test illuminant

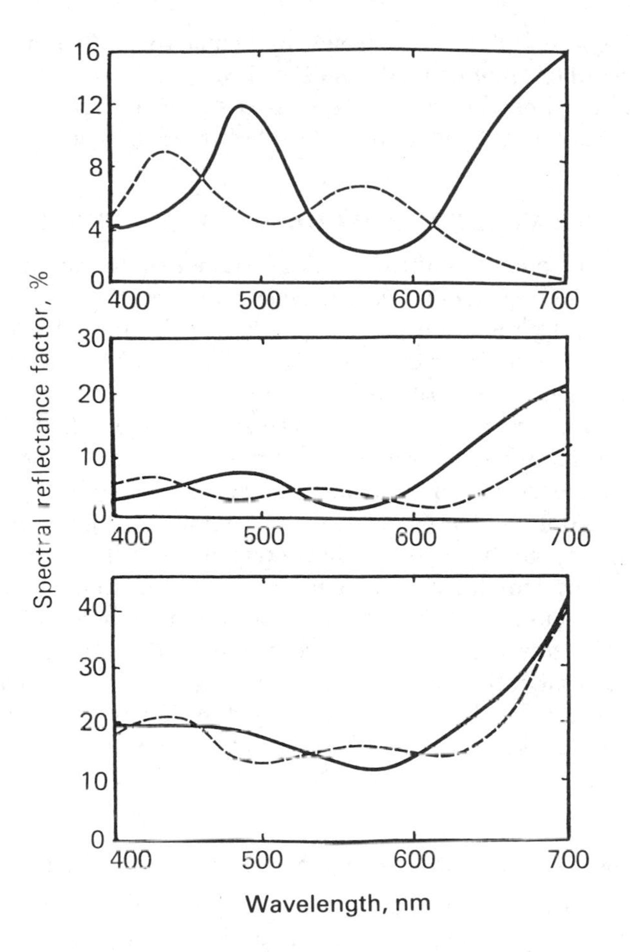

Fig. 6.4 — Spectral reflectance factors of three pairs of dyed fabric which are metameric in Standard Illuminant C for the CIE 1931 Standard Colorimetric Observer.

are $X_{1,t}$, $Y_{1,t}$, $Z_{1,t}$ and $X_{2,t}$, $Y_{2,t}$, $Z_{2,t}$, respectively, then the metamerism index is obtained by calculating the corresponding colour difference; this colour difference should preferably be calculated according to one of the recommended CIE colour-difference formulae, but if a different formula is used this should be stated (CIE 1986). It should be borne in mind that, as mentioned in section 3.13, if a colour difference formula is designed for one illuminant, it may not assess colour differences sufficiently accurately in another illuminant unless corrections are made for chromatic adaptation, for instance by one of the methods given in section 3.13 (Berns & Billmeyer 1983).

The preferred reference illuminant is Standard Illuminant D_{65}; if a different reference illuminant is used this should be stated. Suitable test illuminants include CIE Standard Illuminant A to represent tungsten light, and the illuminants listed in

Appendix 5.3 to represent fluorescent lamps, particularly F2, F7, and F11. The most appropriate choice of test illuminant depends on the application, and it may be useful to determine the metamerism index with respect to several test illuminants. The specific test illuminant must be identified as a subscript to M, such as M_A, M_{F1}, etc.

6.7 INDEX OF METAMERISM FOR CHANGE OF OBSERVER

The CIE has recommended that the degree of metamerism for changes of observer be evaluated by calculating an *Observer Metamerism Index*, M_2 or M_{10}, consisting of the size of the colour difference between a metameric pair caused by substituting, in the place of a reference observer, a standard deviate observer (SDO) having different spectral sensitivities (CIE 1989). The reference observer can be either the CIE Standard Colorimetric Observer (the 2° Observer) or the CIE 1964 Supplementary Standard Colorimetric Observer (the 10° Observer); the symbols M_2 and M_{10}, respectively, can be used to denote which observer is being used. The Standard Deviate Observer is obtained by changing the CIE colour-matching functions of the reference observer by applying the modifications given in Table 6.1. The colour difference should be calculated according to one of the recommended CIE colour-difference formulae, but if a different formula is used this should be stated.

The use of the Standard Deviate Observer is intended to generate values of M_2 and M_{10} that are typical of the colour differences that occur when matches are made by different real observers whose colour vision is classified as normal (that is, those who are not colour deficient, see section 1.10). (Ohta 1985, Nayatani, Hoshimoto, Takahama & Sobagaki 1985.)

6.8 INDEX OF METAMERISM FOR CHANGE OF FIELD SIZE

Although the CIE has not recommended an index of metamerism for a change of field size, the availability of the two CIE Standard Colorimetric Observers makes such assessments possible for changes between 2° and 10°. The 1931 Observer defines colour matches for 2° fields, and the 1964 Supplementary Observer for 10° fields. Hence, for a pair of samples that are a metameric match for the 1931 Observer, their colour difference when evaluated by using the 1964 Observer can be regarded as a measure of their metamerism for a change in field size. A similar assessment can be made in the reverse situation, where the samples are a metameric match for the 1964 Observer. In either case, the colour difference can be calculated by using one of the recommended CIE colour difference formulae, but if a different formula is used this should be stated.

6.9 COLOUR MATCHES AND GEOMETRY OF ILLUMINATION AND VIEWING

If two surface reflecting samples match one another in one set of conditions of illuminating and viewing geometry, they will continue to match in a different geometry only if their surface characteristics are such that their gloss properties are the same. But if, for instance, one sample is glossy and the other matt, then a change

Table 6.1 — Modifications of CIE colour-matching functions to obtain a standard deviate observer

Wave-length nm	$\Delta\bar{x}$	$\Delta\bar{y}$	$\Delta\bar{z}$	nm	$\Delta\bar{x}$	$\Delta\bar{y}$	$\Delta\bar{z}$
380	−0.0001	0.0000	−0.0002	580	−0.0600	−0.0126	−0.0013
385	−0.0003	0.0000	−0.0010	585	−0.0637	−0.0162	−0.0011
390	−0.0009	−0.0001	−0.0036	590	−0.0656	−0.0196	−0.0009
395	−0.0026	−0.0004	−0.0110	595	−0.0638	−0.0199	−0.0008
400	−0.0069	−0.0009	−0.0294	600	−0.0595	−0.0187	−0.0006
405	−0.0134	−0.0015	−0.0558	605	−0.0530	−0.0170	−0.0005
410	−0.0197	−0.0019	−0.0820	610	−0.0448	−0.0145	−0.0004
415	−0.0248	−0.0022	−0.1030	615	−0.0346	−0.0112	0.0000
420	−0.0276	−0.0021	−0.1140	620	−0.0242	−0.0077	0.0002
425	−0.0263	−0.0017	−0.1079	625	0.0155	−0.0048	0.0000
430	−0.0216	−0.0009	−0.0872	630	−0.0085	−0.0025	−0.0002
435	−0.0122	0.0005	−0.0455	635	−0.0044	−0.0012	−0.0002
440	−0.0021	0.0015	−0.0027	640	−0.0019	−0.0006	0.0000
445	0.0036	0.0008	0.0171	645	−0.0001	0.0000	0.0000
450	0.0092	−0.0003	0.0342	650	0.0010	0.0003	0.0000
455	0.0186	−0.0005	0.0703	655	0.0016	0.0005	0.0000
460	0.0263	−0.0011	0.0976	660	0.0019	0.0006	0.0000
465	0.0256	−0.0036	0.0859	665	0.0019	0.0006	0.0000
470	0.0225	−0.0060	0.0641	670	0.0017	0.0006	0.0000
475	0.0214	−0.0065	0.0547	675	0.0013	0.0005	0.0000
480	0.0205	−0.0060	0.0475	680	0.0009	0.0003	0.0000
485	0.0197	−0.0045	0.0397	685	0.0006	0.0002	0.0000
490	0.0187	−0.0031	0.0319	690	0.0004	0.0001	0.0000
495	0.0167	−0.0037	0.0228	695	0.0003	0.0001	0.0000
500	0.0146	−0.0047	0.0150	700	0.0002	0.0001	0.0000
505	0.0133	−0.0039	0.0117	705	0.0001	0.0000	0.0000
510	0.0118	−0.0060	0.0096	710	0.0001	0.0000	0.0000
515	0.0094	−0.0025	0.0062	715	0.0001	0.0000	0.0000
520	0.0061	0.0010	0.0029	720	0.0000	0.0000	0.0000
525	0.0017	0.0005	0.0005	725	0.0000	0.0000	0.0000
530	−0.0033	−0.0011	−0.0012	730	0.0000	0.0000	0.0000
535	−0.0085	−0.0020	0.0020	735	0.0000	0.0000	0.0000
540	−0.0139	−0.0028	−0.0022	740	0.0000	0.0000	0.0000
545	−0.0194	−0.0039	−0.0024	745	0.0000	0.0000	0.0000
550	−0.0247	−0.0044	−0.0024	750	0.0000	0.0000	0.0000
555	−0.0286	−0.0027	−0.0021	755	0.0000	0.0000	0.0000
560	−0.0334	−0.0022	−0.0017	760	0.0000	0.0000	0.0000
565	−0.0426	−0.0073	−0.0015	765	0.0000	0.0000	0.0000
570	−0.0517	−0.0127	−0.0014	770	0.0000	0.0000	0.0000
575	−0.0566	−0.0129	−0.0013	775	0.0000	0.0000	0.0000
				780	0.0000	0.0000	0.0000

in geometry must be expected to result in a breakdown of the match. For example, if the two colours match in diffuse illumination and normal viewing (d/0), then changing to 45° illumination and normal viewing (45/0), will tend to make the glossy sample look darker and (unless it is a neutral grey) more saturated. Changes of this type occur even if, in the matching geometry, the samples have identical spectral reflectances, and are therefore a spectral, rather than a metameric, match. Hence, although the term *geometric metamerism* has sometimes been used for this phenomenon, it is not strictly correct, because, in colorimetry, the term *metamerism* implies a difference in spectral composition. However, whatever term is used to denote it, the effect of geometry on the colour matching of samples of different gloss properties can be very considerable, sometimes causing a good match to become a very obvious mismatch. Once again, the resulting colour difference can be calculated by using one of the recommended CIE colour difference formulae, but if a different formula is used this should be stated.

These effects are of considerable importance in the paint industries, particularly because of the widespread use of metallic and pearlescent pigments in automotive finishes; the full evaluation of such coatings requires instruments capable of measuring spectral distributions for various angles of illumination and viewing (goniospectrophotometers). Some *multi-geometry* instruments are arranged to view always at 45°, but to illuminate at angles that differ from the viewing direction by various amounts, such as 105°, 75°, 45° (that is normal), 20°, and 10°.

6.10 CORRECTING FOR INEQUALITIES OF TRISTIMULUS VALUES

The two samples whose metamerism index is to be evaluated may fail to have exactly the same tristimulus values in the reference condition, a phenomenon sometimes referred to as *paramerism* (Kuehni 1983); in this case a suitable means of correcting for this should be used and explained. Three different methods of correction have been suggested.

The first method is to multiply the X tristimulus value of sample 2 in the test condition, by X_{1r}/X_{2r}, where X_{1r} and X_{2r} are the X tristimulus values for samples 1 and 2, respectively, in the reference condition, with analogous corrections for the Y and Z tristimulus values of sample 2 in the test condition (Brockes 1970, McLaren 1986).

The second method is similar to the first but uses additive, instead of multiplicative, corrections; thus the X tristimulus value of sample 2 in the test condition has added to it $X_{1r}-X_{2r}$, with analogous corrections for the Y and Z tristimulus values (Brockes 1970).

The third method involves changing the spectral power distribution of sample 2 so as to obtain an exact match in the reference condition, but making this change so as to retain, as far as possible, the differences in spectral power distribution that are attributable to metamerism. The procedure involved is as follows (Fairman 1987):

Step 1. Regard the three colour-matching functions of the Standard Observer being used as three spectral power distributions $P_X(\lambda)$, $P_Y(\lambda)$, $P_Z(\lambda)$.

Step 2. Find the quantities of $P_X(\lambda)$, $P_Y(\lambda)$, and $P_Z(\lambda)$ that, when added together, produce a spectral power distribution, $M_1(\lambda)$, that is an exact metameric match to the spectral power distribution, $P_1(\lambda)$, of sample 1 in the reference condition.

Step 3. Find the quantities of $P_X(\lambda)$, $P_Y(\lambda)$, and $P_Z(\lambda)$ that, when added together, produce a spectral power distribution, $M_2(\lambda)$, that is an exact metameric match to the spectral power distribution, $P_2(\lambda)$, of sample 2 in the reference condition.

Step 4. From $P_2(\lambda)$ subtract $M_2(\lambda)$ to obtain the spectral power distribution:

$$P_2(\lambda)-M_2(\lambda)$$

This spectral power distribution will have both positive and negative values, and, because $P_2(\lambda)$ and $M_2(\lambda)$ are an exact metameric match, its tristimulus values will all be zero; hence it may be thought of as a 'metameric black'.

Step 5. To the spectral power distribution, $M_1(\lambda)$, found in Step 2, add that of the metameric black obtained in Step 4, to obtain a corrected spectral power distribution, $P_{2c}(\lambda)$, for sample 2 in the reference condition:

$$P_{2c}(\lambda)=M_1(\lambda)+P_2(\lambda)-M_2(\lambda)$$

Because $M_1(\lambda)$ matches $P_1(\lambda)$ exactly, and $P_2(\lambda)-M_2(\lambda)$ have zero tristimulus values, $P_{2c}(\lambda)$ matches $P_1(\lambda)$ exactly.

Step 6. Use $P_{2c}(\lambda)$ instead of $P_2(\lambda)$ in determining the metamerism index.

6.11 TERMS USED IN CONNECTION WITH METAMERISM

The terms for metamerism included in the 4th edition of the CIE International Lighting Vocabulary (CIE 1987) are as follows:

metameric colour stimuli, *metamers*
: Spectrally different colour stimuli that have the same tristimulus values. The corresponding property is called *metamerism*.

The following additional terms or meanings, although not endorsed by the CIE, are also found in the literature (Fairman 1986):

(perceived) metameric colour stimuli, *(perceived) metamers*
: Spectrally different colour stimuli that are a visual match for a particular real observer under specified viewing conditions. The corresponding property is called *(perceived) metamerism*. (When the meaning is clear from the context the adjective *perceived* can be omitted.)

parameric colour stimuli, paramers

Spectrally different colour stimuli that have nearly the same tristimulus values. The corresponding property is called *paramerism*.

indices of metamerism potential

Indices indicating the degree to which two metameric specimens may develop metamerism, derived solely from their different spectral characteristics.

special indices of metamerism

Indices of degree of metamerism associated with specific changes in illuminating or viewing conditions, such as change of illuminant, change of observer, or change of field size.

REFERENCES

Berns, R. S. & Billmeyer, F. W. *Color Res. Appl.*, **8**, 186 (1983).
Berns, R. S. & Kuehni, R. G. *Color Res. Appl.*, **15**, 23 (1990).
Brockes, A. *Die Farbe*, **19**, 135 (1970).
CIE Publication No. 15.2, *Colorimetry* (1986).
CIE Publication No. 17.4, *International lighting vocabulary* (1987).
CIE Publication No. 80, *Special metamerism index: change in observer* (1989).
Fairman, H. S. *Color Res. Appl.*, **11**, 80 (1986).
Fairman, H. S. *Color Res. Appl.*, **12**, 261 (1987).
Hunt, R. W. G. *The Reproduction of Colour* 4th Ed. Fountain Press, Surbiton (1987).
Kuehni, R. G. *Color Res. Appl.*, **8**, 192 (1983).
McLaren, K. *The Colour Science of Dyes and Pigments* 2nd Ed. Adam Hilger, Bristol (1986).
Ohta, N. *Color Res. Appl.*, **10**, 156 (1985).
Nayatani, Y., Hashimoto, K., Takahama, K. & Sobagaki, H. *Color Res. Appl.*, **10**, 147 (1985).
Nimeroff, I. *Color Eng.*, **6**(Issue 6), 44 (1968).
Nimeroff, I. & Yurow, J. A. *J. Opt. Soc. Amer.*, **55**, 185 (1965).
Pinney, J. E. & DeMarsh, L. E. *J. Phot. Sci.*, **11**, 249 (1963).

7

Colour order systems

7.1 INTRODUCTION

Collections of colour samples are often used to provide examples of colour products. These may be in the form of patches of paint, swatches of cloth, pads of papers, or printings of inks, for example, according to the type of product involved. Such examples are very useful if the number of colours required is fairly limited. But, if a very large number of colours is necessary, and, if intermediate colours lying between samples are to be considered, a system is required in which interpolation can be made between samples in an unambiguous way. Such systems are referred to as *colour order systems*.

7.2 VARIABLES

To facilitate interpolation between samples, it is usually helpful to arrange them according to major perceptual attributes of colour. If an observer is given a random collection of colour samples and asked to systematize them, it is likely that the first step would be to sort them according to their hues. Thus, all the reddish colours, for example, could be separated into one group, all the yellowish into another, and so on for the greenish, bluish, and purplish colours; and another group would have to be provided for all the colours that did not exhibit any hue at all, the white, grey, and black colours. The hue groups could then be further subdivided into intermediate hues, such as yellow–reds (that is, orange colours), green–yellows, blue–greens, purple–blues, and red–purples. Even finer subdivisions can be made, it being possible to identify constancy of hue for samples with considerable precision, particularly for saturated colours, even though different colour names may not suggest themselves for each sub-group.

Having classified the colours according to hue, the next most obvious step would probably be to classify them according to lightness. Thus, for each hue group, the observer might put all the light colours at the top of the arrangement, all the dark colours at the bottom, with the colours of medium lightness in between. The distance of a colour from the bottom could then be made to represent the lightness of that colour. The whites, greys, and blacks could be arranged as a separate group in a single column, in the same way, to form a grey scale with white at the top and black at the bottom.

The observer would then most likely notice that some colours were more colourful than others, and could arrange the least colourful, or pale, samples near one side of the array, say the left-hand side, and the most colourful, or vivid, samples towards the right-hand side. In this case, the distance of the colour from the left-hand side could be made to represent the colourfulness, as shown in Fig. 7.1. However, the

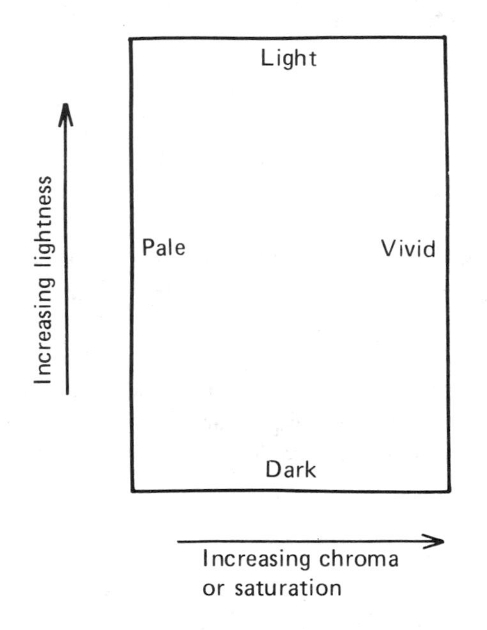

Fig. 7.1 — Two-dimensional array of lightness, and chroma or saturation.

colourfulness could be judged either as chroma (that is, in proportion to the brightness of a reference white for the samples), or as saturation (that is, in proportion to the brightness of the sample itself), as discussed in section 1.9.

If now the observer were to attach the samples to suitable pages, the pages could be arranged in a three-dimensional array, with the grey scale as a central vertical axis, and the different hue groups joining this axis at their left-hand sides, as shown in Fig. 7.2. The different hue groups could be arranged in the same order as the colours of the spectrum, with the purple colours completing a circle between blue and red. This three-dimensional array of colours then provides a systematic arrangement in which interpolation is possible in any direction, so that all colours can be represented that lie within the limits imposed by the colours lying at the extreme edges of the array.

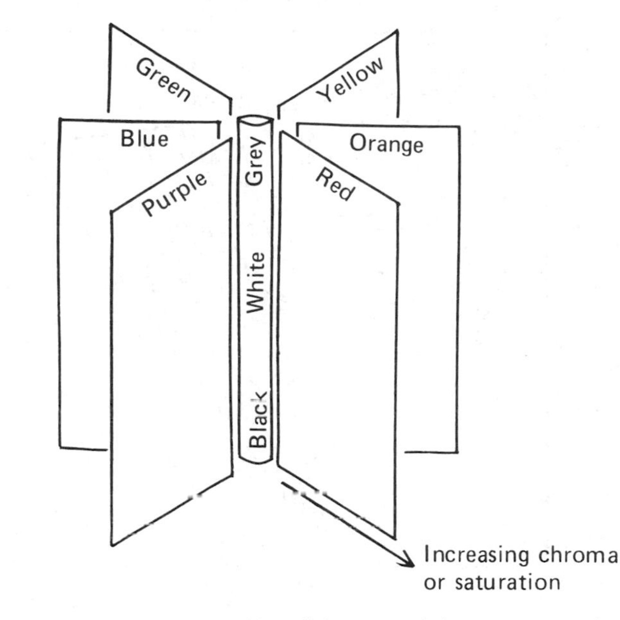

Fig. 7.2 — Three-dimensional array of hue, lightness, and chroma or saturation.

7.3 OPTIMAL COLOURS

There are theoretical limits to the extent of colour arrays of the type described in the previous section. To produce a chromaticity different from that of the illuminant, non-self-luminous samples must absorb some of the light. Assuming, for the moment, that no absorbed radiant power is reradiated by fluorescence, the absorption must result in a reduction in luminance factor; this reduction is minimal when, at each wavelength, the sample either absorbs all, or none, of the light (Rösch 1928, MacAdam 1937). Colours having this property are called *optimal colours*. In Fig. 7.3, the loci of these optimal colours are shown for various luminance factors in the u',v' diagram for Standard Illuminant C, (S_C). If the luminance factor is high, the area within the optimal colour locus is small; if the luminance factor is low, the area becomes much larger. Thus, high values of u,v saturation, s_{uv}, can be attained only if the luminance factor is low. However, there is an exception in the case of yellow colours, which can have both high s_{uv} and high luminance factors; this is because the absorption of the blue light, required to produce the yellow colours, results in only a slight reduction in luminance factor, on account of the small contribution of the β cones to the achromatic signal (as discussed in section 1.6).

Real surface and transmitting colours (that do not fluoresce) are even more restricted in chromaticity than indicated in Fig. 7.3, because, at some wavelengths, they always have absorptions that are intermediate between total and zero. Surface colours are further restricted, because light reflected from their topmost surface results in their minimum possible reflections usually being a few per cent instead of zero, at any wavelength. This greatly reduces the range of chromaticities possible for dark colours, and, in Fig. 7.4, loci for optimal surface colours having a top-surface

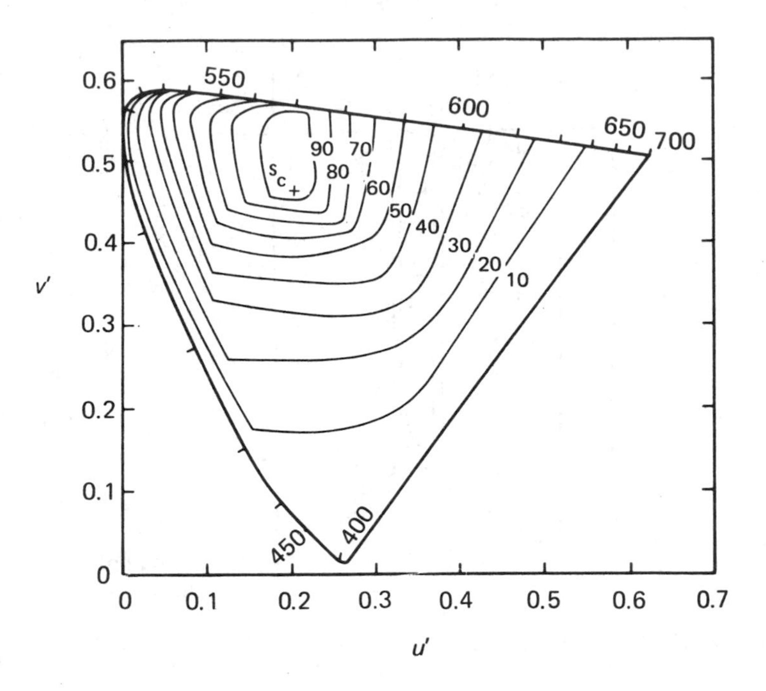

Fig. 7.3 — Limits of optimal colours on u',v' chromaticity diagram for a daylight illuminant (Standard Illuminant C).

reflection factor of 0.56% are shown (by the full lines) in plots of L^* against C^*_{uv} for several hues for a daylight illuminant (S_C). Also shown in this figure (by the broken lines) are the maximum gamuts of surface colours found in practice in a study carried out by Pointer (Pointer 1980); it is clear that the gamuts of these real colours are substantially smaller than those of the optimal colours, except for yellows.

If colours fluoresce, they can lie outside the limits defined by the optimal colours; and if they do lie outside, they usually have a different quality of appearance which Evans termed *fluorent* (Evans & Swenholt 1967, 1969), a term applicable whether physical fluorescence is occurring or not.

7.4 THE MUNSELL SYSTEM

One of the most widely used colour order systems is the *Munsell system*, originated by the artist A. H. Munsell in 1905, and extended and refined in various ways since (Nickerson 1976). The general arrangement follows broadly that depicted in Figs 7.1 and 7.2. An important feature of the Munsell system is that the colours are arranged so that, for each perceptual attribute used, as nearly as possible the perceptual difference between any two neighbouring samples is constant (Berns & Billmeyer 1985). (See Plate 4, pages 112, 113.)

In the case of the grey scale, there are ten main steps, with white designated 10, and black zero, the greys having values from 1 to 9 as they become lighter. The

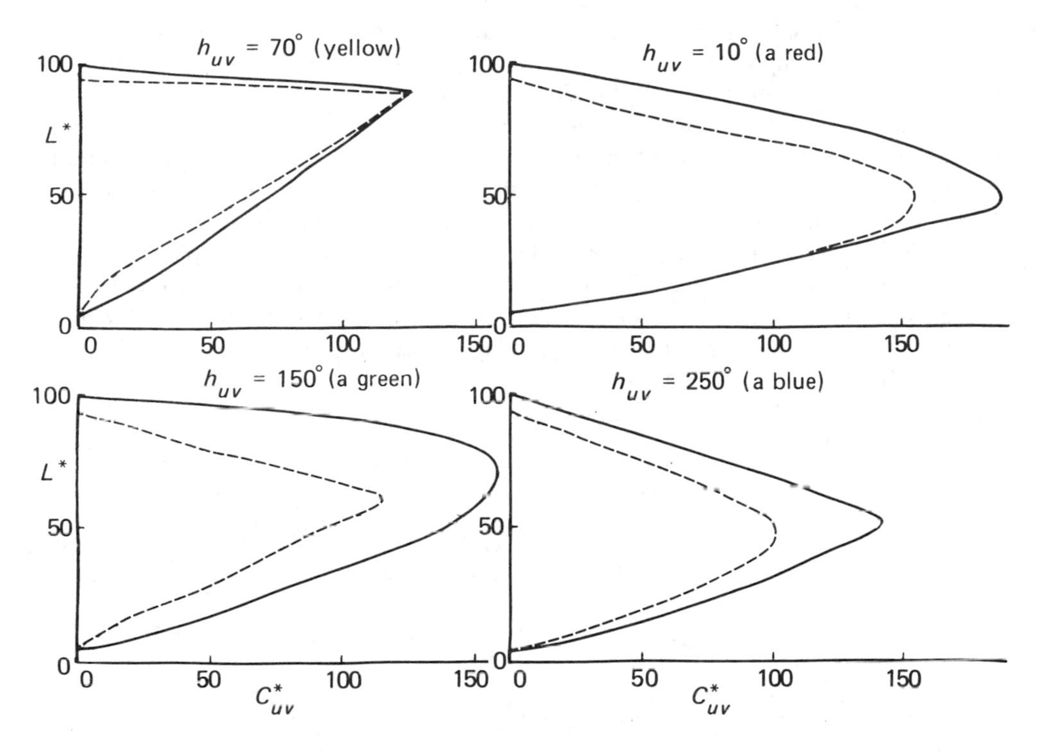

Fig. 7.4 — Limits of optimal colours (full lines) and of real surface colours (broken lines), for four hues on plots of L^* against C^*_{uv} for a daylight illuminant (S_C). The hues have values of h_{uv} equal to 10° (a red), 70° (a yellow), 150° (a green), and 250° (a blue).

difference in lightness between any neighbouring pair of samples, say 3 and 4, is then intended to be perceptually as great as between any other neighbouring pair, say 7 and 8. The luminance factors that the grey samples need to have to achieve this result are affected to some extent by the background against which the samples are viewed. In the case of the Munsell system, the judgements used in deriving it were based on viewing the samples against a medium grey background having a luminance factor of about 20%. The relationship between luminance factor and number on this scale is used for all the coloured samples as well as for the greys, and is referred to as *Munsell Value*. The Munsell Value, V, of a sample can be intermediate between the whole numbers of the eleven basic samples, and it is designated by using decimals; for example, a Value of 7.5 would be intended to be perceptually midway in lightness between samples having Values of 7 and 8.

The spacing of the hues around the grey scale axis is also intended to represent uniform differences in perceived hue between neighbouring hue pages. There are five principal hues, Red, Yellow, Green, Blue, and Purple, and they are designated: 5R, 5Y, 5G, 5B, and 5P, respectively. The intermediate hues are designated: 5YR, 5GY, 5BG, 5PB, and 5RP. Finer divisions between 5R and 5YR are designated: 6R, 7R, 8R, 9R, 10R, 1YR, 2YR, 3YR, and 4YR, with similar designations between other hues, as shown in Fig. 7.5. Once again, finer divisions are represented by decimals; for example, a Munsell Hue of 2.5YR is intended to be perceptually midway between samples having Munsell Hues of 2YR and 3YR.

The distances of the samples from the grey scale axis are intended to represent uniform differences in perceived chroma (not saturation) and are given numbers that are typically as small as 4 or less for weak colours, and 10 or more for strong colours. Munsell Chroma is indicated by an oblique line preceding the numerical value, for example /8. Once again decimals are used to indicate intermediate samples between integers; thus, for example, /8.5 indicates that the sample is intended to be perceptually midway in chroma between samples having Munsell Chromas of /8 and /9. In Fig. 7.5 the dots represent samples having Munsell Chromas of /2, /4, /6, /8, and /10, for Munsell Hues at the 5 and 10 positions.

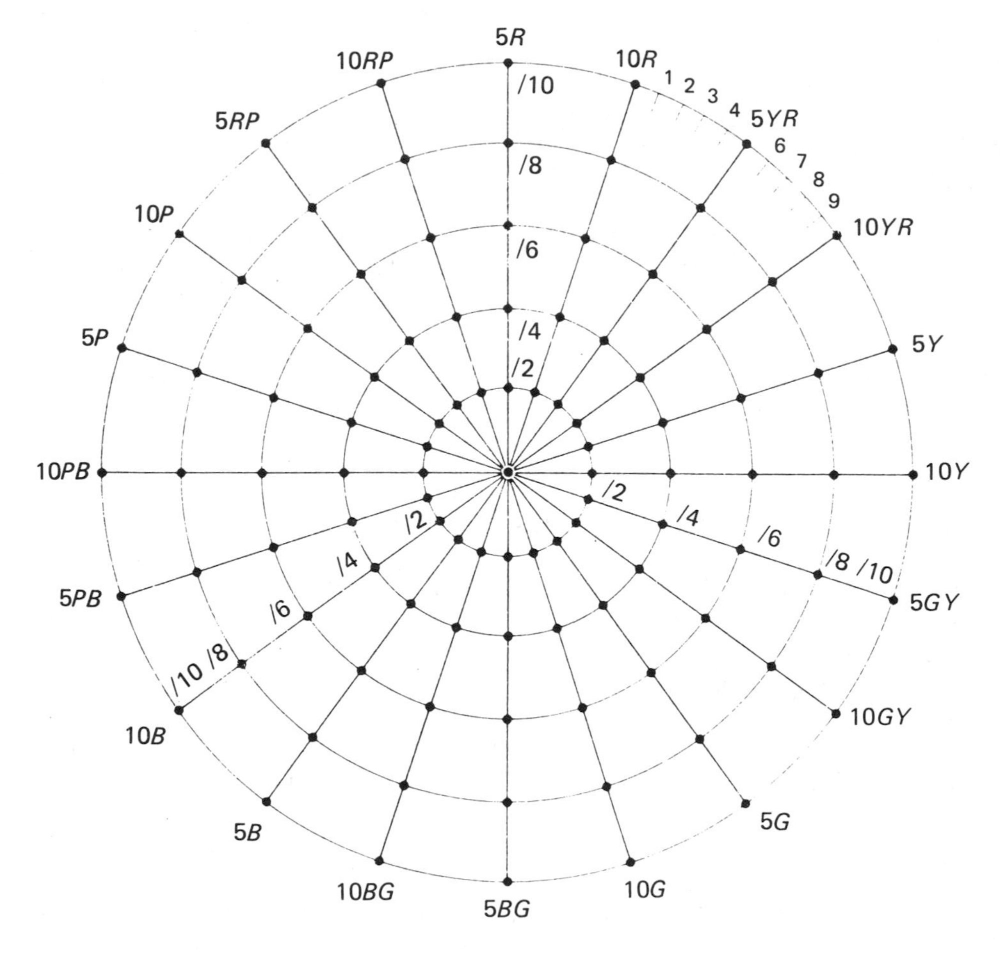

Fig. 7.5 — Arrangement of Hue circle in the Munsell system.

The full Munsell specification is always given in the order Hue, Value, Chroma (sometimes referred to as HVC), for example, 2.5Y6/8 indicates that the Hue is half way between 10YR and 5Y (so that it is a slightly orange yellow, 10YR being a yellowish orange and 5Y a yellow), that the lightness is slightly lighter than a medium grey (which would have a Value of 5), and that it has a fairly strong Chroma.

The maximum designation possible for Munsell Chroma depends on the hue and lightness of the colour; for dark blues it is larger than for light blues (this is because of the small contribution of the β cones to the achromatic signal); and for medium-lightness reds it is larger than for medium-lightness greens (this is because of the overlapping of the cone spectral sensitivity curves in the green part of the spectrum). The outer boundary of the Munsell Colour space is therefore not symmetrical.

If two samples of the same Munsell Hue and Chroma, but differing by 1 unit in Value, are compared with two samples of the same Hue and Value, but differing by 2 units in Chroma, it is found that the two differences are perceptually similar in magnitude; in other words, a difference of 1 unit of Value is similar in perceptual magnitude to a difference of 2 units of Chroma. This difference is also similar in perceptual magnitude to that between two samples having the same Value, and both having a Chroma of 5, but differing by 3 units of Hue. For samples of higher Chroma, the perceptual magnitude of the hue difference would be greater, and for samples of lower Chroma it would be less; this is because of the angular arrangement of the Hue parameter in the Munsell System.

7.5 THE MUNSELL BOOK OF COLOR

So far we have considered only the Munsell *system*, the perceptual basis on which the ordering of the colours depends. We must now consider the actual physical samples, and the way they are arranged to form an atlas, in this case the *Munsell Book of Color*.

The samples in the *Book* consist of painted paper *chips*, typically about 17×20 mm in size; larger size pieces are also available, if required for special purposes. Most of the pages in the *Book* are laid out in the manner shown (for two hues) in Fig. 7.6; each page shows samples of a single Munsell Hue, there being 40 such pages, those included being for Munsell Hues 2.5YR, 5YR, 7.5YR, 10YR, 2.5Y, and so on right round the hue circle to 10R, and so back to 2.5YR.

In 1943, a report was issued (Newhall, Nickerson & Judd 1943) in which, for Standard Illuminant C (S_C), the X, Y, Z, tristimulus values of the samples in the *Munsell Book of Color* were carefully studied. It was found that, although the transitions in the tristimulus values from one sample to a neighbouring one should be smooth, there were some irregularities. A system for which the tristimulus values were smooth was then produced, known as *Munsell Renotation*. Chips made to these specifications were used for the glossy sample version of the *Book*, introduced in 1957. For the original *Book* of matt samples, the chips still used the original notation for some years, but since 1957 all the *Books* produced, both matt and glossy, have been based on the renotation, which is now referred to simply as *Munsell Notation* (Davidson, Godlove & Hemmendinger 1957).

Tables of the x and y chromaticity co-ordinates and the Y tristimulus values (for Standard Illuminant C), corresponding to Munsell Notations can be found in various

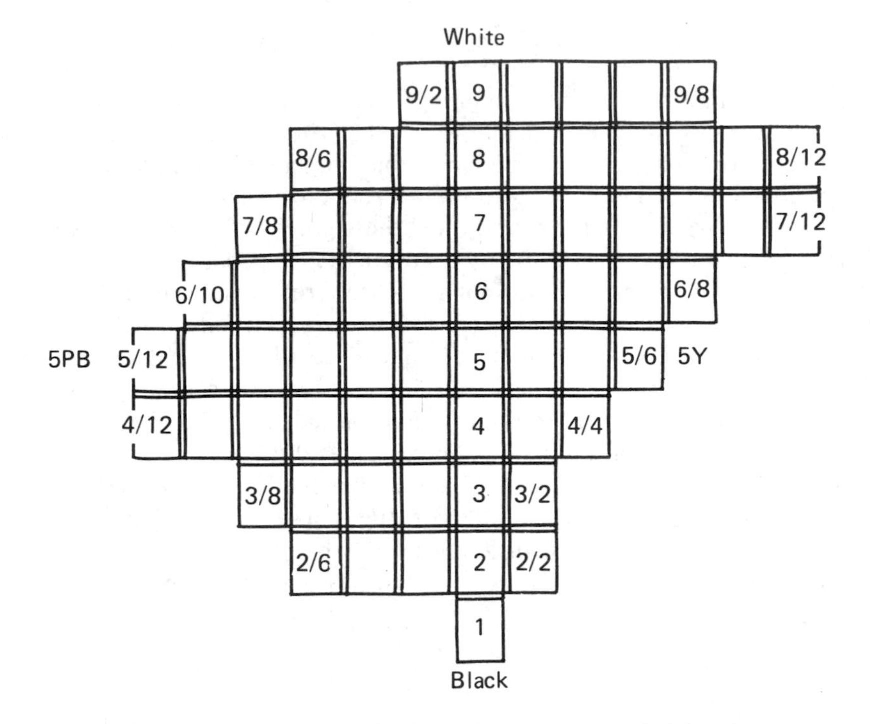

Fig. 7.6 — Arrangement of colours of constant Hue in the Munsell system.

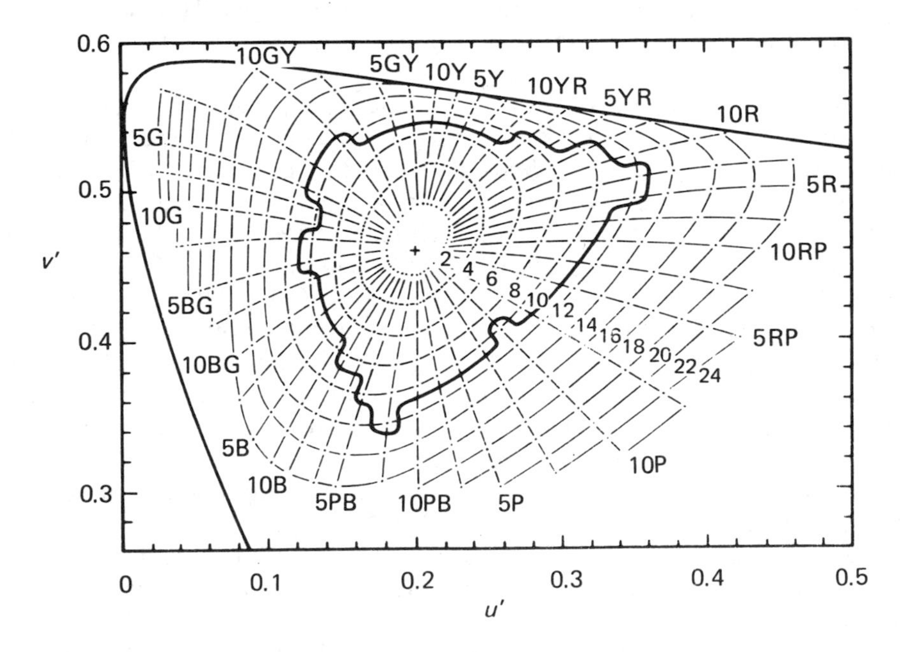

Fig. 7.7 — Chromaticities of colours of Munsell Value 5 shown on the u',v' diagram, for Standard Illuminant C (shown as +).

(a)

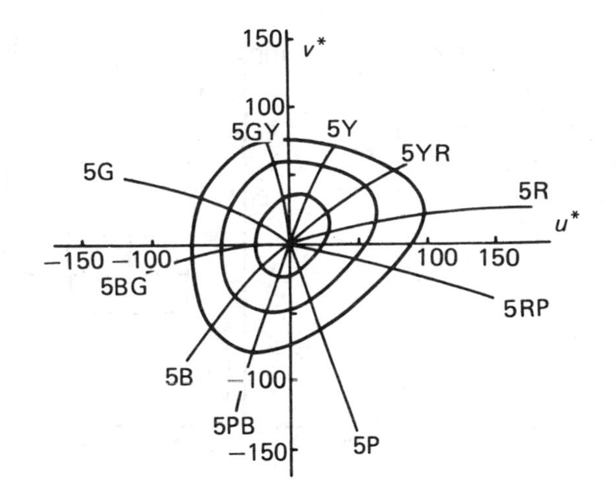

(b)

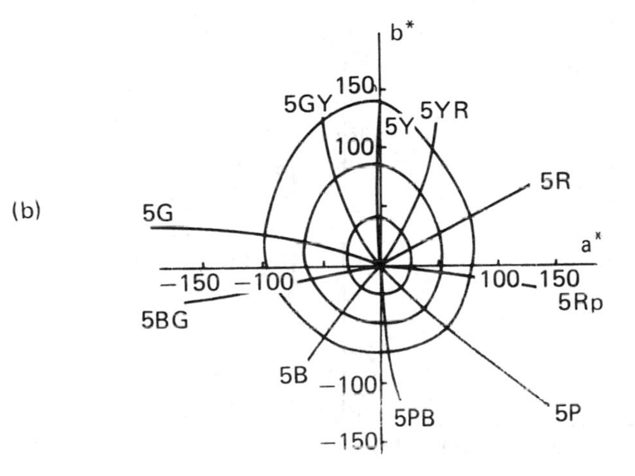

Fig. 7.8 — Loci of colours of constant Munsell Hues, and Munsell Chromas (of 4, 8, and 12) for Munsell Value 5 in u^*,v^* and a^*,b^* diagrams for Standard Illuminant C.

places (see, for instance, Wyszecki & Stiles, pages 488 to 500, 1967, or pages 840 to 852, 1982).

We must now compare the parameters of the Munsell system with some of the CIE measures that were described in Chapter 3. There is a simple approximate relationship between Munsell Value, V, and CIE 1976 lightness, L^*: L^* is approximately 10 times V. However, a Munsell Value of 10 corresponds to the perfect diffuser; hence, if L^* is evaluated using a reference white for which Y_n is less than 100, its value for the perfect diffuser will be more than 100, and thus more than 10 times the Munsell Value.

In Fig. 7.7, the u',v' chromaticities corresponding to Munsell Notation are shown for Munsell Value 5, at Munsell Hue intervals of 2.5, and at Munsell Chroma

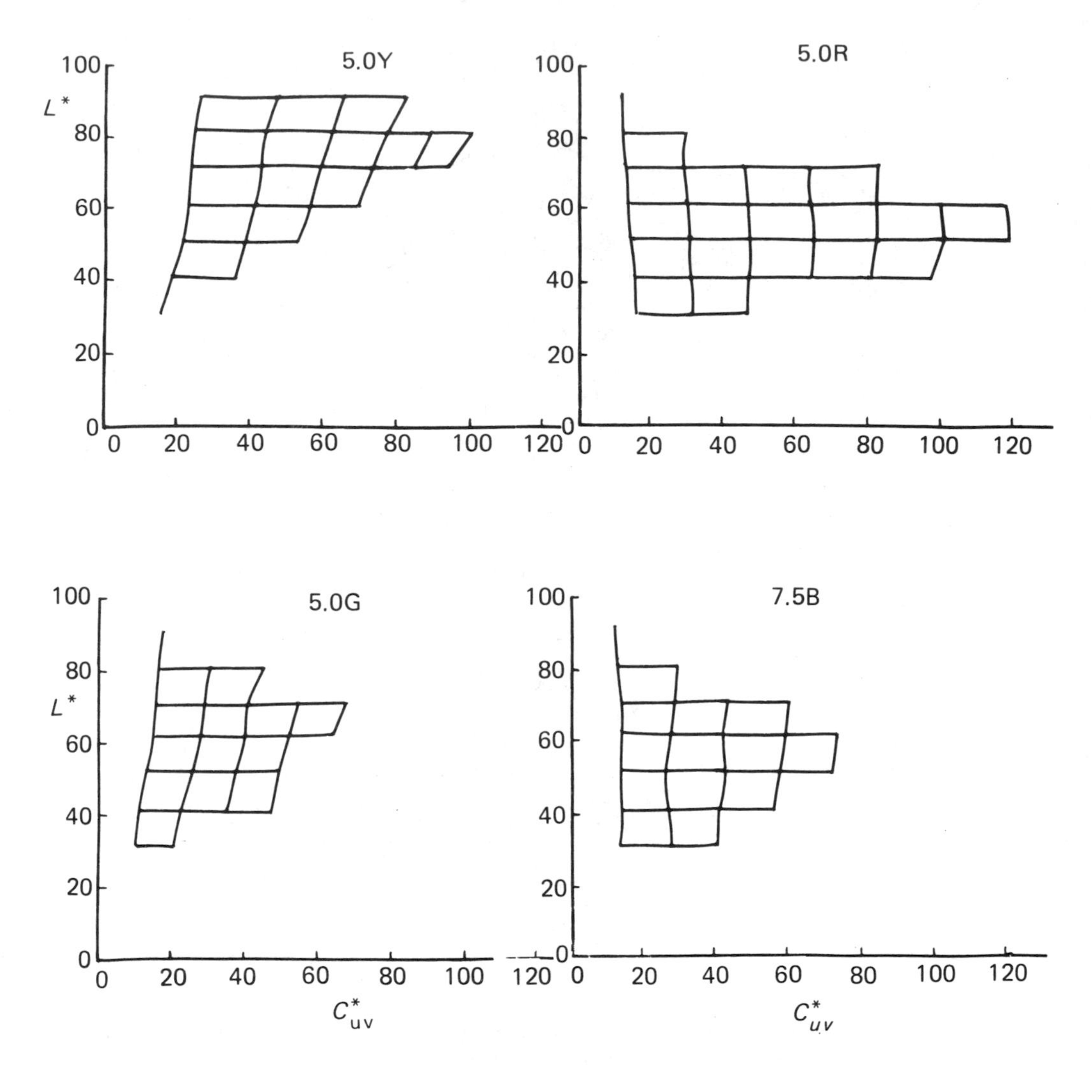

Fig. 7.9 — Positions of colours of four Munsell Hues plotted in L^*, C^*_{uv} diagrams for Standard Illuminant C.

intervals of 2. Only the colours plotting within the unbroken irregular line inside the spectral locus are included in the *Munsell Book of Color* (matt version); those lying outside have been extrapolated and are not based on perceptual experiments on actual chips. If the spacing of the colours in the u', v' diagram were the same as that in the Munsell system at Munsell Value 5, then, in Fig. 7.7, the Munsell Hues would lie on straight lines radiating from the illuminant point, separated by equal angles all round the hue circuit, and the Munsell Chromas would lie on concentric circles, centred on the illuminant point and separated by equal distances, so as to form a grid of the same general shape as that of Fig. 7.5. It is clear from Fig. 7.7 that there is a tendency for this to be the case, but the conformance is only approximate. The points

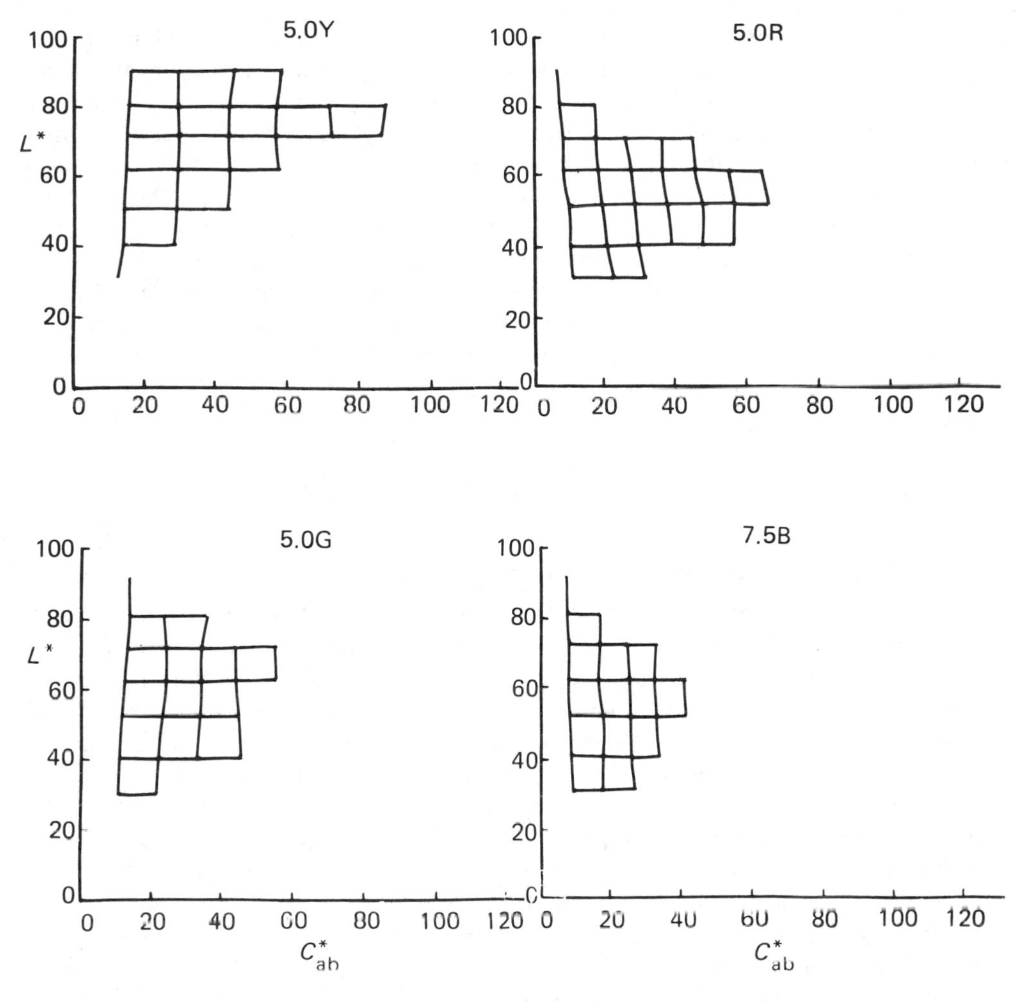

Fig. 7.10 — Same as Fig. 7.9, but for L^*,C^*_{ab} diagrams.

representing single Munsell Hues lie on lines that are slightly curved, especially for some hues; the points representing single Munsell Chromas lie on roughly circular contours which are stretched in the yellow and blue directions, and also, for the higher Chromas, in the red direction.

In Fig. 7.8 plots are shown of the loci of constant Munsell Hue and Chroma for Munsell Value 5 in u^*,v^* and a^*,b^* diagrams. Again, conformity between the Munsell system and the CIELUV and CIELAB systems would be shown by regular patterns of straight lines and concentric circles in this figure, and there is an approximate tendency for this to be so. But the circles are compressed in the purple region in the u^*,v^* diagram, and extended in the yellow region in the a^*,b^* diagram, with various irregularities of hue spacing, particularly in the purple region of the u^*,v^* diagram, and in the yellow–green region of the a^*,b^* diagram.

In Fig. 7.9, plots are shown of Munsell samples in the L^*, C^*_{uv} planes of the CIELUV system. In this case, if the spacing were the same as in the Munsell system, neighbouring points should form a series of squares (because the samples have a Munsell Value spacing of 1, and a Munsell Chroma spacing of /2). It is clear that, again, there is a tendency for this spacing to occur but the conformance is only very approximate. If C^*_{ab} of the CIELAB system is used instead of C^*_{uv}, the resulting plots are broadly similar, but different in detail, as can be seen from Fig. 7.10. The values of C^* are between about 5 and about 12 times those of Munsell Chroma.

The CIELUV and CIELAB systems have a general similarity to the Munsell system because they were designed with this intention. However, it is known that the Munsell system is not perfectly uniform, so that Figs 7.7, 7.8, 7.9, and 7.10 can only be used with caution to estimate the uniformity of the CIELUV and CIELAB systems.

7.6 UNIQUE HUES AND COLOUR OPPONENCY

Towards the end of the last century, the work of Young, Helmholtz, and Maxwell, amongst others, had helped to emphasize the basically *trichromatic* nature of colour vision. But Hering drew attention to the fact that the number of *unique hues* is not three, but four: red, yellow, green, and blue. Today, the trichromatic features of colour vision are associated with the three different types of cone, ρ, γ, and β, and the unique hues with the colour-difference signals, as described in sections 1.6 and 1.7. The hues red, yellow, green, and blue are said to be unique because they cannot be described in terms of any combinations of other colour names. Thus, for instance, although orange can be described as a yellowish red or a reddish yellow, red cannot be described as a yellowish blue or a bluish yellow. In fact the four unique hues comprise two pairs, red and green, and yellow and blue; the colours in each of these pairs are *opponent*, in the sense that they cannot both be perceived simultaneously as component parts of any one colour. That is, it is impossible to have a reddish green, or a greenish red, or a yellowish blue, or a bluish yellow. But yellowish reds, reddish yellows, greenish yellows, yellowish greens, bluish greens, greenish blues, reddish blues, and bluish reds are all possible. Of course, other colour *names* are used, such as orange, turquoise, purple, mauve, or magenta; but, perceptually, these can all be described by using combinations of the unique hue names, yellow and red in the case of orange, green and blue in the case of turquoise, and blue and red in the case of purple, mauve, and magenta. (Some people may think in terms of green being a mixture of blue and yellow, but this comes from an association with a way of producing green colours by mixing pigments or dyes; once a green has been produced, perceptually it may be a yellowish green or a bluish green, but it cannot be both yellowish and bluish at the same time, and hence green is not *perceptually* a yellowish blue or a bluish yellow.)

These four unique hues, together with white and black, make six basic colours; and, as the six constitute three pairs (red and green, blue and yellow, white and black) the trichromacy of colour vision has not disappeared, but is merely expressed in a different form. The black–white pair is different in its opponency from the unique-hue pairs, in that blackish whites and whitish blacks are possible, and in fact are so commonly experienced that the colour name *grey* exists to describe them.

7.7 THE NATURAL COLOUR SYSTEM (NCS)

These ideas of Hering were developed in Sweden by Tryggve Johansson and Sven Hesselgren, and, more recently, by Anders Hård and his co-workers so as to produce the *Natural Colour System* or *NCS* (Hård & Sivik 1981). In the NCS, colours are described in terms of the relative amounts of the basic colours that are perceived to be present; these amounts are expressed as percentages. Thus, a medium grey that is perceived to have equal amounts of whiteness and blackness is described as having a whiteness of 50 and a blackness of 50. A colour that is perceived to be a pure red, with no trace of yellowness or blueness or whiteness or blackness, is described as having a redness of 100. Colours having only combinations of redness, blackness, and whiteness are represented in a triangular array, as shown in Fig. 7.11. The three sides

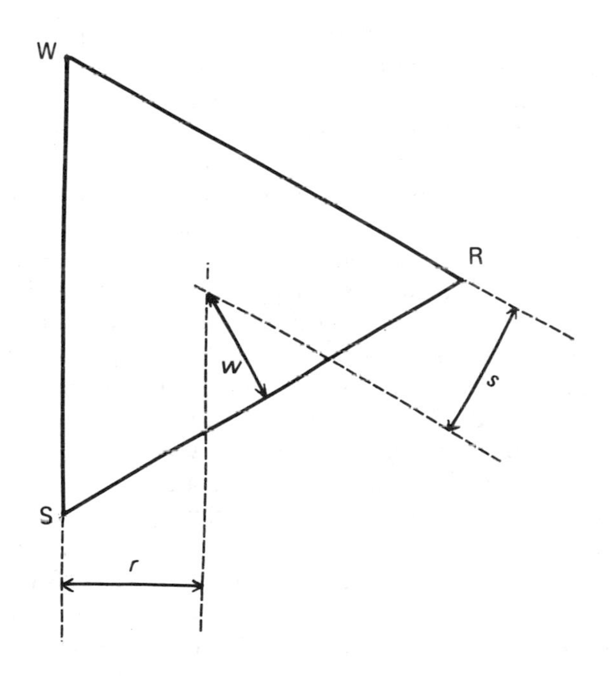

Fig. 7.11 — Arrangement of colours on a white-black-red plane of the NCS.

of the triangle are of equal length, with white represented by the point, W, at the top, black by the point, S, at the bottom, and red by the point, R, to the right and opposite the middle of the line WS. Colours that are perceived to have only redness and whiteness are then situated on the line WR, and their position on this line represents the proportions of redness and whiteness; for instance, a quarter along from W represents 75% whiteness and 25% redness; midway represents 50% of each; and so on. Similarly, points on the line SR represent colours having only blackness and redness; and points on WS, those having only whiteness and blackness (the greys). Points lying within the triangle represent colours having some whiteness, blackness, and redness, the relative amounts being represented by the distances, w, s, and r of the point from the lines SR, WR, and WS, respectively; these distances are expressed as percentages of the maximum possible within the triangle, and this is the distance of a corner of the triangle from the opposite side. The three lengths always add up to this maximum distance, and hence the three numbers, w, s, and r, always add up to 100. Since

$$w+s+r=100$$

it is only necessary to quote two of the three numbers: the convention adopted is that NCS *whiteness*, w, is omitted; thus a colour for which, for example, $w=20$, $s=50$, and $r=30$, would be specified as $s=50$, $r=30$.

Similar triangles exist for the other unique hues, yellow, green, and blue, in which cases the values quoted are s and y, or s and g, or s and b, respectively. Triangles also exist for hues consisting of mixtures of neighbouring pairs of unique hues, and for these intermediate hues the right-hand point of each triangle represents colours of these hues that have no perceived whiteness or blackness content. All the triangles are then fitted in to a three-dimensional array in which the white–black line is vertical and common to all triangles, and the triangles representing different hues occupy different angular positions as shown in Fig. 7.12. The specific hue of any colour is determined by its perceived contents of red and yellow, or yellow and green, or green and blue, or blue and red. Thus, if a colour is perceived to have 80% yellow, 20% red, no whiteness and no blackness, it would occupy the right-hand corner of the triangle at the Y20R position in Fig. 7.12, as indicated by the point C in Fig. 7.13. If another colour had 10% whiteness, 30% blackness, 30% yellowness, and 30% redness, its hue would be determined by the ratio of its two hue contents, that is 30 to 30, which is the same as 50 to 50, so that it would be on the Y50R triangle; its position in this triangle would be 30% of the maximum possible distance from WC, and 60% of the maximum possible from WS as shown by the point A in Fig. 7.13; this 60% is the sum of the two hue contents, and represents the total chromatic content, and this is termed the NCS *chromaticness, c*. The specification of the colour is then $s=30$, $c=60$, with a hue ratio of 50 yellow to 50 red. The NCS specification is then abbreviated to 30, 60, Y50R; the numbers are always given in the order blackness, chromaticness, and hue. The NCS hue is indicated by the initial letter of one unique hue, followed by the percentage of a second unique hue, and then the initial letter of

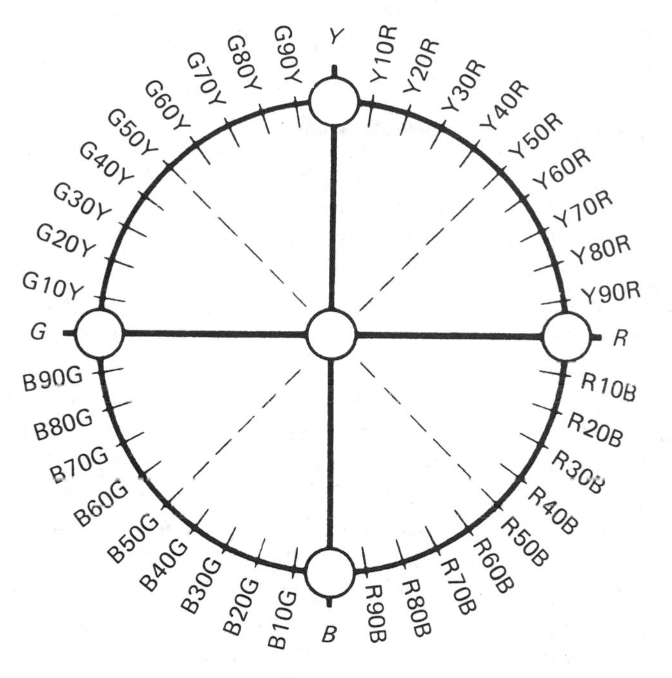

Fig. 7.12 — Arrangement of hues in the NCS.

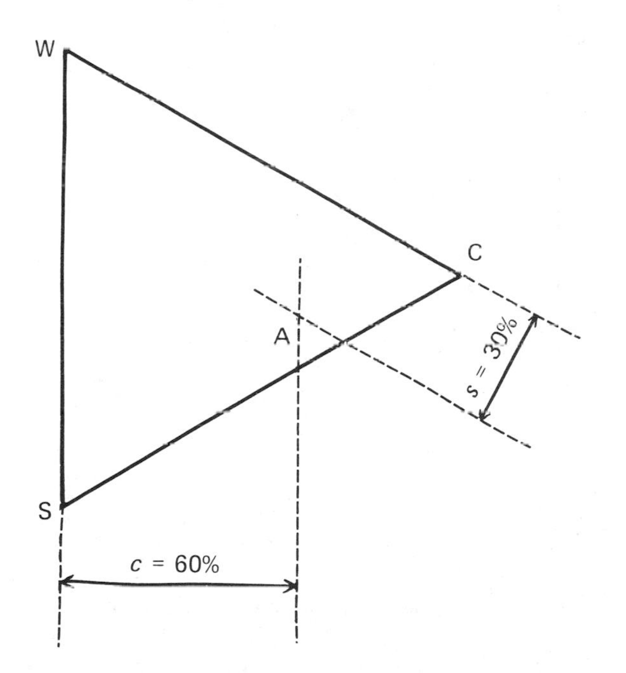

Fig. 7.13 — Same as Fig. 7.11 but for a hue, C.

the second hue, the hues always being given in the order, Y, R, B, G, Y (see Fig. 7.12). (See Plate 5, pages 112, 113.)

7.8 NATURAL COLOUR SYSTEM ATLAS

To produce an atlas based on the above concepts, over 60 000 observations were made on samples of coloured papers viewed in daylight. Based on these observations, colour samples were produced for a prototype atlas having about 1200 samples. The relative spectral power distributions of these samples were measured, for Standard Illuminant C (S_C), from which their X, Y, Z tristimulus values were calculated. Comparison of the observations and the measurements was then used to obtain smoothed and adjusted X, Y, Z tristimulus values for about 16 000 NCS notations (for details, see Swedish Standard, 1982). The final atlas was intended to portray samples at every tenth percentage step in blackness and chromaticness for forty different hues. This would have provided over 2000 samples (55 on each hue page, plus the grey scale), but pigments do not exist to produce some of the colours, so that not all the points can be illustrated. The production atlas has 1530 samples (earlier editions had 1412), and this number includes a few illustrating some positions additional to the tenth percentage steps. The size of the samples in the NCS atlas is 12×15 mm; larger size pieces are also available, if required for special purposes.

In Fig. 7.14 are shown the loci of the u',v' chromaticities for the unique hues of

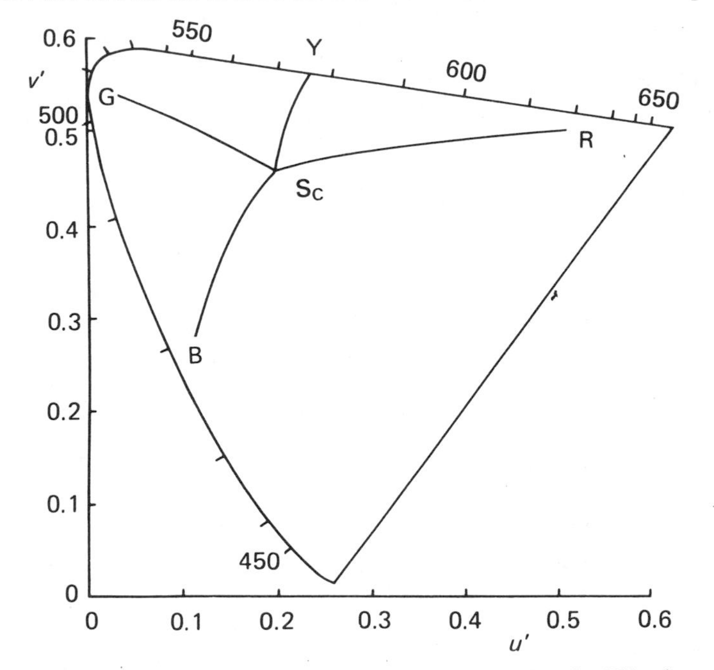

Fig. 7.14 — Loci of NCS unique hues on the u',v' diagram for Standard Illuminant C.

the NCS atlas having no blackness. The curvature of the lines is much the same as that for similar Munsell Hues, as can be seen by comparing Figs 7.14 and 7.7; it can also be seen that NCS unique red corresponds approximately to Munsell 5R, unique

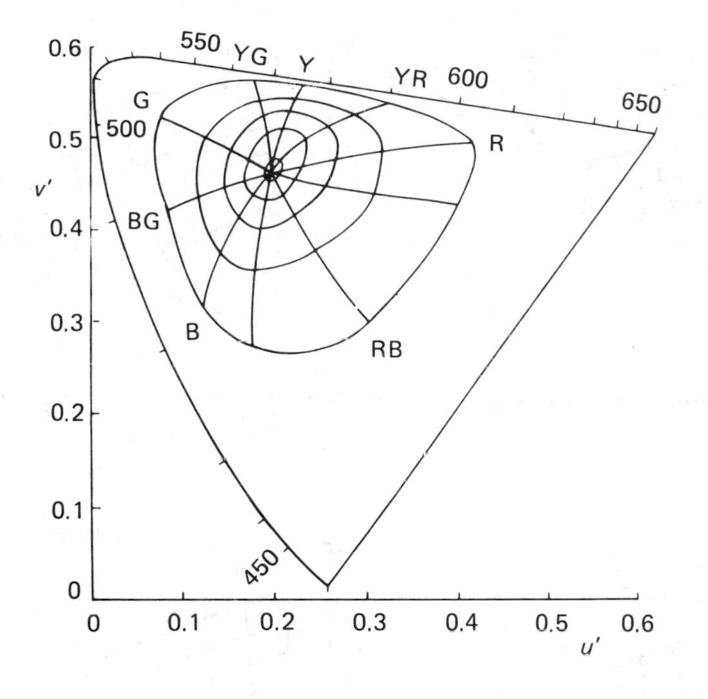

Fig. 7.15 — Loci of colours of zero blackness, for various hues, and for chromaticnesses of 30, 50, 70, and 90, for the NCS, for Standard Illuminant C.

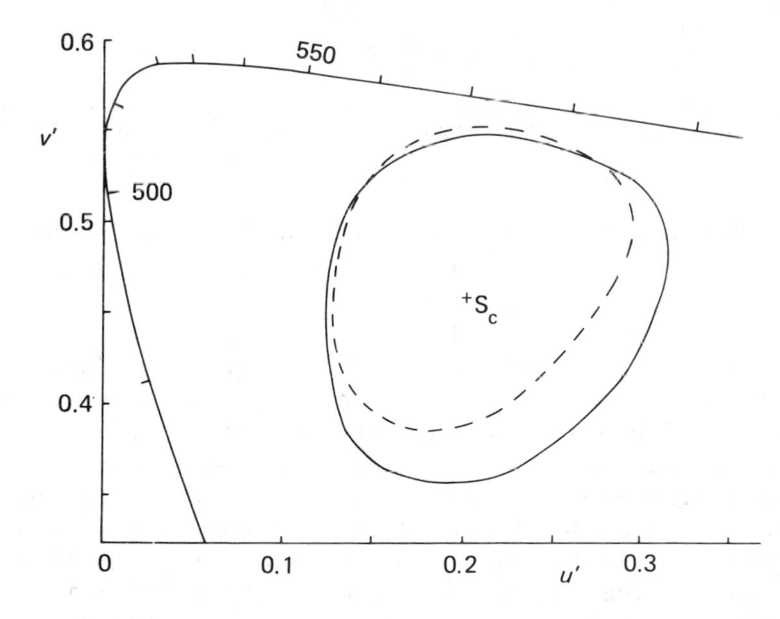

Fig. 7.16 — Loci of colours of a constant Munsell chroma (broken line) and a constant NCS chromaticness (full line) compared for Standard Illuminant C.

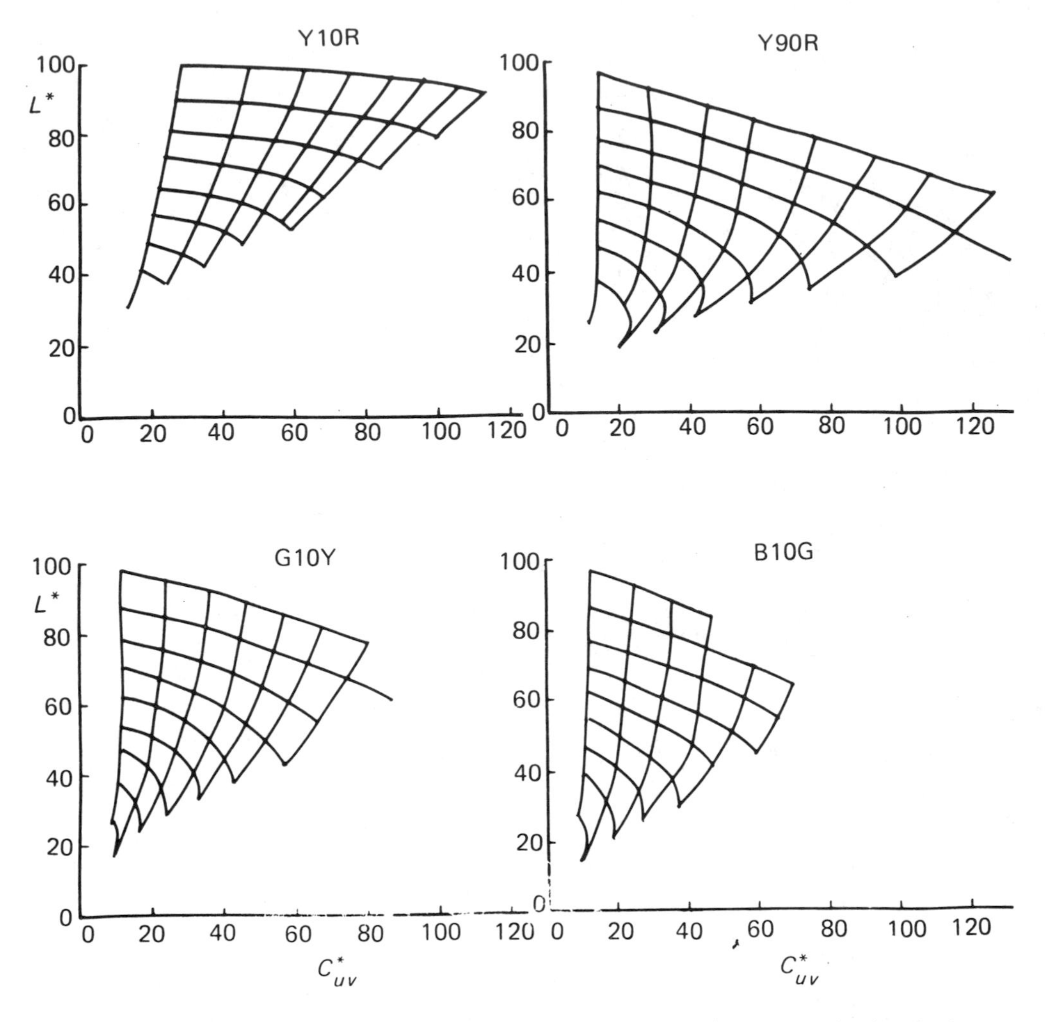

Fig. 7.17 — Positions of colours of four NCS hues plotted in L^*,C^*_{uv} diagrams for Standard Illuminant C.

yellow to Munsell 5Y, unique green to Munsell 5G, and unique blue to Munsell 7.5B. The Munsell system was designed with the aim of having an equal number of perceptual colour differences between its five principal hues 5R, 5Y, 5G, 5B, and 5P, and this is not necessarily compatible with four of them corresponding to unique hues. It is, therefore, not surprising that 5B does not correspond to unique blue. The inclusion of the purple hue in the Munsell system was necessary because of the larger number of perceptual hue differences between blue and red than between red and yellow, or yellow and green, or green and blue. The arrangement of hues in the NCS, as shown in Fig. 7.12, therefore, means that there are many more discriminable hue differences in the blue–red NCS quadrant, than in the other three NCS hue

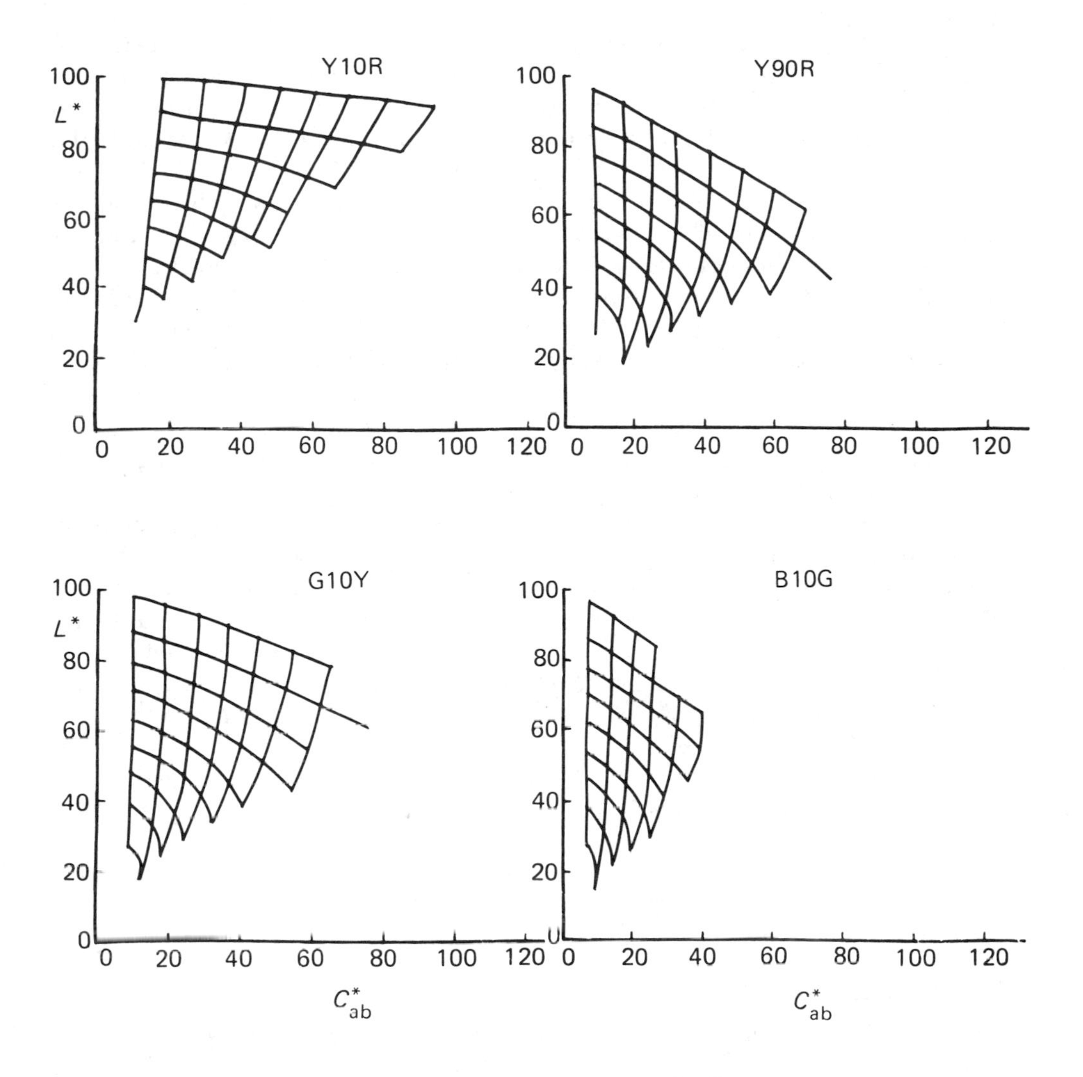

Fig. 7.18 — Same as Fig. 7.17 but for L^*, C^*_{ab} diagrams.

quadrants. This is not necessarily to be thought of as a disadvantage; it is merely a consequence of the basis on which the NCS is formulated.

In Fig. 7.15, u',v' chromaticities are shown for NCS colours of zero blackness, not only for the unique hues, but also for some intermediate hues, and for colours of constant NCS chromaticnesses of 30, 50, 70, and 90. It is clear that the general pattern is not too dissimilar from that of the Munsell system for colours of Munsell Value 5, shown in Fig. 7.7. However, for purple colours, the Munsell Chroma contours tend to be nearer the illuminant point than the NCS chromaticness contours, and this can be seen more clearly in Fig. 7.16, where two contours have been shown for comparison that are very similar in the red–yellow–green–blue quadrants, but appreciably different in the blue–red quadrant. The reason for this difference is not known, but the inclusion of the purple hue in the Munsell system may have had an effect on the observations in this quadrant that were absent in the NCS studies.

In Fig. 7.17, the NCS network is shown for different blackness and chromaticness in each of four NCS hue planes in plots of L^* against C^*_{uv}; in Fig. 7.18 the same is done for L^* plotted against C^*_{ab}. The lines of constant NCS chromaticness are approximately vertical in these plots, but with a tendency to slope in towards the black point, particularly at low lightnesses; this shows that NCS chromaticness is intermediate between CIE 1976 chroma (for which the loci are vertical) and CIE 1976 saturation (for which the loci radiate from the black point in the L^*, C^*_{uv} plots). The lines of constant NCS blackness slope downwards as they radiate from the grey scale axis and are not horizontal, as is the case for lines of constant CIE 1976 lightness, L^*, and also for those of constant Munsell Value.

It is claimed that the attribute of NCS blackness is more readily perceived in colours than the attribute of lightness, with which Munsell Value correlates. There is also evidence that, in the use of colour in architecture, Munsell Value is not an ideal variable, and blackness might be more suitable (Gloag & Gold 1978; Whitfield, O'Connor & Wiltshire 1986). It is interesting in this connection to note that colours of different hues, but having the same NCS blackness and chromaticness, are described by some designers as having a certain equivalence, sometimes referred to as equality of *nuance* or *weight*.

Variables that are more similar to NCS blackness than to Munsell Value are used in other colour systems intended as aids to colour designers; one example is the DIN system.

7.9 THE DIN SYSTEM

The *DIN* (Deusches Institut für Normung) system was developed by M. Richter and his co-workers in Germany (Richter & Witt 1986). The three variables used are hue, T, saturation, S, and darkness, D, given in that order. Thus for a DIN specification of 16:6:4, for instance, $T=16$, $S=6$, and $D=4$. The system uses Standard Illuminant D_{65}, and CIE X, Y, Z tristimulus values for the samples are available (see Wyszecki & Stiles, pages 503 to 505, 1967, or pages 863 to 865, 1982).

Colours of constant dominant (or complementary) wavelength are regarded as being of constant hue. There are 24 principal hues, having values of $T=1$ for a yellow, proceeding via reds, purples, blues, and greens, to a yellow-green of $T=24$, and back to $T=25=1$. The 24 principal hues were chosen to represent equal hue differences between adjacent pairs, all round the hue circle.

The second of the three variables, saturation, S, is a function of distance from the point representing the reference white on a chromaticity diagram, and for colours of the same luminance factor (not the same darkness) equal saturation represents equal perceptual differences from the grey of the same luminance factor.

The third variable is related to darkness rather than to lightness, and is not related in a simple manner to luminance factor; in this respect it is similar to blackness in the NCS. DIN darkness, D, is evaluated as:

$$D=10-6.1723\log[40.7(Y/Y_0)+1]$$

where Y is the luminous reflectance of the colour, and Y_0 is the luminous reflectance of the optimal colour (see section 7.3) having the same chromaticity, the subscript 0 indicating zero darkness (when $Y=Y_0$, $D=0$).

The colour solid associated with the DIN system is formed by having the grey scale as a vertical axis, with white at the top, for which $D=0$, and black at the bottom, for which $D=10$. The top surface of the solid is then a portion of a sphere having the black point as its centre, as shown in Fig. 7.19; this surface represents the optimal

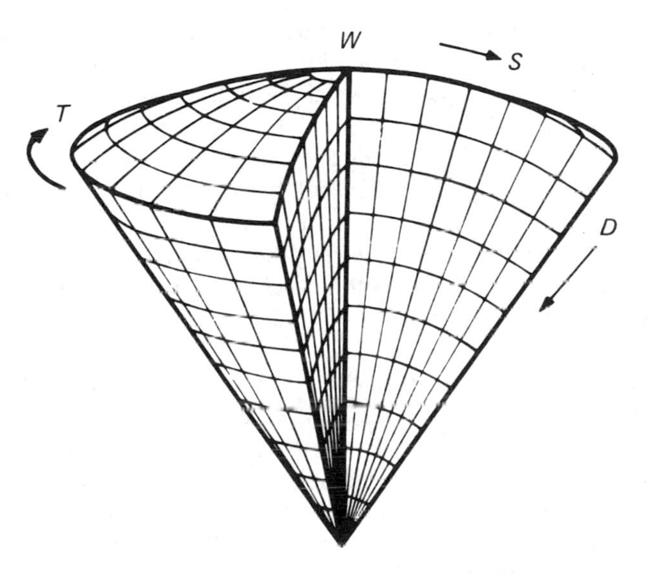

Fig. 7.19 — Colour solid of the DIN system.

colours. Different hues, T, are then situated in 24 different planes with one edge coincident with the grey scale axis, and the same angle, 15°, between adjacent planes. DIN darkness, D, is represented by distance down from the top surface towards the black point. DIN saturation, S, is represented by angular distance out from the grey scale axis, evaluated from the black point; the angle of one saturation step is equal to 5°. Although, for simplicity, the solid is shown in Fig. 7.19 as being a symmetrical cone, the maximum value of S is different for different hues, so that the edge of the top surface is not circular, and the sides of the cone are at different angles for different hues. S has a maximum value of about 16, and this occurs for violet colours.

An atlas for the DIN system is available, known as the *DIN Colour Chart*. In the atlas, for each hue page, colours of equal DIN saturation are arranged in columns parallel to the grey scale axis, instead of in lines radiating from the black point. Samples having the same hue and darkness, but differing in DIN saturation by the same number of units, exhibit a colour difference that becomes progressively smaller as darkness increases. Hence, the dark samples on a DIN Colour Chart are more alike than the light samples. This result is a consequence of choosing saturation as the perceptual attribute with which one of the variables of the system correlates, instead of chroma. A column of colours of constant DIN saturation is a *shadow series* in which the colours all have the same chromaticity. (See Plate 8, page 113.)

The size of the samples in the DIN Colour Chart is 20×28 mm; larger size pieces are also available, if required for special purposes.

7.10 THE COLOROID SYSTEM

A colour order system designed particularly for use by architects is the *Coloroid* system, developed in Hungary by Nemcsics and his co-workers (Nemcsics 1980, 1987). The aim is to provide a system in which the colours are spaced evenly in terms of their aesthetic effects, rather than in terms of colour differences as in the Munsell system, or perceptual content as in the NCS. One interesting feature of the Coloroid system is that notations in it can be calculated from CIE x,y chromaticity co-ordinates and luminance factors, Y, using fairly simple formulae and one look-up table. Standard Illuminant C (S_C) is used (but the system is also applicable to D_{65}).

The general arrangement is conventional in having a central vertical grey scale, with white at the top and black at the bottom, and hue planes that radiate from the grey scale.

A Coloroid specification consists of three numbers, such as 13–22–56. The first number indicates the hue. Colours of equal dominant (or complementary) wavelength are regarded as having equal hue, and the hue number, A, like 13 in the above example, is derived from a table of dominant (or complementary) wavelengths for values of A ranging from 10 at 570.83 nm to 76 at 568.92 nm; increases in A represent increases in dominant (or complementary) wavelength. The spacing of the values of A on the wavelength scale is intended to be aesthetically even. Interpolation between values is used if necessary, using values of ϕ, which are also tabulated against A, where:

$$\tan\phi=(y-y_w)/(x-x_w) \ ,$$

where x_w, y_w are the values of x,y for the reference white. Values of ϕ are given, for Standard Illuminants C and D_{65}, in Table 7.1.

The second of the three numbers indicates the chromatic content, T. Colours are regarded as having the same value of T if they can be produced by additively mixing the same percentage of saturated colour of the same hue (dominant wavelength) with white and black. (By this is meant that, if the colour is produced by an additive mixture of the saturated colour and the reference white and the reference black, presented on a rapidly rotating disc, then T is constant when the saturated colour occupies a constant percentage of the area of the disc.) By saturated colour is meant a spectral colour in the range from 450 to 625 nm or lying on a purple line joining these two points in a chromaticity diagram. The values of T range from zero for the reference white and greys having the same chromaticity, to 100 for the saturated colours. The formula for calculating T is given in Table 7.1.

The third of the three numbers indicates the Coloroid lightness, V, which is defined as:

$$V=10Y^{1/2} ,$$

where Y is the percentage luminance factor. For the perfect diffuser, $Y=100$ and hence $V=100$. For a perfect black, $Y=0$, and $V=0$. The V scale thus extends from 0 to 100.

For a given value of A (hue) and V (lightness), the chromatic content T is related to Munsell Chroma by the expression:

$$T=k_{AV}C^{2/3} ,$$

but k_{AV} varies with A and V. Hence, in the Coloroid system, while the variable A is closely correlated with hue, and V with lightness, the variable T is not related simply to either chroma or saturation; it is intended to provide an aesthetically even representation of chromatic content.

Atlasses based on the Coloroid system are available for practical use by architects and other colour designers. The size of the samples is 15×28 mm.

7.11 THE OPTICAL SOCIETY OF AMERICA (OSA) SYSTEM

A colour order system that is different from the others considered so far is the *OSA* (Optical Society of America) system. It is dedicated primarily to producing a system in which, as nearly as possible, the distance between any two points represents the perceptual size of the colour difference between the two samples represented by them. There is, as a consequence, no attempt to include a correlate of hue, or of chroma or saturation. The OSA system was the culmination of a long series of investigations which started in 1947 and ended in 1977, in which a leading part was taken by Deane B. Judd (Nickerson 1981).

An OSA specification consists of three numbers, such as 3:1:5. The first represents the lightness, L. When $L=0$, the colour has the same lightness as a medium grey of 30% reflectance factor. But, for chromatic colours having $L=0$, the reflectance factor is usually less than 30% because of the contribution of colourfulness to brightness, and therefore to lightness (as mentioned in sections 2.3 and 3.2). L is negative for darker colours, and positive for lighter colours. Values of L range from about -7 to $+5$.

The second number in an OSA specification, j, represents the yellowness of the colour; j is positive for yellowish colours and negative for bluish colours. Values of j range from about -6 to $+11$.

Table 7.1 — Formulae for calculations in the coloroid system

A is derived from the values listed below for ϕ, where

$$\arctan \phi = (y - y_w)/(x - x_w).$$

$$V = 10Y^{1/2}.$$

$$T = \frac{100Ye_w(x_0 - x)}{100e_\lambda(x - x_\lambda) + Y_\lambda e_w(x_0 - x)}$$

or

$$T = \frac{100Y(1 - ye_w)}{100e_\lambda(y - y_\lambda) + Y_\lambda(1 - ye_w)} \ .$$

The formula for T giving the most convenient numbers is used. The subscript w indicates that the value is for the reference white, the subscript 0 that it is for the reference black, the subscript λ that it is for the saturated colour (for which values of x_λ, y_λ, e_λ, and Y_λ are given opposite and in Nemscics 1980; the values of Y_λ are the same as for the $\bar{y}(\lambda)$ function, but multiplied by 100), and

$$e_w = (1/100)(X_w + Y_w + Z_w)$$

where X_w, Y_w, and Z_w are tristimulus values for which the Y tristimulus value is the luminance factor expressed as a percentage.

The reverse equations for deriving x, y, Y from A, T, V are as follows:

$$x = \frac{e_w x_0(V^2 - TY_\lambda) + 100Te_\lambda x_\lambda}{e_w(V^2 - TY_\lambda) + 100Te_\lambda}$$

$$y = \frac{V^2 + 100Te_\lambda y_\lambda - TY_\lambda}{e_w(V^2 - TY_\lambda) + 100Te_\lambda}$$

$$Y = (V/10)^2 \ .$$

In all the evaluations, linear interpolation may be carried out between successive values (using the variable ϕ in the case of A; ϕ_C is used for Standard Illuminant C, and ϕ_{65} for Standard Illuminant D_{65}).

Data for the saturated colours of the Coloroid system

A	ϕ_C	ϕ_{65}	x_λ	y_λ	e_λ	Y_λ
10	59.0	58.05	0.44987	0.54895	1.724349	94.6572
11	55.3	54.17	0.46248	0.53641	1.740845	93.3804
12	51.7	50.38	0.47451	0.52444	1.754986	92.0395
13	48.2	46.70	0.48601	0.51298	1.767088	90.6482
14	44.8	43.58	0.49578	0.50325	1.775953	89.3741
15	41.5	39.61	0.50790	0.49052	1.785074	87.6749
16	38.2	36.30	0.51874	0.43035	1.791104	86.0368
20	34.9	32.88	0.52980	0.46934	1.794831	84.2391
21	31.5	29.39	0.54137	0.45783	1.798665	82.4779
22	28.0	25.82	0.55367	0.44559	1.794822	79.9758
23	24.4	22.17	0.56680	0.43253	1.789610	77.4090
24	20.6	18.35	0.58128	0.41811	1.779484	77.4014
25	16.6	14.32	0.59766	0.40176	1.760984	70.7496
26	12.3	10.08	0.61653	0.38300	1.723444	66.0001
30	7.7	5.53	0.63896	0.36061	1.652892	59.6070
31	2.8	0.74	0.66619	0.33358	1.502608	50.1245
32	−2.5	−4.38	0.70061	0.29930	1.072500	32.1000
33	−8.4	−10.66	0.63925	0.26753	1.136638	30.4093
34	−19.8	−24.35	0.53962	0.22631	1.232286	27.8886
35	−31.6	−34.65	0.50340	0.19721	1.310122	25.8373
40	−43.2	−46.21	0.46041	0.17495	1.376610	24.0851
41	−54.6	−57.28	0.42386	0.15603	1.438692	22.4490
42	−65.8	−67.95	0.38991	0.13846	1.501583	20.7915
43	−76.8	−78.29	0.35586	0.12083	1.570447	18.9767
44	−86.8	−87.66	0.32195	0.10328	1.645584	16.9965
45	−95.8	−96.11	0.28657	0.08496	1.732085	14.7168
46	−108.4	−108.10	0.22202	0.05155	1.915754	9.8764
50	−117.2	−116.63	0.15664	0.01771	2.146310	3.8000
51	−124.7	−123.81	0.12736	0.05227	1.649940	8.6198
52	−131.8	−130.59	0.10813	0.09020	1.273415	11.4770
53	−138.5	−136.98	0.09414	0.12506	1.080809	13.5067
54	−145.1	−143.30	0.08249	0.15741	0.957577	15.0709
55	−152.0	−149.91	0.07206	0.18958	0.868977	16.4626
56	−163.4	−160.96	0.05787	0.24109	0.771732	18.5949
60	−177.2	−174.64	0.04353	0.30378	0.697110	21.1659
61	171.6	174.30	0.03291	0.35696	0.655804	23.4022
62	160.2	162.65	0.02240	0.41971	0.623969	26.1843
63	148.4	150.45	0.01196	0.49954	0.596037	30.1137
64	136.8	138.37	0.00425	0.60321	0.607414	36.6425
65	125.4	126.59	0.01099	0.73542	0.659924	48.5346
66	114.2	114.70	0.08050	0.83391	0.859523	71.7274
70	103.2	103.88	0.20259	0.77474	1.195684	92.6325
71	93.2	93.75	0.28807	0.70460	1.410097	99.0587
72	84.2	84.43	0.34422	0.65230	1.532830	99.9862
73	77.3	77.27	0.37838	0.61930	1.603793	99.3224
74	71.6	71.29	0.40290	0.59533	1.649449	98.1981
75	66.9	66.35	0.42141	0.57716	1.681081	97.0252
76	62.8	62.05	0.43647	0.56222	1.704981	95.8592

The third number, g, represents the greenness of the colour; g is positive for greenish colours and negative for reddish colours. Values of g range from about -10 to $+6$.

The geometrical arrangement of the points representing the colours is such that each point is surrounded by twelve equidistant points, representing twelve nearest-neighbour colours all equally different perceptually from the colour considered. The arrangement of these points is the same as that of the centres of spheres packed as closely together as possible, each sphere touching twelve others. This geometry may be referred to as cubo-octahedral (Billmeyer 1981), and is illustrated in Fig. 7.20.

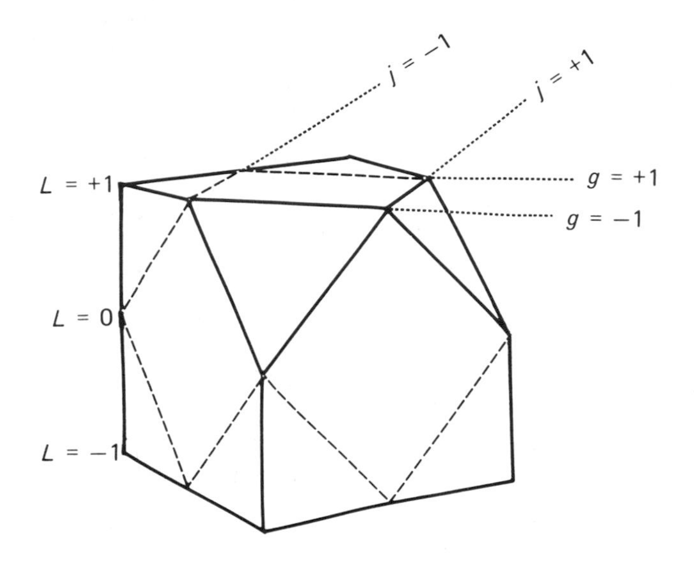

Fig. 7.20 — The cubo-octahedral basis of the OSA system.

A cube has eight corners. If each of these eight corners is cut off down to the mid-point of the edges of the cube, a figure is formed having twelve corners. In Fig. 7.20 these are situated at the four corners of the square formed by the mid-points of the edges of the top surface of the original cube; at the four corners of the square formed by the mid-points of the vertical edges of the original cube; and at the four corners of the square formed by the mid-points of the edges of the bottom surface of the original cube. These twelve points are all equidistant from the point lying at the centre of the original cube. In the OSA system, successive values of L are located on successive horizontal planes, so that, if the five points on the centre plane of the cube of Fig. 7.20 have $L=0$, the four points on the plane above have $L=1$, and the four on the plane below have $L=-1$. On the $L=1$ and $L=-1$ planes, there is no point directly above or below the centre point, but such points occur on planes $L=2, 4$, etc., and $L=-2, -4$, etc. On each horizontal plane, closest neighbouring points in one direction represent values of g differing by + or − 2 units, and, at right angles to that direction, values of j differing by + or − 2 units. If the centre point of Fig. 7.20 represents $L=0, g=0, j=0$, then the four points surrounding it on the same plane all have $L=0$, and have $g=2$, $j=0$, or $g=-2, j=0$, or $g=0, j=2$, or $g=0, j=-2$. The four points on the plane above, all have $L=1$, and have $g=1, j=1$, or $g=1, j=-1$, or $g=-1, j=1$, or $g=-1, j=-1$. If the distance between closest pairs of these colours is regarded as 2 units, the distance between successive horizontal planes differing in value of L by 1, is $2^{1/2}$ units. Thus, if two colours differ in their values of L, g, and j by ΔL, Δg, and Δj, respectively, then the total colour difference is given by:

$$[2(\Delta L)^2+(\Delta g)^2+(\Delta j)^2]^{1/2} .$$

Although this is a colour difference formula, the judgements on which the OSA system was based were not on pairs of colours exhibiting very small differences, so that it is not necessarily applicable to colours that almost match.

Samples of OSA colours, 2×2 inches (about 50×50 mm) square, are available; but they are not pre-arranged to represent any particular planes of the OSA system. This has the advantage that the user can select samples and arrange them on any planes passing through sample points in the OSA space. There are nine different families of parallel planes that include all of the sample points; of these, in one family all the planes are horizontal; in four families all the planes are vertical; and in the other four families all the planes are parallel to one of the four pairs of parallel cut-off corners of the original cube (see Fig. 7. 20). These planes represent unusual combinations of colours and are claimed to be useful in helping designers to find novel patterns for their work. (See Plates 6 and 7, pages 112, 113.)

The OSA samples are all specified in terms of X, Y, Z tristimulus values for Standard Illuminant D_{65}, for the 1964 Standard Observer (the 50 mm square samples subtending an angle of about 10° when viewed at about 300 mm) (MacAdam 1978, Wyszecki & Stiles, pages 866 to 884, 1982), but there is no simple relationship between OSA specifications and CIE tristimulus values. Equivalent Munsell specifications have also been published for OSA specifications (Nickerson 1981). Colours lying between the sample points are represented by decimals, for example, 3.4, −5.7.

7.12 THE HUNTER LAB SYSTEM

In the Hunter Lab system (Hunter 1958; Billmeyer & Saltzman 1981), the general arrangement is similar to that of the OSA system, but the emphasis is on a simple relationship to X, Y, Z tristimulus values, rather than on the closest approximation possible to uniform spacing. The variables used in the Hunter Lab system are L, as the correlate of lightness, a as the correlate of redness or greenness, and b as the correlate of yellowness or blueness. These variables are defined, for Standard Illuminant C′ (S_C), as follows:

$$L=10Y^{1/2}$$
$$a=[17.5(1.02X-Y)]/Y^{1/2}$$
$$b=[7.0(Y-0.847Z)]/Y^{1/2} .$$

There are no samples available to represent this system, but its simple relationship to X, Y, Z tristimulus values has made it readily applicable to colour measuring instruments; as a result, many colorimeters have been marketed that produce these L, a, b values as immediate displays or print-outs from instrumental measurements.

7.13 THE TINTOMETER

A rather different kind of colour order system is provided by an instrument called *The Tintometer*, originally developed by J. W. Lovibond in 1887 for measuring the colour of beer. The samples provided are of coloured glass, and, by superimposing them in suitable combinations, a very wide range of colours can be produced. The glasses are of three different colours, called *red* (actually a magenta to absorb green light), *yellow* (to absorb blue light), and *blue* (actually a cyan to absorb red light). They are each available in scales ranging from almost colourless to highly saturated red, yellow, or blue, as the case may be. The scale divisions are arbitrary, except for two features: first, that if a red, a yellow, and a blue glass of the same value are superimposed, then the result is visually nearly neutral in Standard Illuminant C (S_C); and second, that, if two glasses of the same colour are superimposed, whose numerical designations on the scale are n_1 and n_2, their combined colour is the same as that of a single glass having the designation n_1+n_2.

The glasses are viewed against a standard white surface illuminated by a tungsten filament lamp. If the sample being measured is a reflecting surface, it is illuminated in a similar manner by the same lamp. In the case of transmitting samples, they are viewed against the standard white surface illuminated by the same lamp. The lamp is run at a specified voltage that results in Standard Illuminant A (S_A). For measurements in daylight, a filter can be inserted in the eyepiece of the instrument to produce Standard Illuminant C (S_C).

The glasses are arranged in slides in groups of ten, the first slide containing glasses designated from 0.1 to 0.9 in steps of 0.1, the second slide from 1.0 to 9.0 in steps of 1.0, and the third from 10.0 to the maximum in steps of 10.0. The sample being measured is seen in one half of the field of view of the instrument, and is compared with the colour of a combination of the glasses, seen in the other half. The total field size is 2° in diameter. The combination of the glasses is adjusted until the two colours match; and if the match required, say, 2.5 of the red glasses, 3.7 of the yellow, and 8.6 of the blue, the Tintometer specification would be given as:

2.5 Red, 3.7 Yellow, 8.6 Blue.

Designations of this type are used in a number of industries to define colour standards for products, such as edible oils. The glasses can also be made up into special combinations that give series of colours against which special tests can be evaluated. For instance, in the control of swimming pool water, indicator dyes are used in samples to determine the pH (acidity or alkalinity); the colour of the dyed water is related to the pH, and, by comparing its colour with those of the special series of combination glasses, which have been previously calibrated in terms of pH, the pH is very easily determined. Such series of glasses are conveniently mounted in discs holding up to about 10 glasses, and mounted in a simple comparator.

The difference in colour between two glasses whose numerical values differ by 0.1, which is the smallest interval on the scales, is very small and similar in size to a just noticeable difference in a 2° field. Altogether about 9 million different colours can be produced.

In an alternative form of the instrument, glasses of only two colours are ever used in combination, and the third variable consists of a calibrated means of reducing the amount of light passing through the glasses (Schofield 1939). With this *Lovibond Schofield Tintometer* it is possible to convert the readings into CIE X, Y, Z tristimulus values by means of charts provided for the purpose.

The gamut of chromaticities covered by the Tintometer instruments is very wide, but there is an area towards the green part of the spectral locus that is not covered; however, by using a supplementary cyan filter, for which a special calibration chart is provided, this area can be included.

7.14 ADVANTAGES OF COLOUR ORDER SYSTEMS

The use of colour order systems to select, measure, and check colours is very widespread. The advantages of using them include the following:

First, they are easy to understand, because they usually have actual samples that can be seen.

Second, they are easy to use; in most circumstances side by side comparisons are made without the need for any instrumentation.

Third, the number and spacing of the samples can be adapted for different applications, and different arrangements of the samples can be used for different purposes.

Fourth, most colour order systems are calibrated in terms of CIE tristimulus values, and these can therefore be obtained for colours if required.

7.15 DISADVANTAGES OF COLOUR ORDER SYSTEMS

There are a number of disadvantages that occur when colour order systems are used.

First, there is not just one colour order system in use, but several, and there is no simple means of transferring results from one system to another.

Second, there are gaps between the samples, and this means that interpolation often has to be used both in matching a colour to the samples, and in determining the corresponding specifications.

Third, the visual comparison between colours and the samples is strictly valid only if it is done using an illuminant having the same spectral power distribution, and geometrical arrangement as was used for the original judgements and calibration adopted for the system (see Chapter 6). However, the errors are not likely to be large in most cases if reasonably typical indoor daylight is used; but, if any other illuminant is used, such as tungsten light, or fluorescent light, serious errors may occur (unless a calibration specific for that illuminant is available, as in the case of tungsten light with the Tintometer, for instance). These difficulties will be more pronounced the more the colour being considered differs in spectral composition and gloss characteristics from the samples in the system. It is therefore always advisable, wherever possible, to use a system having samples as similar as possible in these respects to the colours being matched. In this connection, a version of the Munsell system has been produced on textiles, known as the SCOT–Munsell system (Standard Colors Of Textiles); as a further aid in obtaining similarity, the swatches in this system have one side dull and the other shiny. Keeping to a minimum the differences in spectral composition between the samples and the colours being measured is difficult. To produce colours of high chroma requires the use of colorants that are very spectrally selective; in other words they must absorb some wavelengths very strongly and others very weakly. But, if these colorants are mixed with grey colorants to produce samples of low chroma, these samples will be more spectrally selective than most colours of low chroma met with in practice, because these colours usually depend on colorants that are not very spectrally selective. This poses a real problem for the manufacturer of samples for systems; what is usually done is to use colorants whose spectral selectivity is low for low chroma colours, but high for high chroma colours. This means that very careful control is necessary to avoid discontinuities of colour along a series of colours of increasing chroma.

Fourth, different observers may make slightly different matches on the same colour (this is the case even when colour-defective observers are carefully excluded). This effect (often referred to as *observer metamerism*, see section 6.7) will also be more pronounced the greater the difference in spectral composition between the samples and the colours being matched.

Fifth, there may be differences between the samples in the collection actually being used and those for which the calibration of the system applies. This can arise from manufacturing tolerances in the production of the samples, or from deterioration of the samples as a result of extensive use or fading. Naturally manufacturers of samples for colour order systems are aware of these problems, and high consistency and good permanence of the samples are always sought.

Sixth, some colours may lie outside the gamut of the samples available in the system. This can arise in the case of colours of very high chroma; they are often produced by pigments or dyes that are not of the highest stability, and it may not therefore be possible to make samples of these colours that are suitable for inclusion in systems. Fluorescent colours are often in this category.

Seventh, most colour order systems cannot be used for self-luminous colours, such as light sources, unless ancilliary apparatus is used.

This is quite a long list of potential disadvantages, but they are of varying importance, depending on the application and the particular system being used. Thus, for example, in the Tintometer, the gaps between the samples are small enough to be insignificant in most applications; and, because the samples are made of glass, they have excellent permanence.

As has been mentioned several times, colour order systems are calibrated in terms of their CIE tristimulus values. The calibrations can be used to determine the tristimulus values of samples that have been compared to the colours in the system, but, except in the case of the Tintometer, interpolation is usually necessary, and this is not easy to do.

REFERENCES

Berns, R. S. & Billmeyer, F. W. *Color Res. Appl.* **10**, 246 (1985).
Billmeyer, F. W. *Color Res. Appl.* **6**, 34 (1981).
Billmeyer, F. W. & Saltzman, M. *Principles of Color Technology*, 2nd ed., p. 62, Wiley, New York (1981).
Davidson, H. R., Godlove, M. N. & Hemmendinger, H. *J. Opt. Soc. Amer.* **47**, 336 (1957).
Evans, R. M. & Swenholt, B. K. *J. Opt. Soc. Amer.* **57**, 1319 (1967).
Evans, R. M. & Swenholt, B. K. *J. Opt. Soc. Amer.* **59**, 628 (1969).
Gloag, H. L. & Gold, M. J. *Colour Co-ordination Handbook*, pp. 25-29, Building Research Establishment Report, HMSO London (1978).
Hård, A. & Sivik, L. *Color Res. Appl.*, **6**, 129 (1981).
Hunter, R. S. *J, Opt. Soc. Amer.* **48**, 985 (1958).
MacAdam, D. L. *J. Opt. Soc. Amer.* **27**, 294 (1937).
MacAdam, D. L. *J. Opt. Soc. Amer.* **68**, 121 (1978).
Nemcsics, A. *Color Res. Appl.* **5**, 113 (1980).
Nemcsics, A. *Color Res. Appl.* **12**, 135 (1987).
Newhall, S. M., Nickerson, D. & Judd, D. B. *J. Opt. Soc. Amer.* **33**, 385 (1943).
Nickerson, D. *Color Res. Appl.* **1**, 7, 69, and 121 (1976).
Nickerson, D. *Color Res. Appl.* **6**, 7 (1981).
Pointer, M. R. *Color Res. Appl.* **5**, 145 (1980).
Richter, M. & Witt, K. *Color Res. Appl.* **11**, 138 (1986).
Rösch, S. *Phys. Z.* **29**, 83 (1928).
Schofield, S. K. *J. Sci. Instrum.* **16**, 74 (1939).
Swedish Standard, SS 01 91 03, *CIE tristimulus values and chromaticity co-ordinates for the colour samples in SS 01 91 02* (1982).
Whitfield, T. W. A., O'Connor, M. & Wiltshire, T. J. *Color Res. Appl.* **11**, 215 (1986).
Wyszecki, G. & Stiles, W. S. *Color Science*, 1st. ed., Wiley, New York (1967).
Wyszecki, G. & Stiles, W. S. *Color Science*, 2nd. ed., Wiley, New York (1982).

GENERAL REFERENCES

Color Res. Appl. **9**, Number 4 (Winter, 1984).
Color Res. Appl. **10**, Number 1 (Spring, 1985).

8

Precision and accuracy in colorimetry

8.1 INTRODUCTION

By precision, is meant the consistency with which measurements can be made. By accuracy, is meant the degree to which measurements agree with those made by a standard instrument or procedure in which all possible errors are minimized.

Precision is affected by random errors. The most common sources of random errors in photo-electric colorimeters, spectroradiometers, and spectrophotometers are variations in sensitivity, electronic noise, and sample presentation. With a modern instrument, for a given sample, the mean colour difference from the mean of a set of measurements (MCDM) can usually be expected to be about 0.1 (or less) of a CIELAB or CIELUV colour difference unit. Bearing in mind that, under average conditions, a just noticeable difference is usually regarded as corresponding to approximately one of these units, this level of precision can be considered to be satisfactory. This performance has also been found to apply to the evaluation of colour differences of pairs of samples measured on the same instrument; it also applies both to such measurements made on the same day, and to series of such measurements made over a period of several weeks or more. (Marcus 1978, Billmeyer & Alessi 1981).

Accuracy is affected by systematic errors. Common sources of systematic errors in modern instruments are wavelength calibration, detector linearity, geometry of illumination and viewing, and polarization. These errors may be associated with stray light, wavelength scale, bandwidth, reference-white calibration, thermochromism, and fluorescence (Carter & Billmeyer 1979, Berns & Petersen 1988). The accuracy is usually assessed by comparing results with 'standard' results obtained by national standardizing laboratories, using the best possible instrumentation and procedures. A distinction has to be drawn between the accuracy of results obtained from spectroradiometers or spectrophotometers, and from colorimeters using filtered photocells. With spectroradiometers and spectrophotometers, the mean colour difference from the standard results (MCDS) is generally found to be at least an order of magnitude greater than the precision (MCDM) quoted above: that is, the MCDS may be about 1 or more CIELAB or CIELUV units. (Billmeyer & Alessi 1981). With colorimeters using filtered photocells, the accuracy can be very dependent on the spectral composition of the sample, and errors considerably in excess of one CIELAB or CIELUV unit may sometimes occur; these errors may be reduced by determining the X, Y, Z tristimulus values from a matrix of the three filter readings (Erb, Krystek & Budde 1984), or from more than three filter readings (Wharmby 1975). The importance of accuracy depends on the application. When the same instrument is used to monitor the consistency of a product of nearly constant spectral composition, such as successive batches of nominally the same paint or dye, good precision is vital, but great accuracy is not. Accuracy becomes progressively more and more important, as the measurements involve different instruments of the same type, instruments of different types, and samples of increasingly different spectral compositions.

When colorimetric results are compared, it is essential to ensure that like is being compared to like. Thus, the illuminant (S_A, S_C, or D_{65}, etc.), the standard observer (2° or 10°), and the illuminating and viewing geometry (45/0, diffuse/0, specular included or excluded, etc.) must all be the same for the two sets of data. If they are not, no meaningful comparisons can be made, unless the purpose is solely to demonstrate the effect of some difference in the measurement conditions. For example, if one measurement refers to Standard Illuminant C, and another to D_{65}, then the comparison has no meaning unless the same sample, and the same conditions (except for the illuminant), were involved, and the purpose was only to compare the colorimetry of that sample under these two illuminants.

8.2 SAMPLE PREPARATION

In many applications, colorimetric measurements are made only on selected samples from a larger population of material. It is therefore necessary that the sample be representative of the larger population. If the population involves variation in colour, then it is clearly necessary to use a number of samples sufficiently large to yield a reasonable average, if this is what is wanted, or to indicate the nature of the variability, if this is what is wanted (ASTM 1990b). A properly designed sampling procedure, based on statistical considerations, is a necessity. It is also necessary to avoid samples that have uncharacteristic non-uniformities, such as may be caused by dirt, bubbles, streaks, or variations in flatness, for example. Visual inspection is a very sensitive method of detecting the presence of such defects, and samples should always be checked in this way; failure to do so can lead to very misleading results.

If samples are translucent, then their appearance, and the results of measurements made upon them, will be very dependent on the nature of the material behind them. It is useful, with these materials, to make measurements with both white and black backings; any difference between the two sets of readings then indicates the presence of translucency and its extent. If only one type of backing is used, this should be opaque and of constant colour, such as black (Wightman & Grum 1981).

Some samples, such as woven fabrics, have directional effects, so that their colours are dependent on the directions of illumination and viewing (see section 5.9).

These differences can be explored by orienting the samples at different angles when using the 0/45 or 45/0 geometry; if the 0/diffuse or diffuse/0 geometries are used, the results will tend to be an average for all directions. But, in all cases, for the highest consistency of results, such samples should always be oriented in the same direction.

8.3 THERMOCHROMISM

Some materials change colour with temperature. This *thermochromism* occurs with some red, orange, and yellow glasses, and with some tiles that contain selenium (Carter & Billmeyer 1979). In such cases, it is necessary to standardize the temperature at which the measurements are made. For this purpose, 'room temperature' is often a sufficiently precise definition, because significant changes usually occur only when the sample is quite hot. However, although instruments are normally operated at room temperature, the samples may become quite hot in certain circumstances. If measurements are made on samples illuminated with monochromatic light, there is little chance of any appreciable thermochromism occurring. But, when samples are illuminated with 'white' light, as is necessary, for instance, when fluorescence occurs (as will be discussed in Chapter 9), then there may be a sufficient rise in sample temperature for some significant thermochromism to take place. This can be checked by taking a series of measurements with the sample left in the measuring position in the instrument; if the measurements show a steady drift with time, thermochromism may be taking place.

8.4 GEOMETRY OF ILLUMINATION AND VIEWING

For transmitting samples, unless they are either perfectly non-diffusing, or perfectly diffusing, the geometry of illumination and viewing will affect the results. The amount of diffusion in many transparent samples is sufficiently small for the geometry to be unimportant. For reflecting samples, unless they are perfectly matt, the geometry of illumination and viewing will affect the results, and often does so very markedly (ASTM 1985, 1988, 1990a, 1990d, 1990e). Very few reflecting samples are sufficiently matt to be unaffected by the geometry, and much of the difficulty in obtaining consistent colorimetric results from different instruments stems from this source. There are several factors that contribute to this inconsistency.

First, the standard conditions of illumination and viewing recommended by the CIE, as described in section 5.9, are not sufficiently restrictive to avoid differences. In particular, the $\pm 10°$ permissible variation of illuminating or viewing axis from the normal direction, and the permissible spread of the beam from the axis of $\pm 8°$or $\pm 5°$, can result in significant differences for some types of sample. It has been suggested that these angles should be reduced to $\pm 2°$ from the nominal illuminating and viewing angles, and $\pm 4°$ for beam spread, in all cases (Rich 1988).

Second, the sizes, positions, and effective reflectances, of the gloss traps commonly used in instruments can differ sufficiently to cause appreciable discrepancies (Billmeyer & Alessi 1981).

Third, in instruments using integrating spheres, there are several factors that can affect the results: these include the sizes, positions, and effective areas of the ports; the positions and reflectances of the baffles that are usually included to prevent light travelling directly from the samples to the detector; and the uniformity and reflectance of the coating on the interior of the sphere (Clarke & Compton 1986).

Finally, if the results of measurements are compared with visual inspection of samples, it is unlikely that the measuring and inspection geometry will be exactly the same, and this can have appreciable effects.

8.5 REFERENCE WHITE CALIBRATION

As mentioned in section 5.7, although the primary standard of reflectance, established in 1969, is the perfect diffuser (CIE 1971 and 1986), this is not available as a material standard; but national standardizing laboratories are able to calibrate material standards relative to it. These 'transfer' standards can then be used to calibrate working standards for general use. These working standards are often made of opal glass, or are in the form of a ceramic or porcelain–enamel tile, or a fluorinated polymer (Grum & Saltzman 1976), or a pressed powder, such as barium sulphate (Erb & Budde 1979). White working standards of this type are then used to normalize instruments before measurements are carried out. The normalizing step in filter-type colorimeters ensures that correct tristimulus values are obtained for the working standard; in spectrophotometers, the normalizing step may result in corrected transmittance factors, or reflectance factors, being obtained throughout the spectrum. It has been noted that, when opal glasses are used, because they are translucent, the total amount of light they reflect in an instrument can depend on the ratio of the illuminated area to the sample port area (Billmeyer & Hemmendinger 1981); for this reason, dense opal, or tiles, are preferable. It is possible to correct numerically for some errors in the calibration of the working standard (to be discussed in section 8.12).

8.6 POLARIZATION

In the case of non-metallic glossy materials, the fraction of the light reflected *regularly* (or *specularly*, that is, in the same direction as by a mirror) depends on its plane of polarization. This is shown in Fig. 8.1, where regular reflectance is plotted against angle of incidence for different planes of polarization, for a ratio of refractive indices of 1.5. This can affect measurements in spectroradiometers, spectrophotometers, and colorimeters, because, in these instruments, it is possible for the light incident on the samples to become partly polarized as it passes through, or is reflected by, their optical components.

Thus, if the light incident on a sample is polarized to any appreciable degree, and if that sample exhibits some gloss, then the spectral reflectance factor will depend on the angle at which the light is incident on the sample. For example, from Fig. 8.1, if this angle is 45°, a glossy sample, whose colorant is in a medium of refractive index 1.5, will reflect specularly from its topmost surface about 9% of the light, if it is fully polarized in the plane containing the beam and the normal to the surface, but about 1% if it is fully polarized at right angles to this plane. These effects can be avoided if

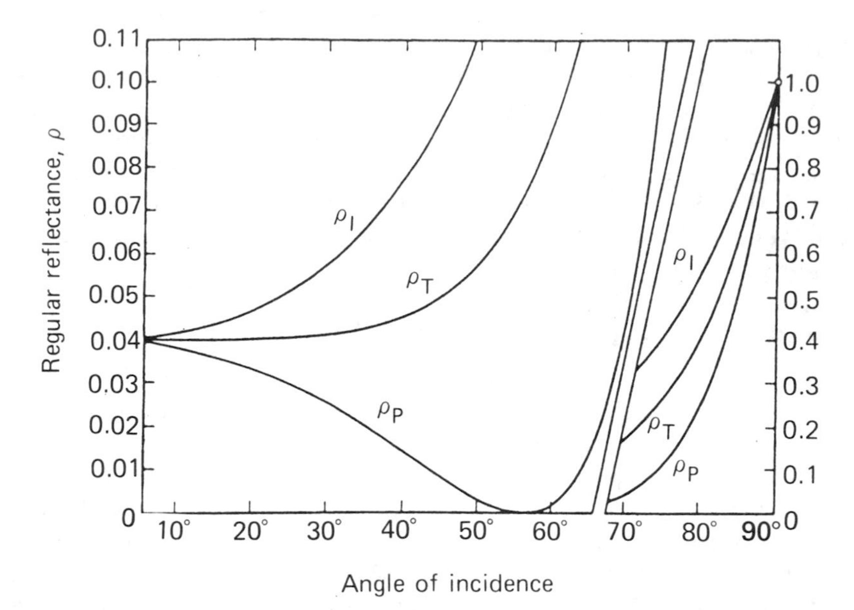

Fig. 8.1 — Regular reflectance from a surface as a function of angle of incidence for light polarized in the plane of the illuminating beam and the normal to the surface, ρ_I, and at right angles to it, ρ_P, and for both together, ρ_T, for ratio of refractive indices 1.5.

the illumination is either normal or diffuse. In the 0/diffuse geometry, the illumination is normal to within 15° (up to 10° for the axis, plus up to 5° for the spread), and because, as can be seen from Fig. 8.1, the polarizing effect is still quite weak up to this angle, any effects of polarization are likely to be quite small. Similarly, in the 0/45 geometry the illumination is normal to within 18° (up to 10° for the axis, and up to 8° for the spread), and again this is likely to cause only small effects. In the diffuse/0 geometry, the effects of polarization should be averaged out because of the many angles of incidence involved. But if the 45/0 geometry is used, then particular attention must be paid to the possibility that polarization may be affecting the results.

Furthermore, when the light *reflected* from a sample is polarized to any appreciable degree, the effective transmittance of any subsequent optical components can depend on the extent and angle of polarization of the light; the effect of this on the measurements depends on the type of instrument used.

In liquid crystal displays light passes through crossed polarizing filters, so that the emerging light is always strongly polarized. Extreme care over polarizing effects is therefore necessary when making colorimetric measurements on these devices. Extruded plastic materials may also polarize the light to some extent; in this case the samples should be remeasured after rotation through a right-angle, and an average taken.

8.7 WAVELENGTH CALIBRATION

In filter-type colorimeters, the spectral sensitivity functions achieved in the instrument are only ever approximations to those theoretically required (ASTM 1990c);

the discrepancies can be regarded as errors in wavelength scale. In spectroradiometers and spectrophotometers, the wavelength scale can be checked by measuring samples of known spectral emission, spectral transmittance factor, or spectral reflectance factor; a very useful sample is a didymium filter, because it has rapid changes of transmittance factor with wavelength in several parts of the spectrum (Alman & Billmeyer 1975). If the instrument has provision for adjusting the wavelength scale, then the results from the known sample can be used to determine the corrections that should be made. Some instruments make no provision for adjusting the wavelength scale, but, in this case, it may be possible to correct the data numerically (to be discussed in section 8.12).

8.8 STRAY LIGHT

In any optical instrument some light is reflected or scattered in a non-imaging way, and this is usually referred to as *stray light*. In spectroradiometers and spectrophotometers this can cause errors. For instance, a yellow sample may have a reflectance factor of about 95% at the long, and only about 3% at the short, wavelength end of the spectrum. If even as little as 2% of the long wavelength light is scattered in such a way as to be present when the short wavelengths are being measured, readings of 5% instead of 3% may result, and this may be seriously misleading. The stray-light condition of these instruments can be checked by measuring samples that have been carefully calibrated by using instruments known to have very low levels of stray light; these samples should have very low values in some parts of the spectrum. It may be possible to reduce the stray light in an instrument by cleaning the optical components, or by improving the performance of baffles near the light paths. If the amount of stray light present is appreciable and can be reliably estimated, the data can again be corrected numerically (to be discussed in section 8.12).

8.9 LINEARITY

It is necessary for the electrical signals from the detector in spectroradiometers and spectrophotometers to be proportional to the incident radiant power, or, in filter-type colorimeters, to be proportional to the integrated spectral power transmitted by each filter. The most common cause of a departure from this condition is an error in the setting of the zero in the instrument. Measurements on dark samples of known transmittance factor or reflectance factor provide the most convenient way of checking this setting. In some instruments this setting can be readjusted, but, in others, it is fixed. If no adjustment is possible, the data can again be corrected numerically (to be discussed in section 8.12).

8.10 USE OF SECONDARY STANDARDS

Reference has already been made several times to the use of known standards to help detect or correct errors (Verrill 1987). In the case of transmission work, a set of calibrated filters should be used. One such set, provided by the (USA) National

Bureau of Standards (now the National Institute for Standards and Technology, NIST), is designated NBS SRM 2101-2105 (Keegan, Schleter & Judd 1962, Eckerle & Venable 1977); but new sets of these filters are no longer available. In the case of reflection work, a set of calibrated reflecting tiles should be used (Clarke 1969). One such set, provided by the (British) National Physical Laboratory, is designated CCSII (Malkin & Verrill 1983). A set of plastic tiles has also been produced (Simon 1991).

8.11 BANDWIDTH

Unless a sample has properties that are constant with wavelength, the width of the band of wavelengths used in spectroradiometers or spectrophotometers will generally affect the results. The bandwidths used in these instruments vary from a few tenths of a nanometre to as much as about 20 nm. The smaller the bandwidth, the more exactly the instrument is able to detect very steep or narrow variations in emission, in reflectance factor, or in transmittance factor, with wavelength; but also the smaller will be the amount of light available for detection. The larger the bandwidth, the more approximate will be the results for specimens that are very spectrally selective, but the more will be the amount of light available for detection. For best accuracy, the measurement interval and the bandpass should be about equal (to be discussed in section 8.13).

Spectroradiometers and spectrophotometers, in effect, filter the white light they use through a spectral 'window', which is equivalent to a filter that has maximum transmittance at, or near, its nominal recording wavelength, and lower transmittances at longer and shorter wavelengths. These sidebands quickly reduce to zero transmittance on either wavelength side, and this variation of transmittance with wavelength, appropriately weighted for any variation in detector spectral sensitivity, is referred to as the *bandwidth function*. If the bandwidth function varies with wavelength, this can introduce further errors. Numerical corrections can also be made for some effects of bandwidth function (to be discussed in section 8.12).

8.12 CORRECTING FOR ERRORS IN THE SPECTRAL DATA

As mentioned in sections 8.5, 8.7, 8.8, 8.9, and 8.11, numerical corrections for known errors in spectral data can be provided, and these can improve accuracy and agreement between instruments. Systematic methods for deriving such corrections have been suggested (Robertson 1987, Berns & Petersen 1988). These methods are applicable to spectral measurements of either reflectance factor or transmittance factor; but, for simplicity, their description will be limited to reflectance factor data. The methods depend on devising equations that are likely to represent the form of certain types of error present in typical instruments. The technique works well for correcting errors in photometric scale and in wavelength. It does not address differences in measurement that are attributable to differences in geometry between instruments, such as in aperture size, illuminating and viewing angles, or sphere design.

Consider, for instance, an instrument with an error in its zero setting, perhaps caused by stray light, by an imperfect black trap, or by an improperly adjusted analog-to-digital converter. This type of error is likely to produce a constant offset throughout the spectrum. Hence, if $R_m(\lambda)$ is the measured value of the reflectance factor at the various wavelengths, λ, the true value $R_t(\lambda)$ is likely to be given by:

$$R_t(\lambda) = R_m(\lambda) + B_0$$

where B_0 is a constant (which, as also in the cases of the constants B_1, B_2, B_3, and B_4, to be introduced later, could be positive or negative).

An error in the assumed value of the reflectance factor of the instrument's white working standard, caused, for instance, by it being miscalibrated or improperly handled or maintained, is likely to produce a constant percentage error throughout the spectrum. Hence, in this case:

$$R_t(\lambda) = R_m(\lambda) + B_1 R_m(\lambda)$$

where B_1 is another constant. By combining the effects represented by both the above equations, photometric scale errors in $R_m(\lambda)$ can be related to $R_t(\lambda)$ by a straight line where B_1 is the slope of the line and B_0 is the intercept.

Nonlinearity in the response of the photo-detector, or in its associated electronic circuits, would result in the photometric scale errors being nonlinear. Such errors can be represented approximately by:

$$R_t(\lambda) = R_m(\lambda) + B_2[R_m(\lambda)]^2$$

where B_2 is another constant.

An error in the wavelength scale consisting of a simple shift of all wavelengths by the same increment in the same direction is likely to produce errors that are proportional to the slope of the curve of $R_m(\lambda)$ plotted against λ: in regions where the curve is flat, there would be no error; small errors would occur where the curve had a low slope; large errors would occur where the curve had a high slope. Hence, in this case:

$$R_t(\lambda) = R_m(\lambda) + B_3 \mathrm{d}R_m/\mathrm{d}\lambda$$

where $\mathrm{d}R_m/\mathrm{d}\lambda$ is the slope of R_m plotted against λ, and where B_3 is another constant. (This is the simplest form of wavelength error, corresponding, for instance, to a misalignment of a diffraction grating in a monochromator, and it may not represent the true form of the error of the instrument; but formulae representing more complicated forms of error can be devised: Berns & Petersen 1988.)

An instrument possessing too large a bandwidth function is likely to produce errors that are proportional to the rate of change with wavelength of the slope of the curve of $R_m(\lambda)$ plotted against λ: in regions where this curve is a straight line, there would be no error, because the extra-long wavelengths would be balanced by the extra short wavelengths; but, where this curve is not straight, this balance is upset, and errors occur to a degree that is proportional to the rate of change of the slope with wavelength. Hence, in this case:

$$R_t(\lambda) = R_m(\lambda) + B_4 \mathrm{d}^2 R_m/\mathrm{d}\lambda^2$$

where $\mathrm{d}^2 R_m/\mathrm{d}\lambda^2$ is the rate at which the slope of the curve of R_m plotted against λ changes with wavelength, and where B_4 is another constant.

If an instrument had all these systematic errors, the true reflectance factor could then be calculated by combining all five of the above equations, thus:

$$R_t(\lambda) = R_m(\lambda) + B_o + B_1 R_m(\lambda) + B_2[R_m(\lambda)]^2 + B_3 \mathrm{d}R_m/\mathrm{d}\lambda + B_4 \mathrm{d}^2 R_m/\mathrm{d}\lambda^2 \ .$$

To make such calculations, it is necessary to determine the values of B_0, B_1, B_2, B_3, and B_4. But, with real instruments, the errors will not usually correspond exactly to those assumed above. Hence, those values of B_0, B_1, B_2, B_3, and B_4 that give the best representation of the real errors are required.

A method of correction can then be formulated that requires the following steps to be carried out.

Step 1. A set of stable calibrated chromatic and neutral reference materials is obtained, together with their spectral reflectance factor data. The British Ceramic Research Association's CCSII tile set is a good example (Malkin & Verrill 1983). The calibration must be for the same geometry of illumination and viewing as in the instrument being used for measurement. The data must be available at whatever wavelength interval is adopted.

Step 2. Each reference material is measured on the measuring instrument. The reported results should be the average of at least three measurements. When making these measurements, care should be taken to avoid raising the temperature of the materials to a level different from the reported temperature applying to the calibration.

Step 3. A chromatic material is selected that has a wide range of reflectance factor values and steep slopes as a function of wavelength. This material will be used to characterize and correct the systematic errors of the instrument. Careful selection of this material is important; because of thermochromism, some materials will change their measured reflectance factor dramatically with temperature changes. Berns and Petersen found that the cyan tile of the CCSII set was a good compromise between spectral properties and little thermochromic change in measured reflectance factor (Berns & Petersen 1988). For the selected material, the differences between the true result, $R_t(\lambda)$, and its measured result, $R_m(\lambda)$, obtained in Step 2, are then calculated:

$$R_t(\lambda) - R_m(\lambda) \ .$$

If the sampling is at an interval of 10 nm from 400 to 700 nm, for example, this yields 31 values.

Step 4. Using the measurements, $R_m(\lambda)$, from the selected material, a table of values, F_3, is calculated that approximates the slope of the curve of $R_m(\lambda)$ plotted against wavelength at the same 31 wavelengths; and a table of values, F_4, is calculated that approximates the rate of change of this slope with wavelength, at the same 31 wavelengths. The values F_3 and F_4 are calculated as follows:

$$F_{3i} = [R(\lambda_{i+1}) - R(\lambda_{i-1})]/[\lambda_{i+1} - \lambda_{i-1}]$$
$$F_{4i} = [R(\lambda_{i+1}) + R(\lambda_{i-1}) - 2R_i(\lambda)]/[(\lambda_{i+1} - \lambda_{i-1})/2]^2 \ .$$

The subscript, i, indicates the wavelength considered, and the subscripts $i+1$ and $i-1$ indicate neighbouring sampling wavelengths. (At the ends of the spectrum, the values of F_3 and F_4 for the first and last measured wavelengths can be set equal to those of the second, and second to last, wavelengths, respectively.)

Step 5. If B_0, B_1, B_2, B_3, and B_4 are set equal to some trial values, a set of trial corrected reflectance factors $R_c'(\lambda)$, is obtained for the 31 wavelengths:

$$R_c'(\lambda) = R_m + B_0 + B_1 R_m(\lambda) + B_2[R_m(\lambda)]^2 + B_3 F_3(\lambda) + B_4 F_4(\lambda) \ .$$

These values can be compared with the true values $R_t(\lambda)$ for this material to obtain the value of the error $e(\lambda)$ at each of the 31 wavelengths:

$$e(\lambda) = R_t(\lambda) - R_c'(\lambda) \ .$$

The values of B_0, B_1, B_2, B_3, and B_4 are now obtained that result in the sum of the squares of the values of $e(\lambda)$ at the 31 wavelengths being the minimum possible. (This *multiple linear regression* procedure is available in the form of readily applicable computer programs.)

Step 6. Optimum corrected values, $R_c(\lambda)$, of the data for the selected material are then obtained as:

$$R_c(\lambda) = R_m(\lambda) + B_0 + B_1 R_m(\lambda) + B_2[R_m(\lambda)]^2 + B_3 F_3(\lambda) + B_4 F_4(\lambda)$$

where B_0, B_1, B_2, B_3, and B_4 have the values that minimize $e(\lambda)$.

Step 7. Step 6 is repeated for the other materials for which the true values are known, and it should be found that the errors have been reduced. (If there are still important errors, different formulae, or a larger number of formulae, for representing the errors may be needed; see Berns & Petersen 1988.)

Step 8. Step 6 is used to correct all measured data obtained from the instrument, using the same geometry.

8.13 COMPUTATIONS

As described in section 2.6, CIE tristimulus values, X, Y, Z, are obtained by weighting spectral power (or other radiant quantity) distributions with the CIE colour-matching functions, $\bar{x}(\lambda)$, $\bar{y}(\lambda)$, $\bar{z}(\lambda)$. This is expressed mathematically by the summations:

$$X = K[P_1\bar{x}_1 + P_2\bar{x}_2 + \ldots + P_n\bar{x}_n]$$

$$Y = K[P_1\bar{y}_1 + P_2\bar{y}_2 + \ldots + P_n\bar{y}_n]$$

$$Z = K[P_1\bar{z}_1 + P_2\bar{z}_2 + \ldots + P_n\bar{z}_n]$$

where the subscripts 1, 2, ... n indicate the values of the quantities indicated, at a series of equally spaced wavelength intervals throughout the spectrum from wavelength 1 to wavelength n; P are appropriate measures of spectral power, such as the colour stimulus function (the absolute measure of a radiant quantity per small constant-width wavelength interval throughout the spectrum), $\varphi_\lambda(\lambda)$, or the equivalent relative measure, the relative colour stimulus function, $\varphi(\lambda)$, which is often derived as the product of a relative spectral power distribution, $S(\lambda)$, for a source, and the spectral transmittance factor or the spectral reflectance factor, $R(\lambda)$, for a sample; and K is a suitably chosen constant. For transmitting or reflecting samples, K is given a value that results in Y being equal to 100 for the perfect transmitter or the perfect reflecting diffuser; hence, in this case:

$$K = 100/[S_1\bar{y}_1 + S_2\bar{y}_2 + \ldots + S_n\bar{y}_\mathrm{n}] \; .$$

The results obtained by such summations differ according to the size of the wavelength interval used, and the CIE regards the correct result as that obtained when the interval is infinitesimally small. This is expressed mathematically by replacing the summations by the equivalent integrals:

$$X = K\int P(\lambda)\bar{x}(\lambda)\mathrm{d}\lambda$$

$$Y = K\int P(\lambda)\bar{y}(\lambda)\mathrm{d}\lambda$$

$$Z = K\int P(\lambda)\bar{z}(\lambda)\mathrm{d}\lambda$$

$$K = 100 / \int S(\lambda)\bar{y}(\lambda)\mathrm{d}\lambda$$

The process of integration requires that the quantities being integrated have values that are defined continuously throughout the spectrum. For quantities defined by the CIE at 1 nm intervals (this includes $\bar{x}(\lambda)$, $\bar{y}(\lambda)$, and $\bar{z}(\lambda)$, and $S(\lambda)$ for Standard Illuminants D_{65} and A), this is achieved by linear interpolation of the values within each 1 nm interval. For quantities only available at larger intervals, a more elaborate method of interpolation, such as one of those given in Table 8.1, is preferable.

Table 8.1 — Interpolation formulae

The known values of the spectral quantity are designated:

F_1, F_2, F_3, F_4 at wavelengths
$\lambda_1, \lambda_2, \lambda_3, \lambda_4$.

It is required to estimate the value of F at a wavelength λ between the wavelengths λ_2 and λ_3.

Third degree polynomial formula

This formula can be used only when the known values of F are at regular wavelength intervals. The required value of F is given by:

$$F = F_2 + p\Delta F + p(p-1)\Delta_2 F/2 + p(p-1)(p-2)\Delta_3 F/6$$

where

ΔF is the first difference, $F_2 - F_3$
$\Delta_2 F$ is the second difference, the average of
$(F_1 - F_2) - (F_2 - F_3)$ and
$(F_2 - F_3) - (F_3 - F_4)$,
$\Delta_3 F$ is the third difference,
$(F_1 - F_2) + (F_3 - F_4)$.

and

$$p = (\lambda - \lambda_2)/(\lambda_3 - \lambda_2)$$

Lagrange formula

This formula can be used when the known values of F occur at any wavelengths. The required value of F is given by:

$$\begin{aligned} F = {} & F_1(\lambda-\lambda_2)(\lambda-\lambda_3)(\lambda-\lambda_4)/(\lambda_1-\lambda_2)(\lambda_1-\lambda_3)(\lambda_1-\lambda_4) \\ & + F_2(\lambda-\lambda_1)(\lambda-\lambda_3)(\lambda-\lambda_4)/(\lambda_2-\lambda_1)(\lambda_2-\lambda_3)(\lambda_2-\lambda_4) \\ & + F_3(\lambda-\lambda_1)(\lambda-\lambda_2)(\lambda-\lambda_4)/(\lambda_3-\lambda_1)(\lambda_3-\lambda_2)(\lambda_3-\lambda_4) \\ & + F_4(\lambda-\lambda_1)(\lambda-\lambda_2)(\lambda-\lambda_3)/(\lambda_4-\lambda_1)(\lambda_4-\lambda_2)(\lambda_4-\lambda_3) \end{aligned}$$

A colorimeter that uses filtered photocells performs this type of integration by virtue of the fact that the response of the photocell is the result of the additive effects of the filtered radiation continuously throughout the part of the spectrum transmitted by the filter. The difficulty with this type of instrument is that it is usually impossible to find filters that, when combined with the spectral sensitivity of the unfiltered photocell, result in a perfect match to the $\bar{x}(\lambda)$, $\bar{y}(\lambda)$, $\bar{z}(\lambda)$ functions. Thus, although integration is carried out, errors are introduced on account of the incorrect spectral sensitivities achieved.

The alternative method of obtaining tristimulus values, computation from spectral data, has the advantage that correct spectral sensitivities can be simulated in the computation, but true integration is not carried out because the computations always consist of summations at discrete wavelength intervals (Billmeyer & Fairman 1987).

The rate at which the $\bar{x}(\lambda)$, $\bar{y}(\lambda)$, $\bar{z}(\lambda)$ functions, and the vast majority of spectral power distributions, change with wavelength is sufficiently slow that summations of data at 1 nm intervals are usually considered to be equivalent to integration (Fairman 1983). The only exceptions are light sources emitting very sharp spectral lines, as, for example, in the case of low pressure sodium lamps; in such cases, allocation of the spectral power in the line to the two closest 1 nm wavelengths in proportion to their proximity to the line, is usually considered to be equivalent to integration.

However, even with summations at 1 nm intervals, there is the possibility of error. If the data for the spectral power distribution are obtained by sampling the spectrum through a spectral 'window' that is wider than 1 nm, then the data do not usually correctly represent the value of the spectral power distribution at each 1 nm wavelength (Stearns 1981a). Ideally, the sampling of the spectrum should be through a 'triangular window' in which, as illustrated in Fig. 8.2(a), the response is at a maximum at the nominal wavelength, is zero at the two neighbouring wavelengths, and falls linearly from maximum to zero over the intermediate wavelengths; such a bandwidth function has a width, at its half height, equal to the wavelength interval. (As mentioned in section 8.11, the bandwidth function represents the combined effects of both the spectral transmittance of the measuring instrument and the spectral sensitivity of the detector.) For a spectral power distribution that is constant with wavelength, summation of data obtained with a 1 nm triangular bandwidth function, at successive 1 nm wavelengths, results in perfect integration (see Fig. 8.2(b)). For spectral power distributions that are not constant with wavelength, perfect integration does not occur; but, because, in general, spectral power distributions are as likely to increase as to decrease with wavelength, the triangular window is at least impartial in the way in which the data is treated.

Since the CIE has published data for Standard Illuminants D_{65} and A at 1 nm intervals, it is appropriate to use this with sample data obtained with a triangular window having a half-height width of 1 nm. To obtain such data, an instrument with a slit-width corresponding to about 1 nm would have to be used. But such instruments tend to have poor signal-to-noise ratios (especially with dark samples), because their monochromators transmit so little light. It is therefore much more common to use instruments having wider slits. These wider slits usually still result in approximately triangular windows of bandwidth, but of course with greater half-height width.

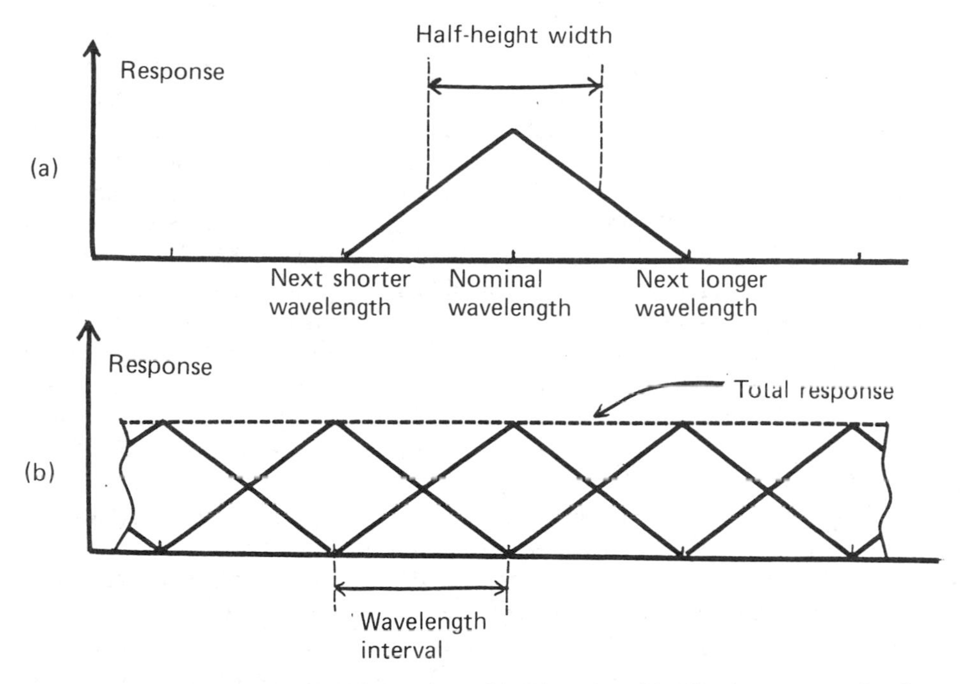

Fig. 8.2 — Spectral bandwidth functions. (a) Triangular. (b) Effective response for the summation of triangular bandwidth functions, separated by a wavelength interval equal to their half-height width, for a spectral power distribution that is constant with wavelength.

Another problem with using data at 1 nm intervals is the very large number of values that have to be included in the summations. For these reasons, most determinations of tristimulus values by the summation of spectral data use wavelength intervals of 5, 10, or 20 nm instead of 1 nm.

When the bandwidth function is triangular with a half-height equal to the wavelength interval, and the summations are obtained by using as weights the entries for the 1 nm $\bar{x}(\lambda)$, $\bar{y}(\lambda)$, $\bar{z}(\lambda)$ functions at each nominal wavelength, then, for a 5 nm interval, the errors are usually less than about $0.2\,\Delta E^*{}_{ab}$; but, if a 10 nm bandwidth and interval is used, errors of up to about $0.7\,\Delta E^*{}_{ab}$ may be encountered; and with a 20 nm bandwidth and interval, errors of up to about $3\,\Delta E^*{}_{ab}$ may occur (Venable 1989). Even larger errors are possible if data corresponding to the wrong bandwidth function are used. Attempts to reduce these errors have been made in various ways.

First, the measured data can be interpolated (using one of the methods given in Table 8.1) to provide values at every 5 nm; but this procedure does not usually improve the accuracy of the results (Venable 1989).

Second, deconvolution procedures can be used that are designed to make the data more like those that would have been obtained with a 1 nm window (to which the entries in the tables of data refer). Instead of using the measured data, P, deconvoluted data, P_c, are used, where:

$$P_{c\lambda} = 1.2P_{\lambda} - 0.1P_{\lambda-w} - 0.1P_{\lambda+w}$$

where w is the half-height bandwidth and interval (Stearns & Stearns 1988). Thus each measurement of P is affected by two neighbouring measurements separated along the wavelength scale by $\pm w$. This changes the values of $P(\lambda)$ from those of the broad window to those that approximate equivalent values for a 1 nm window at the same nominal wavelengths. With 1 nm window values for the nominal wavelengths available, 1 nm window values for all the intermediate wavelengths can be obtained by interpolation (see Table 8.1). These 1 nm values can then be used with the 1 nm $\bar{x}(\lambda)$, $\bar{y}(\lambda)$, $\bar{z}(\lambda)$ data to approximate integration (Stearns 1981b, Stearns 1981c, Stearns & Stearns 1981). But it has been found (Venable 1989) that, if the estimated 1 nm values are used just at the nominal wavelengths (at 5, 10, or 20 nm intervals as the case may be), then the errors are similar, so that the interpolation procedure can be omitted. Thus, data deconvoluted to a 1 nm bandpass can be used with weights at the nominal wavelengths that are correct for 1 nm data, such as those given in Appendix 3, or in similar sets of tables (for example, in ASTM Document E 308 (ASTM 1985; Fairman 1985)).

Third, optimized weights can be used (Venable 1989). The principle of optimized weights can be illustrated by considering, first, a filter-type colorimeter that uses two different filters for approximating the major and minor lobes of the $\bar{x}(\lambda)$ function. Pairs of readings through the two filters have to be added together, and the best weights to use for the two contributions to the final result are those that tend to minimize the errors from the correct result, represented by integration with the true $\bar{x}(\lambda)$ function. Because, in general, spectral data of samples is as likely to increase as to decrease with wavelength, a reasonable approach is to optimize the weights for a non-selective neutral sample with the illuminant being used. This amounts to selecting the weights so that the difference between the computed observer–illuminant product, and the true observer–illuminant product, D_{nm}, evaluated at successive 1 nm wavelengths, sums to zero over the entire spectrum. Data obtained from a spectroradiometer or spectrophotometer can be thought of as having been obtained through a number of filters, N, where N is the number of nominal wavelengths used, each filter having a transmittance equal to the bandwidth function of the instrument; N is usually not less than 16 (20 nm intervals from 400 to 700 nm). The best weights to use for combining these N contributions to the final result will be those that tend to minimize the errors from the true result. Again, it is reasonable to optimize the weights for a non-selective neutral sample with the illuminant being used. Optimization of the N weights is carried out by considering D_{nm} in wavelength intervals equal to the wavelength separation of adjacent spectrophotometric readings. Each interval is used with its centre at successive 1 nm positions along its wavelength range. The sum of the values of D_{nm} in each interval, ΣD_{nm}, is then calculated. The weights are then chosen, by an iterative procedure, so that ΣD_{nm} is as near zero as possible in all the intervals at all the positions (minimum root-mean-square). This ensures that there is no appreciable wavelength range where the differences, D_{nm}, are all of the same sign.

The advantage of using optimized weights can be illustrated by considering the case where the equi-energy stimulus, S_E, is the illuminant, and where the bandwidth function, instead of being triangular, is rectangular, and covers the bandwidth bounded by adjacent mid-points between neighbouring nominal wavelengths, as shown in Fig. 8.3(a). Again, we consider the $\bar{x}(\lambda)$ function as an example. In this case, because the spectral power for S_E is constant with wavelength and the bandwidth is rectangular, the computed observer–illuminant product is constant over each bandwidth, and has the shape of a histogram, as shown in Fig. 8.3(b), with the heights of the steps being equal to the weights used; and the true observer–illuminant product is the same as the colour-matching function. For a non-selective

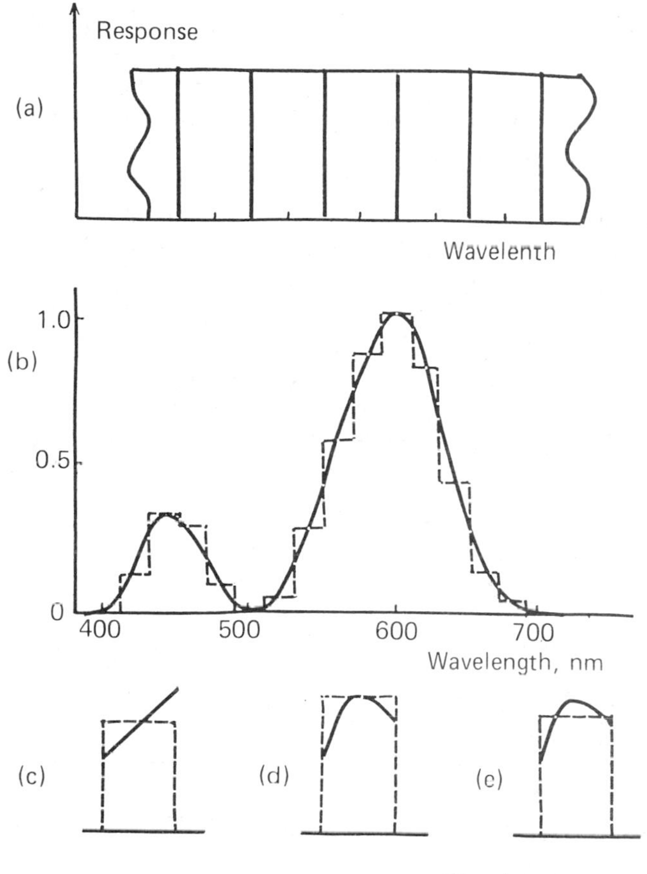

Fig. 8.3 — (a) Rectangular bandwidth functions that cover the whole of the spectrum without any overlapping. (b) An observer–illuminant function approximated by a histogram. (c) Element of a histogram in a single bandwidth, using a step height equal to the value of a function at the mid-wavelength, the function being straight throughout the bandwidth considered. (d) Same as (c) but with the function being curved. (e) Same as (d) but using a step height adjusted to equalize the areas under the step and under the curve.

neutral, the difference between the computed and true tristimulus values will then be equal to the difference between the area under the histogram and the area under the true colour-matching function. In a single bandwidth, the difference between the computed and true observer–illuminant product will be equal to the difference in area under the step considered and under the colour-matching function in the same bandwidth. If weights are used that are simply the 1 nm entries, the height of each step will always be that of the true colour-matching function at the nominal wavelength at the centre of the bandwidth; in this case, the differences between the areas will be zero only if the colour-matching function is a straight line over the bandwidth, as shown in Fig. 8.3(c). When the colour-matching function has curvature, the areas will no longer be equal, as shown in Fig. 8.3(d). But, if the height of the step is altered, the areas can be made equal again, as shown in Fig. 8.3(e). These altered heights are the optimized weights; they are obtained by an iterative procedure that minimizes the differences in area for bandwidths having their centres at successive 1 nm positions throughout the spectrum.

It is clear from the above discussion that the optimum weights will be different for each combination of observer (1931 or 1964), wavelength interval, bandwidth function, and illuminant, and that they are not easy to calculate; but they can be very effective in reducing errors.

Table 8.2 shows typical maximum errors for three cases as found in the study by Venable (Venable 1989): using measured data with table entries (MT); using deconvoluted data with table entries (DT); and using measured data with optimized weights (MO); in all cases the data are assumed to have been measured with triangular windows having the same half-height bandwidth as the wavelength interval, and the table entries are selected from the CIE 1 nm tables. For the MT

Table 8.2 — Typical maximum errors for different types of summation

Half-height band-width	Wave-length interval	Data used	Weights used	Case maximum	Typical errors (ΔE^*_{ab})
5 nm	5 nm	Measured	Table entries	MT	0.2
5 nm	5 nm	Deconvoluted	Table entries	DT	0.02
5 nm	5 nm	Measured	Optimized	MO	0.002
10 nm	10 nm	Measured	Table entries	MT	0.7
10 nm	10 nm	Deconvoluted	Table entries	DT	0.06
10 nm	10 nm	Measured	Optimized	MO	0.01
20 nm	20 nm	Measured	Table entries	MT	3.0
20 nm	20 nm	Deconvoluted	Table entries	DT	0.7
20 nm	20 nm	Measured	Optimized	MO	0.2

case, the errors range from small ($0.2\,\Delta E^*_{ab}$) at 5 nm to considerable ($3\,\Delta E^*_{ab}$) at 20 nm. For the DT case, the errors range from very small ($0.02\,\Delta E^*_{ab}$) at 5 nm to significant ($0.7\,\Delta E^*_{ab}$) at 20 nm. For the MO case, the errors range from negligible

($0.002\,\Delta E^*_{ab}$) at 5 nm to small ($0.2\,\Delta E^*_{ab}$) at 20 nm. The samples used in this study were the NBS SRM 2101-2104 red, yellow, green, and blue glass filters (see section 8.10) together with a didymium glass filter; and, for each filter, the curve of spectral transmittance factor against wavelength was shifted in the long wavelength direction by nineteen 1 nm increments to generate nineteen extra test samples. Some other samples might give larger errors.

The data given in Appendix 3 are the 1 nm entries selected at 5 nm intervals, rounded to four places of decimals. The results of Table 8.2 indicate that, if weights corresponding to these table entries are used with data obtained with a half-height triangular bandwidth of 5 nm, the typical maximum error is only $0.2\,\Delta E^*_{ab}$. This indicates that the 5 nm table entries in Appendix 3 can be used with reasonable confidence in conjunction with spectral data obtained with half-height triangular bandwidths of 5 nm (or less than 5 nm, because the deconvoluted data used with the 5 nm entries have a typical maximum error of only $0.02\,\Delta E^*_{ab}$). But, if these table entries are used at only 10 or 20 nm intervals, or with data obtained with half-height bandwidths of more than 5 nm, appreciable errors may occur.

As mentioned in section 5.12, if data at the two ends of the spectrum are not available, they should be provided by setting the missing values equal to the nearest measured values of the appropriate quantities. To ignore these wavelengths is equivalent to assuming zero values for these missing quantities and this is likely to lead to greater errors (Erb & Krystek 1983). The full range of wavelengths for which the 1 nm values are available for the $\bar{x}(\lambda)$, $\bar{y}(\lambda)$, $\bar{z}(\lambda)$ functions is from 360 to 830 nm; but, when, as is often the case in practice, the computations are made to only 4 places of decimals, it is sufficient to use a range of wavelengths from 380 to 780, because the values above 780 are then all zero, and those below 380 are then nearly zero and are often combined with low values of the spectral power distribution.

In view of these inaccuracies that can arise from the computation of tristimulus values, it is very important that any results that are to undergo critical comparison should have been derived by using the same bandwidth function, wavelength interval, wavelength range, and method of computation, including the choice of weights; this applies both to the evaluation of colour differences, and to the evaluation of the variables in the CIELAB and CIELUV systems, which depend on the relationships between samples and reference whites. This was illustrated in Table 5.3, where the the table entries for the $\bar{x}(\lambda)$, $\bar{y}(\lambda)$, and $\bar{z}(\lambda)$ functions, and for $S(\lambda)$ for Standard Illuminants A and D_{65}, were taken at 1 nm intervals from 360 to 830 nm, at 5 nm intervals from 380 to 780 nm, at 10 nm intervals from 380 to 780 nm and from 400 to 700 nm, and at 20 nm intervals from 400 to 700 nm; both the wavelength interval and the wavelength range affected the results obtained, with differences from the 1 nm results varying from about 0.01 to $1.4\,\Delta E^*_{uv}$ or ΔE^*_{ab}.

8.14 PRECAUTIONS TO BE TAKEN IN PRACTICE

It is clear from the topics discussed in this chapter that there are many sources of imprecision and error in colorimetry, spectroradiometry, and spectrophotometry. Useful precautions that can be taken in practice include the following.

(1) Ensure by visual inspection that all samples are properly selected, and are clean, uniform, and not misshapen.

(2) Store all working standards in protective containers when not in use, and ensure that they are clean and in good condition whenever used.

(3) Ensure that equipment is maintained and operated according to the manufacturer's instructions.

(4) Mount all reflecting samples on standard backings, and measure all textured samples at the same orientation.

(5) When samples are illuminated with white light in instruments, check for thermochromism by making successive measurements over a suitable period.

(6) Select the appropriate geometry of illumination and viewing, and use it without change through each set of measurements.

(7) If polarization is suspected, do not use 45° illumination.

(8) When computing tristimulus values:
   (a) if possible use a 5 nm wavelength interval with a bandwidth half-height of 5 nm or less for the spectral measurements, and use the 5 nm entries in the 1 nm tables of $\bar{x}(\lambda)$, $\bar{y}(\lambda)$, and $\bar{z}(\lambda)$, such as the table entries given in Appendix 3 or in similar sets of tables (for instance, those in ASTM Document E 308 (ASTM 1985)).
   (b) if a wavelength interval greater than 5 nm has to be used, make the spectral measurements with a bandwidth function half-height equal to the wavelength interval, and use the measured data with entries at the greater interval in the 1 nm tables of $\bar{x}(\lambda)$, $\bar{y}(\lambda)$, and $\bar{z}(\lambda)$; if desired, the accuracy can be improved by deconvoluting the spectral data to simulate 1 nm bandwidth data at the measured wavelengths; alternatively, the measured data can be used with optimized weights if these are available.
   (c) When using data with a range of wavelengths less than from 380 to 780 nm, do not set the missing spectral powers equal to zero, but equal to the nearest available data.

(9) When comparing tristimulus values, or any measures derived from them, use the same measuring and computing procedures for all the data; this is particularly important when parameters in colour solids or colour differences are being calculated.

(10) When measuring fluorescent samples (to be discussed in Chapter 9), always illuminate them with white light.

(11) Make repeated measurements on working standards to check precision.

(12) Measure calibrated samples to check accuracy.

(13) When quoting results list the following conditions under which they were obtained:

Geometry of illumination and viewing.

Measuring source (used for fluorescent samples).

Type of spectrophotometer or spectroradiometer used (or bandwidth) for any spectral measurements, or type of photo-electric colorimeter used.

Illuminant and Standard Observer used for tristimulus values.

Method of computation used for deriving tristimulus values from spectral data (including wavelength range and interval, and the use of any correction procedures or optimized weights).

REFERENCES

Alman, D. H. & Billmeyer, F. W. *J. Chem. Educ.* **52**, A281 and A313 (1975).

ASTM Document E 308. *Standard method for computing the colors of objects by using the CIE system*. American Society for Testing Materials, Philadelphia (1985).

ASTM Document E 1164. *Standard practice for obtaining spectrophotometric data for object-color evaluation*. American Society for Testing Materials, Philadelphia (1987, revised 1988).

ASTM Document E 1331. *Standard test method for reflectance factor and color by spectrophotometry using hemispherical geometry*. American Society for Testing Materials, Philadelphia (1990a).

ASTM Document E 1345. *Standard practice for reducing the effect of variability of color measurement by use of multiple measurements*. American Society for Testing Materials, Philadelphia (1990b).

ASTM Document E 1347. *Standard test method for color and color-difference measurement by tristimulus (filter) colorimetry*. American Society for Testing Materials, Philadelphia (1990c).

ASTM Document E 1348. *Standard test method for transmittance and color by spectrophotometry using hemispherical geometry*. American Society for Testing Materials, Philadelphia (1990d).

ASTM Document E 1349. *Standard test method for reflectance factor and color by spectrophotometry using bidirectional geometry*. American Society for Testing Materials, Philadelphia (1990e).

Berns, R. S. & Petersen, K. H. *Color Res. Appl.* **13**, 243 (1988).

Billmeyer, F. W. & Alessi, P. J. *Color Res. Appl.* **6**, 195 (1981).

Billmeyer, F. W. & Fairman, H. S. *Color Res. Appl.* **12**, 27 (1987).

Billmeyer, F. W. & Hemmendinger, H. in: *Golden jubilee of colour in the CIE*, 98–112, Society of Dyers & Colorists, Bradford (1981).

Carter, E. C. & Billmeyer, F. W. *Color Res. Appl.* **4**, 96 (1979).

CIE Publication No. 15, *Colorimetry* (1971).

CIE Publication No. 15.2 *Colorimetry*, 2nd ed. (1986).

Clarke, F. J. J. *Printing Technology*, **13**, 101 (1969).

Clarke, F. J. J. & Compton, J. A. *Color Res. Appl.* **11**, 253 (1986).

Eckerle, K. L. & Venable, W. H., Jr. *Color Res. Appl.* **2**, 137 (1977).
Erb, W. & Krystek, M. *Color Res. Appl.* **8**, 17 (1983).
Erb, W., Krystek, M. & Budde, W. *Color Res. Appl.* **9**, 84 (1984).
Erb, W. & Budde, W. *Color Res. Appl.* **4**, 113 (1979).
Fairman, H. S. *Color Res. Appl.* **8**, 245 (1983).
Fairman, H. S. *Color Res. Appl.* **10**, 199 (1985).
Grum, F. & Saltzman, M. in: CIE Publication No. 36, *Proc. CIE 18th Session, London* 91–98 (1976).
Keegan, H. J., Schleter, J. B. & Judd, D. B. *J. Res. Natl. Bur. Stand.* **66A**, 203 (1962).
Malkin, F. & Verill, J. F. in: CIE Publication No. 56, *Proc. CIE 20th Session, Amsterdam Paper* E37 (1983).
Marcus, R. T. *Color Res. Appl.* **3**, 29 (1978).
Rich, D. C. *Color Res. Appl.* **13**, 113 (1988).
Robertson, A. R. In: C. Burgess & K. D. Mielenz, (eds), *Advances in standards and methodology in spectrophotometry*, Elsevier, New York, (1987).
Simon, F. T. *Color Res. Appl.* **16**, 67 (1991).
Stearns, E. I. *Color Res. Appl.* **6**, 78 (1981a).
Stearns, E. I. *Color Res. Appl.* **6**, 203 (1981b).
Stearns, E. I. *Color Res. Appl.* **6**, 210 (1981c).
Stearns, E. I. & Stearns, R. E. *Color Res. Appl.* **6**, 207 (1981).
Stearns, E. I. & Stearns, R. E. *Color Res. Appl.* **13**, 257 (1988).
Venable, W. H. *Color Res. Appl.* **14**, 260 (1989).
Verrill, J. *CIE J.* **6**, 23 (1987).
Wharmby, D. O. *J. Phys. E: Scientific Instrum.* **8**, 41 (1975).
Wightman, T. E. & Grum, F. *Color Res. Appl.* **6**, 139 (1981).

9

Fluorescent colours

9.1 INTRODUCTION

When a sample fluoresces, some of the power incident on it is re-emitted with a change of wavelength. Therefore, at each wavelength, the total light re-emitted consists of the sum of that due to reflection or transmission (where there has been no change of wavelength) and that due to fluorescence (where there has been a change of wavelength). For the sake of simplifying what is inevitably a complex matter, we will consider only opaque samples; transmitting samples can be dealt with in similar ways.

9.2 TERMINOLOGY

To facilitate the discussion of the effects of fluorescence the following terms are used:

Spectral reflected radiance factor, $\beta_S(\lambda)$
Ratio, at a given wavelength, of the radiance produced by reflection by a sample to that produced by the perfect reflecting diffuser identically irradiated.

Spectral luminescent radiance factor, $\beta_L(\lambda)$
Ratio, at a given wavelength, of the radiance produced by luminescence by a sample to that produced by reflection by the perfect reflecting diffuser identically irradiated.

Spectral total radiance factor, $\beta_T(\lambda)$
The sum of the spectral reflected radiance factor and the spectral luminescent radiance factor.

$$\beta_T(\lambda) = \beta_S(\lambda) + \beta_L(\lambda)$$

Spectral conventional reflectometer value, $\rho_c(\lambda)$
The apparent reflectance factor obtained when a fluorescent sample is measured relative to a non-fluorescent standard white sample, using monochromatic illumination and heterochromatic detection.

The term *radiance factor* has been used in the above definitions, because *reflectance factor* applies only to reflected light and not to fluorescent light, which is produced by photoluminescence. The CIE recommends *spectral radiance factor*, β_e, instead of *spectral total radiance factor*, β_T, but the latter term is used in this chapter to emphasize its composite nature when fluorescence occurs. To measure radiance factors, the cone of light collected should be negligibly small, but with real instruments this may not be so; however, the CIE has not defined terms for luminescent and total radiation collected within a defined cone, hence, in the following discussions, such instrumental measurements are regarded as approximations to radiance factors.

9.3 USE OF DOUBLE MONOCHROMATORS

The most fundamental way of evaluating the colour of fluorescent samples is to irradiate them with monochromatic light and then, for each wavelength of the spectrum, to record the amount of light re-emitted at every wavelength. For an illuminant of known spectral power distribution (such as D_{65}) these data can then be used to calculate the total spectral radiance factor throughout the spectrum, from which tristimulus values can be evaluated in the usual manner (Donaldson 1954, Grum 1971, Minato, Nanjo & Nayatani 1985). Although this procedure is correct, it requires elaborate apparatus, and entails many computations. Simpler methods are therefore needed for practical use.

9.4 ILLUMINATION WITH WHITE LIGHT

The most useful practical approach is to use an instrument in which the sample is irradiated with a suitably chosen 'white' light, after which it is dispersed into the spectrum for the usual evaluation against the spectrum produced similarly by the perfect diffuser. It is essential that the sample is irradiated with white light, the dispersion of the light into the spectrum taking place subsequently. If this is not done, and the light is first dispersed into the spectrum and then used to irradiate the sample, the light produced by the fluorescence, although of various wavelengths, will all be treated as if it were of the same wavelength as the irradiating light. Errors typical of those that are consequently produced are shown in Fig. 9.1 by the curve marked 'conventional'.

If the illuminant used is tungsten light, then Standard Illuminant A can be used in the form of Standard Source A, a tungsten filament lamp having a colour temperature of 2856 K. But tungsten light contains only a small proportion of the ultraviolet radiation that usually plays a very important part in the excitation of fluorescence in samples; so the main interest is in the evaluation of fluorescent materials under daylight illuminants. This requirement, however, poses a difficult problem regarding the choice of white light to use. Standard Illuminant C is realizable as a source, but is deficient in ultraviolet radiation present. On the other hand, Standard Illuminant D_{65}, which has a more correct ultraviolet content, is not realizable as a source.

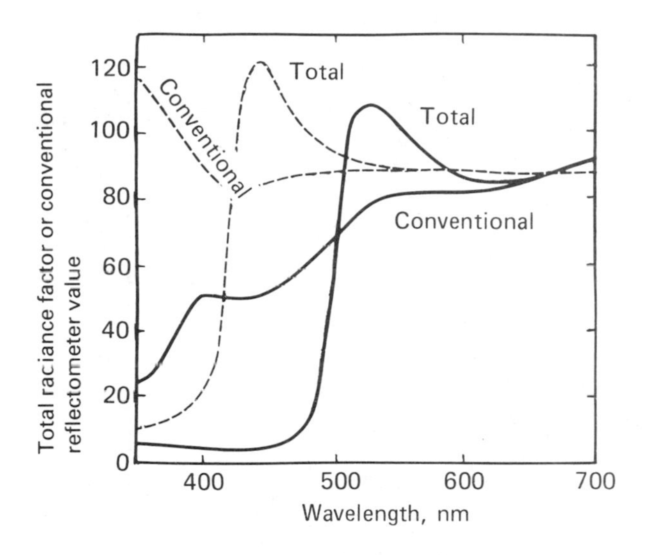

Fig. 9.1 — An example of the difference between total radiance factor ('total'), and conventional reflectometer value ('conventional'), when the sample fluoresces. Full lines: yellow sample. Broken lines: white sample. (After Grum & Bartleson 1980, page 246.)

However, it is possible to use an illuminant that approximates a daylight distribution (such as that of D_{65}), and then to make corrections that give results that are satisfactory for many practical purposes (Grum & Costa 1977). An example of a source that approximates D_{65} is a xenon arc with suitably chosen filters; in Fig. 9.2 relative spectral power distributions are given for such a source and for D_{65} for comparison. To correct for the effects of such differences in a source, it is necessary to know how the measured spectral total radiance factor is composed of the contributions from the spectral reflected radiance factor and the spectral luminescent radiance factor, and the range of wavelengths over which the fluorescing agent is excited. Four methods of obtaining this information will be described later. Assuming that it has been obtained satisfactorily, the procedure for making the corrections is then as follows (Billmeyer & Chong 1980, Grum & Bartleson 1980, Billmeyer 1988).

9.5 CORRECTING FOR DIFFERENCES BETWEEN MEASURING AND DESIRED SOURCES

The use of the measuring source instead of the desired source does not affect the values obtained for the spectral reflected radiance factor. By its definition, it is given by:

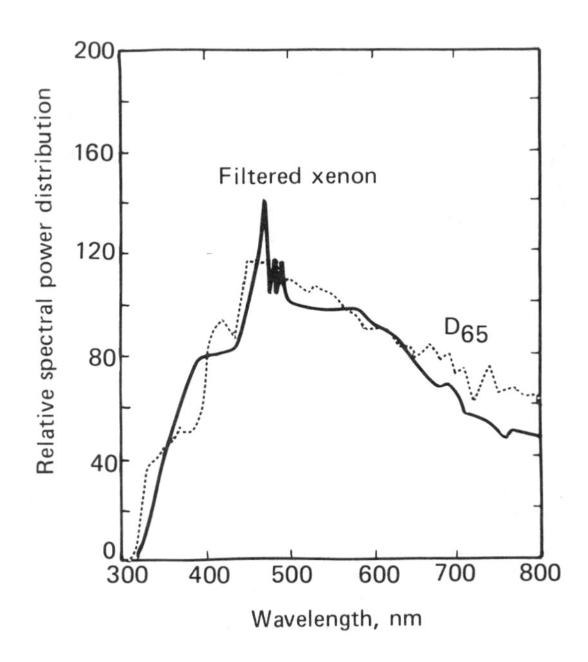

Fig. 9.2 — Relative spectral power distribution of a filtered xenon source intended to approximate D_{65}, whose relative spectral power distribution is shown by the broken line. (After Grum & Bartleson 1980, page 268.)

$$\beta_S(\lambda) = [\beta_R(\lambda).k_aS(\lambda)]/[\beta_P(\lambda).k_aS(\lambda)]$$

where $\beta_R(\lambda)$ is the fraction of the radiance produced by reflection by the sample relative to that produced by the perfect diffuser identically irradiated, $S(\lambda)$ is the relative spectral power distribution of the source, k_a is a constant that converts $S(\lambda)$ to absolute radiance values, and $\beta_P(\lambda)$ is the fraction of the radiance reflected by the perfect diffuser relative to itself. Because the $k_aS(\lambda)$ terms cancel out, $\beta_S(\lambda)$ is independent of the spectral power distribution of the source; and because $\beta_P(\lambda)$ is equal to unity at all wavelengths, $\beta_S(\lambda)$ reduces to $\beta_R(\lambda)$, as is to be expected.

However, the spectral luminescent radiance factor, $\beta_L(\lambda)$ certainly is affected by the spectral power distribution of the source. By its definition, it is given by:

$$\beta_L(\lambda) = F(\lambda)/[\beta_P(\lambda).k_aS(\lambda)]$$

where $F(\lambda)$ is the spectral distribution of the radiance produced by fluorescence. Because $\beta_P(\lambda) = 1$, we have:

$$\beta_L(\lambda) = F(\lambda)/[k_aS(\lambda)]$$

and thus $\beta_L(\lambda)$ is dependent on $S(\lambda)$. If the source contains too great a proportion of ultraviolet radiation, $\beta_L(\lambda)$ will tend to be too high, and if this proportion is too small, $\beta_L(\lambda)$ will tend to be too low.

$F(\lambda)$ is usually at longer wavelengths than those of the irradiating source, and proportional to the number of quanta absorbed from it (Stokes' Law). Hence:

$$F(\lambda) = k_b Q$$

where k_b is a constant, and Q represents the number of quanta from the source absorbed per second by the fluorescing agent; Q can be calculated as follows:

$$Q = k_c \int \alpha(\lambda').k_d S(\lambda').\lambda'.\mathrm{d}\lambda'$$

where k_c is another constant, $\alpha(\lambda')$ is the spectral absorptance of the fluorescing agent, $S(\lambda')$ is the relative spectral power distribution of the irradiating source over the range of wavelengths, λ', that excite the fluorescing agent, and k_d is a constant that converts $S(\lambda')$ to absolute power values. $S(\lambda')$ is multiplied by λ' to allow for the fact that, for a given amount of power of a monochromatic radiation, the number of quanta per second is proportional to the wavelength.

To simplify the following steps, it is now assumed that $\alpha(\lambda')$ is constant with wavelength; although this is unlikely to be strictly true, if, as is often the case, the fluorescing agent is present in high concentration, $\alpha(\lambda')$ will be nearly equal to unity at most wavelengths and hence will not vary much. The above equation can then be rewritten as:

$$Q = k_c k_d \alpha \int S(\lambda').\lambda'.\mathrm{d}\lambda' \ .$$

Approximate numerical values of Q can be obtained from:

$$Q = k_c k_d \alpha [S_1.\lambda_1' + S_2.\lambda_2' + \ldots\ldots S_n.\lambda_n']$$

where the subscript 1 indicates the shortest wavelength of the fluorescent excitation region, the subscripts 2, etc., indicate wavelengths at regular wavelength intervals within that region, and the subscript n indicates the longest wavelength of that region.

Because $\beta_L(\lambda) = F(\lambda)/[k_a S(\lambda)]$ and $F(\lambda) = k_b Q$, we have:

$$\beta_L(\lambda) = k_b Q / k_a S(\lambda)$$

It follows from the above that, for two sources, indicated by the subscripts d denoting the desired source, and m denoting the measuring source:

$$\frac{\beta_{Ld}(\lambda)}{\beta_{Lm}(\lambda)} = \frac{[S_1\lambda_1' + S_2\lambda_2' + \ldots\ldots S_n\lambda_n']_d/S_d(\lambda)}{[S_1\lambda_1' + S_2\lambda_2' + \ldots\ldots S_n\lambda_n']_m/S_m(\lambda)}$$

the constants, k_a, k_b, k_c, k_d, and α all cancelling out. This expression is then used to obtain the desired spectral luminescent radiance factor $\beta_{Ld}(\lambda)$ from the measured spectral luminescent radiance factor $\beta_{Lm}(\lambda)$.

The spectral total radiance factor for the sample illuminated by the desired illuminant, $\beta_{Td}(\lambda)$, is then calculated as $\beta_{Ld}(\lambda)$ plus the spectral reflected radiance factor, $\beta_S(\lambda)$.

$$\beta_{Td}(\lambda) = \beta_{Ld}(\lambda) + \beta_S(\lambda)$$

The four methods that will now be described for separating the contributions of the spectral reflected radiance factor, β_S, and the spectral luminescent radiance factor, β_L, to the spectral total radiance factor, β_T, and for determining the range of wavelengths exciting fluorescence, may be referred to as the *two-monochromator* method, the *two-mode* method, the *filter-reduction* method, and the *luminescence-weakening* method.

9.6 TWO-MONOCHROMATOR METHOD

In the two-monochromator method, the sample is illuminated with monochromatic light by means of a first monochromator, and then a second monochromator is used to transmit light only of the same wavelength; in this way, any fluorescent light generated by irradiance will, because of its wavelength shift, not pass through the second monochromator, and β_S alone is determined. By scanning the wavelength range of interest with the two monochromators in step for wavelength and by using a narrow bandpass for each, β_S can be measured with high accuracy. This is a simpler procedure than measuring the complete spectrum of the light emitted by the sample for each wavelength of irradiation, and is therefore a useful technique. The steps necessary for this method are given in Table 9.1.

However, if, as is often the case, only single monochromators are available, then one of the other methods has to be used, although they tend to be less accurate. In one study of the three remaining methods (Alman, Billmeyer & Phillips 1976) it was found that the filter-reduction method was preferable for samples with small amounts of fluorescence, and the luminescence-weakening method for samples with large amounts of fluorescence; the two-mode method was found to be less accurate, but it is a useful diagnostic tool, and is therefore now described first.

9.7 TWO-MODE METHOD

In the two-mode method (Simon 1972) an instrument is required that can be used in both the white-irradiation and the monochromatic-irradiation modes; the latter mode is referred to as a *conventional reflectometer*. Such an instrument provides a

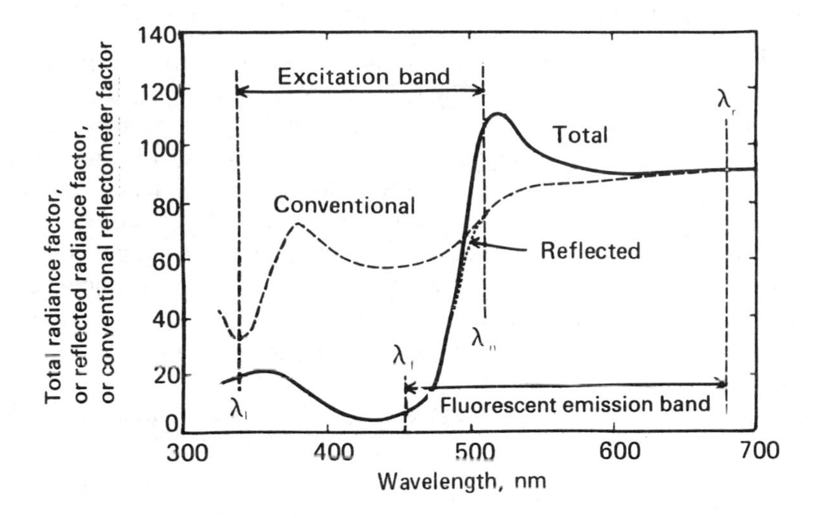

Fig. 9.3 — Total radiance factor ('total'), reflected radiance factor ('reflected'), and conventional reflectometer value ('conventional') for a green fluorescent sample. The excitation band of wavelengths is from λ_1 to λ_n. The fluorescent emission band of wavelengths is from λ_f to λ_r. (After Grum & Bartleson 1980, page 276.)

good means of showing whether any fluorescence is occurring: this is illustrated by the differences between the curves for spectral total radiance factor and for spectral conventional reflectometer value shown in Figs 9.1 and 9.3. The steps necessary for this method are given in Table 9.2.

9.8 FILTER-REDUCTION METHOD

In the filter-reduction method (Eitle & Ganz 1968) a two-mode instrument is not necessary, and the work can be carried out by using only a white-light-irradiating monochromatic-light-detecting instrument. The white light irradiating the sample is used both unfiltered and with a series of filters each of which sharply cuts out all radiation of wavelengths shorter than a chosen one. These chosen wavelengths cover the region where both excitation and fluorescent emission occur (see Fig. 9.3). The steps necessary for this method are given in Table 9.3.

9.9 LUMINESCENCE-WEAKENING METHOD

In the luminescence-weakening method (Allen 1973), as was the case for the filter-reduction method, a two-mode instrument is not necessary, and the work can be carried out by using only a white-light-irradiating monochromatic-light-detecting instrument. The white light irradiating the sample is used both with and without two filters, one of which cuts out all the radiation that causes fluorescence, and the other that reduces that radiation appreciably. The steps necessary for this method are given in Table 9.4.

9.10 PRACTICAL CONSIDERATIONS

If one of these four methods, or a similar procedure, is not carried out, the best that can then be done is to use the white-light irradiation mode with a source that approximates D_{65} as closely as possible, and to record the exact nature of the source used. Various sources have been suggested for this purpose, and some are much more appropriate than others (Terstiege 1989). One method is to use a spectrophotometer with an ultraviolet absorbing filter which can be inserted to a varying extent into the white-light irradiating beam, so as to adjust its ultraviolet content. A setting for the position of the filter is then chosen that most nearly produces spectral radiance measurements that are the same as known true values for a set of standard samples for the desired illuminant, such as D_{65}. A set of fluorescing whites are sometimes used as the standard samples.

In evaluating fluorescent samples, the following are important areas to keep under control:

(1) The illuminating system in the measuring instrument must be fully defined and standardized.
(2) The samples must be stable under the measurement conditions. In the white-light irradiating mode the amount of radiant power falling on them must not affect their colour by heating or fading.
(3) The appropriate illuminating and viewing geometry must be used. If integrating spheres are used, then the specimen modifies the spectral power distribution of the illumination in the white-light irradiating mode (Alman & Billmeyer 1976), but this effect can be negligible if a large sphere with small ports is used.
(4) The coating on the interior surface of any integrating sphere that is used must be non-fluorescent.

REFERENCES

Allen, E. *Applied Optics* **12**, 289 (1973).
Alman, D. H. & Billmeyer, F. W. *Color Res. Appl.* **1**, 141 (1976).
Alman, D. H., Billmeyer, F. W., & Phillips, D. G. in CIE Publication No. 36, *Proc. CIE 18th Session, London* 237-244 (1976).
Billmeyer, F. W. *Color Res. Appl.* **13**, 318 (1988).
Billmeyer, F. W. & Chong, T.-F. *Color Res. Appl.* **5**, 156 (1980).
Donaldson, R. *Brit. J. Appl. Phys.* **5**, 20 (1954).
Eitle, D. & Ganz, E. *Textilveredlung* **3**, 389 (1968).
Grum, F. in CIE Publication No. 21 *Proc. CIE 17th Session, Barcelona* Paper No. P71.22 (1971).
Grum, F. & Bartleson, C.J. *Color Measurement* Academic Press, New York (1980).
Grum, F. & Costa, L. F. *Tappi* **60**, 119 (1977).
Minato, H., Nanjo, M. & Nayatani, Y. *Color Res. Appl.* **10**, 84 (1985).
Simon, F. T. *J. Color Appearance* **1** (Issue 5), 28 (1972).
Terstiege, H., *Color Res. Appl.* **14**, 131 (1989).

Table 9.1 — Steps for the two-monochromator method

Step 1. The spectral total radiance factor, $\beta_T(\lambda)$, is measured with the sample irradiated by a stable readily available source that approximates D_{65} as closely as possible.

Step 2. The spectral reflected radiance factor, $\beta_S(\lambda)$, is measured by irradiating the sample with monochromatic light and using the second monochromator to isolate light only of the same wavelength.

Step 3. The spectral luminescent radiance factor, $\beta_L(\lambda)$, is evaluated as the difference between the spectral total radiance factor and the spectral reflected radiance factor.

Step 4. The range of wavelengths over which the fluorescing agent is excited is estimated as follows. The longest wavelength is taken as that at which the 'total' curve is at its greatest height above the 'reflected' curve. The shortest wavelength is taken as the shortest (on the short wavelength side of the fluorescent emission band) at which the spectral reflected radiance curve is similar to that of a version of the material being studied that contains no fluorescing agent (this indicating that the fluorescing agent is not absorbing any radiation).

Table 9.2 — Steps for the two-mode method

Step 1. The spectral total radiance factor, $\beta_T(\lambda)$, is measured with the sample irradiated by a stable readily available source that approximates D_{65} as closely as possible.

Step 2. The spectral conventional reflectometer value, $\rho_C(\lambda)$, is measured, preferably using the same instrument, but operating it in a monochromatic-irradiating mode.

Step 3. The spectral reflected radiance factor, $\beta_S(\lambda)$, is deduced. This is given by the spectral total radiance factor at wavelengths shorter than those of the fluorescent emission band (the shortest wavelength of which is usually about 50 nm below that at which the 'total' and 'conventional' curves cross), and by the conventional spectral reflectometer value at wavelengths longer than those of the region of excitation for fluorescence (the longest wavelength of which is usually about 15 nm above that at which the 'total' and 'conventional' curves cross). For wavelengths within both these regions the results have to be estimated by interpolation. (See Fig. 9.3.)

Step 4. The spectral luminescent radiance factor, $\beta_L(\lambda)$, is evaluated as the difference between the spectral total radiance factor and the spectral reflected radiance factor.

Step 5. The range of wavelengths over which the fluorescent agent is excited is estimated as follows. The longest wavelength is the point of intercept of the values for the spectral reflected radiance factor and the spectral conventional reflectometer value; the shortest wavelength is that at which the total spectral radiance factor and the spectral conventional reflectometer value are closest. This is illustrated in Fig. 9.3.

Table 9.3 — Steps for the filter-reduction method

Step 1. The spectral total radiance factor, $\beta_T(\lambda)$, is measured with the sample irradiated by a stable readily available source that approximates D_{65} as closely as possible. At wavelengths shorter than the short-wavelength boundary, λ_f, of the fluorescent emission band, the curve obtained provides the spectral reflected radiance factor. The wavelength, λ_f, is usually about 80 nm below the wavelength, λ_T, at which $\beta_T(\lambda)$ is maximum in the fluorescent emission band.

Step 2. A sharp cut-off filter is introduced into the incident beam that completely eliminates all wavelengths below λ_T. At wavelengths above the cut-off wavelength of the filter, this curve provides the spectral reflected radiance factor. It is now necessary to determine the spectral reflected radiance factor at the intermediate wavelengths, between λ_f and λ_T.

Step 3. A cut-off filter is introduced into the incident beam that eliminates all wavelengths below a wavelength in the region between λ_f and λ_T. This reduces the fluorescent emission to give an approximation to the spectral reflected radiance factor.

Step 4. Step 3 is repeated, using filters with different chosen cut-off wavelengths (between λ_f and λ_T) to obtain further approximations to the spectral reflected radiance factor. These approximations are then used to obtain a best estimate of the true spectral reflected radiance factor, the most accurate estimate at each wavelength always being the lowest of those available.

Step 5. The spectral luminescent radiance factor, $\beta_L(\lambda)$, is evaluated as the difference between the spectral total radiance factor and the spectral reflected radiance factor.

Step 6. The range of wavelengths over which the fluorescing agent is excited is estimated as follows. The longest wavelength is taken as that at which the 'total' curve is at its greatest height above the 'Reflected' curve. The shortest wavelength is taken as the shortest (on the short wavelength side of the fluorescent emission band) at which the spectral reflected radiance curve is similar to that of a version of the material being studied that contains no fluorescing agent (this indicating that the fluorescing agent is not absorbing any radiation).

Table 9.4 — Steps for the luminescent-weakening method

Step 1. The spectral total radiance factor, $\beta_T(\lambda)$, is measured with the sample irradiated by a stable readily available source that approximates D_{65} as closely as possible. At wavelengths shorter than the short-wavelength boundary, λ_f, of the fluorescent emission band, the curve obtained provides the spectral reflected radiance factor. The wavelength, λ_f, is usually about 80 nm below the wavelength, λ_T, at which $\beta_T(\lambda)$ is maximum in the fluorescent emission band.

Step 2. A sharp cut-off filter is introduced into the incident beam that completely eliminates all wavelengths below λ_T, and a 'completely filtered (CF)' version of $\beta_T(\lambda)$ is obtained, $\beta_{TCF}(\lambda)$. At wavelengths above the cut-off wavelength of the filter, this curve provides the spectral reflected radiance factor. It is now necessary to determine the spectral reflected radiance factor at the intermediate wavelengths between λ_f and λ_T.

Step 3. Instead of the sharp cut-off filter, a partial cut-off filter is introduced into the incident beam that eliminates all wavelengths below a wavelength just above the minimum of the unfiltered $\beta_T(\lambda)$ curve, and a 'partly filtered (PF)' version of $\beta_T(\lambda)$ is obtained, $\beta_{TPF}(\lambda)$. The spectral transmittance curve, $T(\lambda)$, of this filter has to be known, and can be measured on the spectrophotometer being used.

Step 4. The curves obtained with and without the partial cut-off filter are now used to determine the spectral reflected radiance factor, $\beta_S(\lambda)$, from the equation:

$$\beta_S(\lambda) = [\beta_{TPF}(\lambda).T(\lambda) - \beta_T(\lambda).k]/[T(\lambda) - k]$$

the value of the constant k being given by:

$$k = T(w)[\beta_{TPF}(w) - \beta_S(w)]/[\beta_T(w) - \beta_S(w)]$$

where w is the lowest-wavelength at which readings were obtained with the sharp cut-off filter; the values of $\beta_{TPF}(w)$, $\beta_T(w)$, and $T(w)$ are read off their curves at the wavelength w, and the value of β_S at this wavelength is the same as that of β_{TCF} and can thus be obtained from the curve of β_{TCF}.

Step 5. The spectral luminescent radiance factor, $\beta_L(\lambda)$, is evaluated as the difference between the spectral total radiance factor and the spectral reflected radiance factor.

Step 6. The range of wavelengths over which the fluorescing agent is excited is estimated as follows. The longest wavelength is taken as that at which the 'total' curve is at its greatest height above the 'reflected' curve. The shortest wavelength is taken as the shortest (on the short wavelength side of the fluorescent emission band) at which the spectral reflected radiance curve is similar to that of a version of the material being studied that contains no fluorescing agent (this indicating that the fluorescing agent is not absorbing any radiation).

Justification for the procedure in Step 4. That the procedure in Step 4 is correct can be shown as follows. Using the subscripts PF to indicate 'partly filtered' and CF to indicate 'completely filtered', we can write:

$$\beta_T(\lambda) = \beta_S(\lambda) + F(\lambda)/[k_a S(\lambda)]$$
$$\beta_{TPF}(\lambda) = \beta_S(\lambda) + f_{PF} F(\lambda)/[k_{aPF} S_{PF}(\lambda)]$$

where f_{PF} is a constant that allows for the different level of fluorescent excitation produced by the partly filtered source. If the spectral transmittance of the partial cut-off filter is $T(\lambda)$, then

$$T(\lambda).S(\lambda) = S_{PF}(\lambda) \ .$$

Hence:

$$\beta_{TPF}(\lambda) = \beta_S(\lambda) + f_{PF} F(\lambda)/[k_{aPF} T(\lambda).S(\lambda)] \ .$$

Eliminating $F(\lambda)/S(\lambda)$ we obtain:

$$\beta_S(\lambda) = [\beta_{TPF}(\lambda).T(\lambda).k_{aPF} - \beta_T(\lambda) k_a f_{PF}]/[k_{aPF} T(\lambda) - k_a f_{PF}] \ .$$

Changing the power of the partial cut-off light source would change f_{PF} and k_{aPF} in the same proportion; and f_{PF} is inversely proportional to k_a, because, if k_a were increased $F(\lambda)$ would increase in proportion and f_{PF} would have to decrease in the same proportion to keep $f_{PF} F(\lambda)$ in the same relationship to the power of the partial cut-off light source $k_{aPF} S_{PF}(\lambda)$. Therefore, where k is a new constant:

$$f_{PF} = k k_{aPF}/k_a \ .$$

Substituting this expression for f_{PF} we obtain:

$$\beta_S(\lambda) = [\beta_{TPF}(\lambda).T(\lambda) - \beta_T(\lambda).k]/[T(\lambda) - k]$$

If w represents a particular wavelength, we can write the above equation for w, solve for w, and obtain:

$$k = T(w)[\beta_{TPF}(w) - \beta_S(w)]/[\beta_T(w) - \beta_S(w)] \ .$$

10

RGB Colorimetry

10.1 INTRODUCTION

In the years immediately following the introduction of the CIE system of colorimetry in 1931, visual instruments providing additive mixtures of controllable beams of red, green, and blue light were used to obtain colour matches on samples, from which CIE tristimulus values were calculated. The skill required to make such matches, and particularly the length of time that it took even an experienced observer to match large numbers of samples, have resulted in such instruments now being obsolete. Visual colour matches are still made when colour atlasses are used (see Chapter 7), and approximate CIE specifications can be obtained from such matches; and, if visual colour matches are made on the Lovibond-Schofield Tintometer, CIE specifications that are accurate can be derived (see section 7.13). But, today, most colorimetry depends on responses from photocells, either in colorimeters employing filtered photocells, or in spectroradiometers or spectrophotometers to provide spectral data for computation.

However, there remain two areas where the production and specification of colours by additive mixtures of beams of red, green, and blue light are still carried out. In visual research work, this is still a very useful way of providing colour stimuli in a continuously variable form; and, in television and VDU displays, the colours are also produced by such additive mixtures. In this chapter, we shall therefore discuss the colorimetric procedures involved.

10.2 CHOICE AND SPECIFICATION OF MATCHING STIMULI

The gamut of chromaticities that can be matched with a set of three colour matching stimuli is defined by the triangle formed by the three points representing them in a chromaticity diagram. In Fig. 10.1, such a triangle is shown, in the x,y and in the u',v' diagrams, for the three phosphors (*E.B.U. phosphors*) chosen as standard for

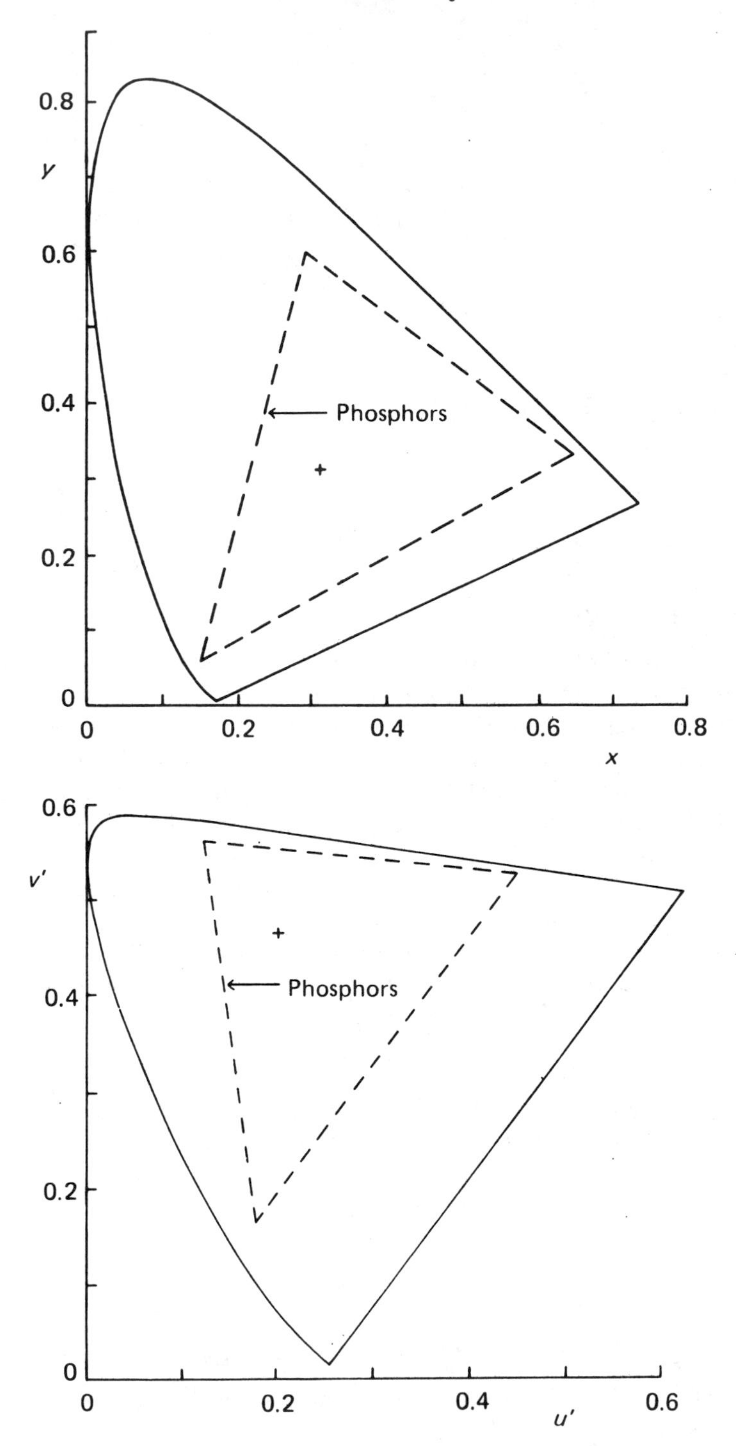

Fig. 10.1 — Gamut of E.B.U. phosphors in (a) the x,y chromaticity diagram, and (b) the u',v' chromaticity diagram.

European colour television (BREMA 1969, Sproson 1978); the chromaticities of these *primaries*, as they are usually called, are:

Red	x=0.64	y=0.33	u'=0.451	v'=0.523
Green	x=0.29	y=0.60	u'=0.121	v'=0.561
Blue	x=0.15	y=0.06	u'=0.175	v'=0.158

It is clear that the triangles in Fig. 10.1 cover less than half the domain of chromaticities lying within the spectral locus and purple boundary. If more saturated matching stimuli, lying on the spectral locus, had been chosen, then more of the domain of chromaticities would have been included in the triangles; but, even if phosphors for television display devices were available having such chromaticities, their luminous efficacies would be very low, so that the luminances attainable with such phosphors would also be very low. Low luminances result in poor precision in colour matching, and in low brightness and colourfulness in displayed colours. The choice of primaries nearly always requires a compromise between covering as much of the chromaticity domain as possible and having adequate luminance; the primaries whose chromaticities are shown in Fig. 10.1 were the result of such a compromise for television.

The triangle for the primaries appears to be particularly lacking in its coverage of greenish colours in the x,y diagram, whereas in the u',v' diagram a greater lack is apparent for purplish colours. This provides a good illustration of the way in which the x,y diagram can be misleading; as already shown in Fig. 3.5, colour differences of a given size are represented by much larger distances in the greenish than in the purplish regions of this diagram. Such distortions are much reduced in the u',v' diagram, so that its use is more appropriate, although it is not without some distortions itself, as can be seen in Fig. 3.6.

10.3 CHOICE OF UNITS

As already explained in section 2.4, if the amounts of red, green, and blue are measured in photometric units, such as units of luminance, a match on a white is represented by three very unequal numbers, the value for the blue being particularly low. It is, therefore, common practice to use units defined as resulting in equal amounts of the three primaries being required to match a suitably chosen white. In television displays this white is usually D_{65} in European systems, and S_C in American systems. The luminances of these units are denoted by L_R, L_G, and L_B.

10.4 CHROMATICITY DIAGRAMS USING *r* AND *g*

Having defined the chromaticities of the three matching primaries, and the units in which they are to be measured, the colorimetric system is complete, and the amounts of red, green, and blue light needed to match any colour are the tristimulus values, ***R***,

G, B, in the system. From these can be obtained corresponding chromaticity co-ordinates:

$$r = R/(R+G+B)$$

$$g = G/(R+G+B)$$

$$b = B/(R+G+B)\ .$$

Since $r+g+b=1$, if two of these values are known the third can be deduced by subtracting their sum from 1. A chromaticity diagram in which one of the co-ordinates is plotted against one of the others can therefore be used, and it is customary to use r and g for this purpose, as shown in Fig. 10.2.

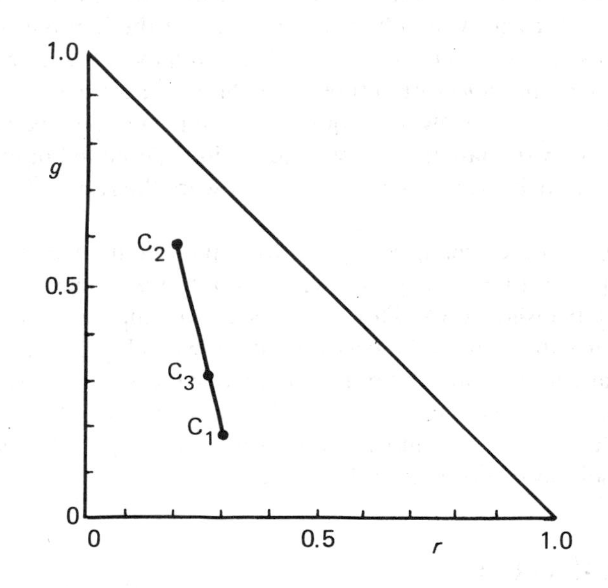

Fig. 10.2 — r,g chromaticity diagram. The colour C_3 can be matched by an additive mixture of the colours C_1 and C_2.

Additive mixtures of colours on r,g chromaticity diagrams, such as that shown in Fig. 10.2, lie on the straight line joining the two points representing the constituent colours in the mixture. If the calculation given in section 3.5 for mixtures on the x,y diagram is reworked for mixtures on an r,g diagram, it is found that the chromaticity, r_3,g_3 of the colour C_3, produced by the additive mixture of m_1 luminance units of colour C_1, and m_2 luminance units of colour C_2, having chromaticities r_1,g_1 and r_2,g_2, respectively, is the same as the centre of gravity of weights, W_1 and W_2 placed at the points representing C_1 and C_2, respectively, where:

$$W_1 = m_1/(L_R r_1 + L_G g_1 + L_B b_1)$$
$$W_2 = m_2/(L_R r_2 + L_G g_2 + L_B b_2)$$

This means that the point C_3 divides the line C_1C_2 in the ratio

$$\frac{C_1C_3}{C_2C_3} = \frac{W_2}{W_1} \, .$$

The centre of gravity law of colour mixture thus applies to mixtures in the *r,g* diagram, but is a little more complicated than in the *x,y* diagram where the weights used are m_1/y_1 and m_2/y_2, or in the u',v' diagram where the weights used are m_1/v'_1 and m_2/v'_2.

10.5 COLOUR-MATCHING FUNCTIONS IN R,G,B, SYSTEMS

If the red, green, and blue primaries are used to match equal amounts of power per small constant-width wavelength interval throughout the spectrum, the results constitute a set of colour-matching functions. If the results are plotted against wavelength, a set of curves similar to those shown in Fig. 10.3 is obtained. If these curves are compared with those shown in Fig. 2.4, it can be seen that they are broadly of the same type but differ in details. These differences result from the differences in chromaticities between the television phosphors used as the primaries in Fig. 10.3,

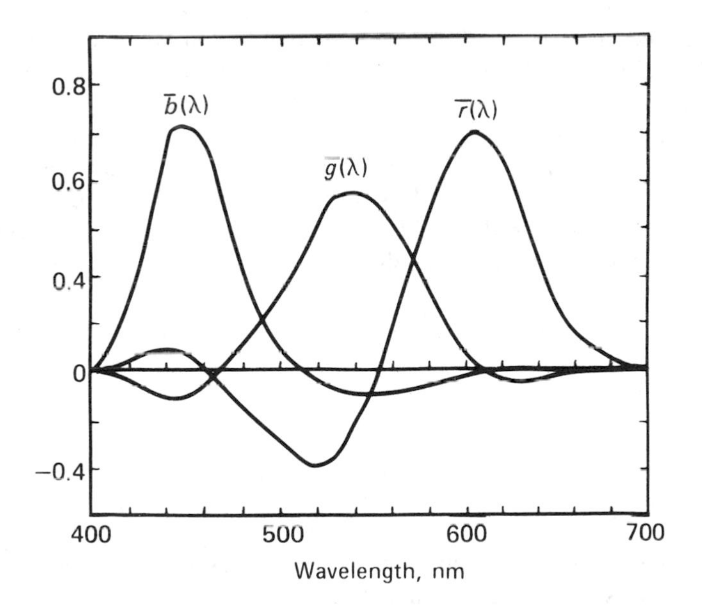

Fig. 10.3 — Colour-matching functions for E.B.U. phosphors.

and the monochromatic wavelengths of 700, 546.1, and 435.8 nm used as the matching stimuli in Fig. 2.4.

10.6 DERIVATION OF X,Y,Z FROM R,G,B TRISTIMULUS VALUES

If it is required to obtain CIE X,Y,Z tristimulus values from the R,G,B tristimulus values of a red, green, and blue system, then a set of *transformation equations* is required. These are usually in the form:

$$X=A_1R+A_2G+A_3B$$
$$Y=A_4R+A_5G+A_6B$$
$$Z=A_7R+A_8G+A_9B\ ,$$

with the corresponding reverse equations in the form:

$$R=B_1X+B_2Y+B_3Z$$
$$G=B_4X+B_5Y+B_6Z$$
$$B=B_7X+B_8Y+B_9Z\ ,$$

where A_1 to A_9 and B_1 to B_9 are constants. The reverse equations can be used to transform the CIE colour matching functions $\bar{x}(\lambda),\bar{y}(\lambda),\bar{z}(\lambda)$ into $\bar{r}(\lambda),\bar{g}(\lambda),\bar{b}(\lambda)$, the corresponding functions for the CIE Standard Observer in the red, green, and blue system being used. The curves of Fig. 10.3 were obtained in this way, and therefore represent the colour-matching properties of the CIE 1931 Standard Colorimetric Observer in this system.

The calculation of the constants A_1 to A_9 and B_1 to B_9 in the transformation equations proceeds as follows. The initial data are usually in the form of the CIE chromaticities of the primaries, and of the white used for equalizing their units. These chromaticities are best derived by computation in the usual way from spectral power data. We can denote them as follows:

	x	y	z
Red	a_1	a_2	a_3
Green	a_4	a_5	a_6
Blue	a_7	a_8	a_9
White	j_1	j_2	j_3 .

It is convenient now to use equations to represent colour matches. When this is done, the equations are written in the form:

$$C(\mathrm{C})\equiv R(\mathrm{R})+G(\mathrm{G})+B(\mathrm{B})\ .$$

The sign ≡ in these equations means 'matches', the italicized symbols C, R, G, B represent the amounts of the colours in the match, and the bracketed symbols (C),(R),(G),(B) indicate the colours to which these amounts refer; these bracketed symbols are only labels, and there is no sense in which C(C) is to be regarded as C being multiplied by (C), for instance. Although these labels are unnecessary in the above equations because the symbols themselves indicate the colour concerned, this is not generally so, particularly when actual numbers are used for the amounts. The same convention is used for the X,Y,Z system, in equations of the type:

$$C(\mathrm{C}) \equiv X(\mathrm{X}) + Y(\mathrm{Y}) + Z(\mathrm{Z}) \ .$$

The chromaticity co-ordinates can be regarded as tristimulus values for a certain amount, k_1 of (R), k_2 of (G), and k_3 of (B). We can therefore represent these amounts in colour-matching equations, as follows:

$$k_1(\mathrm{R}) \equiv a_1(\mathrm{X}) + a_2(\mathrm{Y}) + a_3(\mathrm{Z})$$

$$k_2(\mathrm{G}) \equiv a_4(\mathrm{X}) + a_5(\mathrm{Y}) + a_6(\mathrm{Z})$$

$$k_3(\mathrm{B}) \equiv a_7(\mathrm{X}) + a_8(\mathrm{Y}) + a_9(\mathrm{Z}) \ .$$

The matches on the white in the two systems can be represented by the colour-matching equations:

$$k_4(\mathrm{W}) \equiv H_1(\mathrm{R}) + H_2(\mathrm{G}) + H_3(\mathrm{B})$$

$$k_4(\mathrm{W}) \equiv J_1(\mathrm{X}) + J_2(\mathrm{Y}) + J_3(\mathrm{Z}) \ ,$$

where H_1,H_2,H_3 are the amounts of (R),(G),(B), respectively, required to match the white, and J_1,J_2,J_3 are proportional to the x, y, z chromaticity co-ordinates, j_1,j_2,j_3, of the white, but such that J_2 is equal to the luminance factor of the white. It is now necessary to evaluate k_1,k_2,k_3, and to do this it is required to solve the three simultaneous equations for k_1(R),k_2(G),k_3(B) to obtain the equivalent equations for:

$$1.0(\mathrm{X}) \equiv c_1k_1(\mathrm{R}) + c_2k_2(\mathrm{G}) + c_3k_3(\mathrm{B})$$

$$1.0(\mathrm{Y}) \equiv c_4k_1(\mathrm{R}) + c_5k_2(\mathrm{G}) + c_6k_3(\mathrm{B})$$

$$1.0(\mathrm{Z}) \equiv c_7k_1(\mathrm{R}) + c_8k_2(\mathrm{G}) + c_9k_3(\mathrm{B}) \ .$$

Substituting (X),(Y),(Z) in the equation

$$k_4(\mathrm{W}) = J_1(\mathrm{X}) + J_2(\mathrm{Y}) + J_3(\mathrm{Z})$$

and comparing the result with the equation

$$k_4(\mathrm{W}) \equiv H_1(\mathrm{R}) + H_2(\mathrm{G}) + H_3(\mathrm{B})$$

we obtain:

$$k_1 = H_1/(J_1c_1 + J_2c_4 + J_3c_7)$$
$$k_2 = H_2/(J_1c_2 + J_2c_5 + J_3c_8)$$
$$k_3 = H_3/(J_1c_3 + J_2c_6 + J_3c_9) \ .$$

Hence k_1, k_2, k_3 are evaluated, and can be used in equations:

$$1.0(\mathrm{R}) \equiv (a_1/k_1)(\mathrm{X}) + (a_2/k_1)(\mathrm{Y}) + (a_3/k_1)(\mathrm{Z})$$
$$1.0(\mathrm{G}) \equiv (a_4/k_2)(\mathrm{X}) + (a_5/k_2)(\mathrm{Y}) + (a_6/k_2)(\mathrm{Z})$$
$$1.0(\mathrm{B}) \equiv (a_7/k_3)(\mathrm{X}) + (a_8/k_3)(\mathrm{Y}) + (a_9/k_3)(\mathrm{Z}) \ .$$

If then a colour has been produced or measured in the (R),(G),(B) system by amounts R of (R), G of (G), and B of (B), this can be expressed as:

$$C(\mathrm{C}) \equiv R(\mathrm{R}) + G(\mathrm{G}) + B(\mathrm{B}) \ ,$$

and by substituting the expressions for 1.0(R),1.0(G),1.0(B) in this equation we obtain:

$$\begin{aligned} C(\mathrm{C}) \equiv & (Ra_1/k_1 + Ga_4/k_2 + Ba_7/k_3)(\mathrm{X}) + \\ & (Ra_2/k_1 + Ga_5/k_2 + Ba_8/k_3)(\mathrm{Y}) + \\ & (Ra_3/k_1 + Ga_6/k_2 + Ba_9/k_3)(\mathrm{Z}) \ . \end{aligned}$$

So the X,Y,Z tristimulus values are calculated from the R,G,B tristimulus values by the equations:

$$X = (a_1/k_1)R + (a_4/k_2)G + (a_7/k_3)B$$
$$Y = (a_2/k_1)R + (a_5/k_2)G + (a_8/k_3)B$$
$$Z = (a_3/k_1)R + (a_6/k_2)G + (a_9/k_3)B \ .$$

The set of reverse transformation equations enabling R, G, B to be obtained from X, Y, Z are given by the expressions:

$$R = c_1k_1X + c_4k_1Y + c_7k_1Z$$
$$G = c_2k_2X + c_5k_2Y + c_8k_3Z$$
$$B = c_3k_3X + c_6k_3Y + c_9k_3Z \ .$$

These coefficients were obtained in the colour matching equations for 1.0(X), 1.0(Y), and 1.0(Z); because they are now used in tristimulus value equations, their positions are transposed diagonally.

10.7 USING TELEVISION AND VDU DISPLAYS

When television and VDU displays are being used, it is necessary, for critical work, to check the characteristics of the monitor. The chromaticities of the phosphors define the three primaries involved; the chromaticity of the reference white defines the units in which they are expressed; and the relationship between the voltages applied to the electron guns and the amount of light produced by the phosphors enables tristimulus values to be calculated from the three voltages applied to the guns. In an ideal monitor, the displayed colour would have the same tristimulus values as those calculated. However, in a given actual monitor, there may be differences between the displayed and calculated colours; these can arise from at least four sources (Cowan 1983).

First, the monitor may vary with time. A convenient way of checking this is to compare a displayed grey scale with a standard grey scale produced by means that are known to be reasonably stable. A very convenient way of doing this is to have a small fluorescent lamp emitting light of chromaticity the same as that of the reference white, with a series of non-selective neutral grey filters wrapped round it at convenient intervals. By comparing the displayed and standard grey scales on a daily, or if necessary more frequent, basis, the stability of the monitor is checked in quite a comprehensive fashion. The relationship between the applied voltages, and the amounts and colours of the light produced by the phosphors, is checked; although slight small identical changes in all three channels may pass unnoticed, these are not usually very important, but different changes in the channels are readily detected.

Second, the relationship between the applied voltages and the colour produced may vary over the area of the display. This can be checked by displaying a constant signal all over the display. However, because the eye is not very sensitive to changes in colour that are spatially very gradual it is advisable to check the colour at different positions on the display using some form of comparator that enables the colours of separated areas to be seen side by side.

Third, the chromaticities of the phosphors may not be the same as those given in the monitor specification. This can arise in several different ways. The phosphors may not be quite the same chemically or physically as those intended by the manufacturer. They may alter with extensive usage. There may be some minor excitation of a phosphor by electrons from the wrong guns; a phenomenon known as *beam-landing errors*. To check these effects requires spectroradiometry, and this should be done at different positions because beam-landing errors are usually very position-dependent (King & Marshall 1984).

Fourth, there may be some interaction between the strengths of the three electron beams, so that the amount of red light, for instance, produced by the corresponding signal in the red channel, may be affected by changes in the strengths of the other two electron beams. This can only be checked by producing a variety of colours, in which the strengths of all three beams are altered, and comparing the predicted with the displayed colours (Cowan & Rowell 1986).

REFERENCES

BREMA, *Radio and Electronic Engineer* **38**, 201 (1969).
Cowan, W. B. *Computer Graphics* **17**, 315 (1983).
Cowan, W. B. & Rowell, N. *Color Res. Appl.* **11**, S34 (1986)
King, P. A. & Marshall, P. J., I.B.A. Technical Review No. 22, *Light and colour principles*, p. 46 (1984).
Sproson, W.N. *Proc. IEE.* **125**, 603 (1978)

GENERAL REFERENCE

Sproson, W. N. *Colour Science in Television and Display Systems*, Hilger, Bristol (1983).

11

Miscellaneous topics

11.1 INTRODUCTION

In this chapter various items are included that have not been covered in the previous chapters.

11.2 THE EVALUATION OF WHITENESS

Whiteness is a very important attribute of colours in certain industries, such as paper making, textiles, laundering, and paint manufacture. The higher the luminance factor, the whiter a sample will look, although if luminance factors of over 100% can be achieved they may have an appearance of fluorescence rather than of whiteness. But whiteness does not depend only on luminance factor; chromaticity also has an effect. If two whites have the same luminance factor and one is slightly bluer than the other it will look whiter. To promote uniformity of practice in the evaluation of whiteness the CIE has recommended that the formulae for whiteness, W or W_{10}, and for tint, T_W or $T_{W,10}$, given below, be used for comparisons of the whiteness of samples evaluated for CIE Standard Illuminant D_{65}. The application of the formulae should be restricted to samples that are called 'white' commercially, that do not differ much in colour and fluorescence, and that are measured on the same instrument at nearly the same time; within these restrictions, the formulae provide relative, but not absolute, evaluations of whiteness, that are adequate for commercial use, when employing measuring instruments having suitable modern and commercially available facilities.

$$W = Y + 800(x_n - x) + 1700(y_n - y)$$

$$W_{10} = Y_{10} + 800(x_{n,10} - x_{10}) + 1700(y_{n,10} - y_{10})$$

$$T_W = 1000(x_n - x) - 650(y_n - y)$$

$$T_{W,10} = 900(x_{n,10} - x_{10}) - 650(y_{n,10} - y_{10}) \ ,$$

where Y is the Y tristimulus value of the sample, x and y are the x,y chromaticity co-ordinates of the sample, and x_n,y_n are the chromaticity co-ordinates of the perfect reflecting diffuser, all for the CIE 1931 standard colorimetric observer; the subscript 10 indicates similar values for the CIE 1964 supplementary standard colorimetric observer.

The higher the value of W or W_{10}, the greater is the indicated whiteness. The more positive the value of T_W or $T_{W,10}$, the greater is the indicated greenishness; the more negative the value of T_W or $T_{W,10}$, the greater is the indicated reddishness. However, these variables do not provide uniform scales of whiteness, greenishness, or reddishness. For the perfect reflecting diffuser W and W_{10} are equal to 100, and T_W and $T_{W,10}$ are equal to zero.

The formulae are applicable only to samples whose values of W or W_{10} and T_W or $T_{W,10}$ lie within the following limits:

W or W_{10}	greater than 40 and less than $5Y-280$ or $5Y_{10}-280$
T_W or $T_{W,10}$	greater than -3 and less than $+3$.

The formulae for T_W and $T_{W,10}$ are based on the empirical fact that lines of equal tint in whites are approximately parallel to lines of dominant wavelength 466 nm.

11.3 COLORIMETRIC PURITY

As explained in section 3.4, a measure, *excitation purity*, p_e, that correlates approximately with saturation can be obtained from the x,y chromaticity diagram by dividing the distance on this diagram from the white point to the colour considered by the distance from the white point to the point on the spectral locus having the same dominant wavelength. An alternative measure of a similar type is called *colorimetric purity*, p_c. It is defined as:

Colorimetric purity, $p_c = L_d/(L_n + L_d)$

> where L_d and L_n are, respectively, the luminances of the spectral (monochromatic) stimulus and of the reference white that match the colour stimulus in an additive mixture.'

Unlike excitation purity, colorimetric purity is independent of the use of any particular chromaticity diagram. In the case of stimuli characterized by complementary wavelength, suitable mixtures of the light from the two ends of the spectrum are used instead of the monochromatic stimuli. Colorimetric and excitation purities are related by the expression:

$$p_c = p_e y_d / y$$

where y_d and y are the y-chromaticity co-ordinates, respectively, of the monochromatic stimulus and the colour stimulus considered. Similar measures, $p_{c,10}$ and $p_{e,10}$, for the CIE 1964 supplementary standard colorimetric observer can also be calculated and are related by a similar expression.

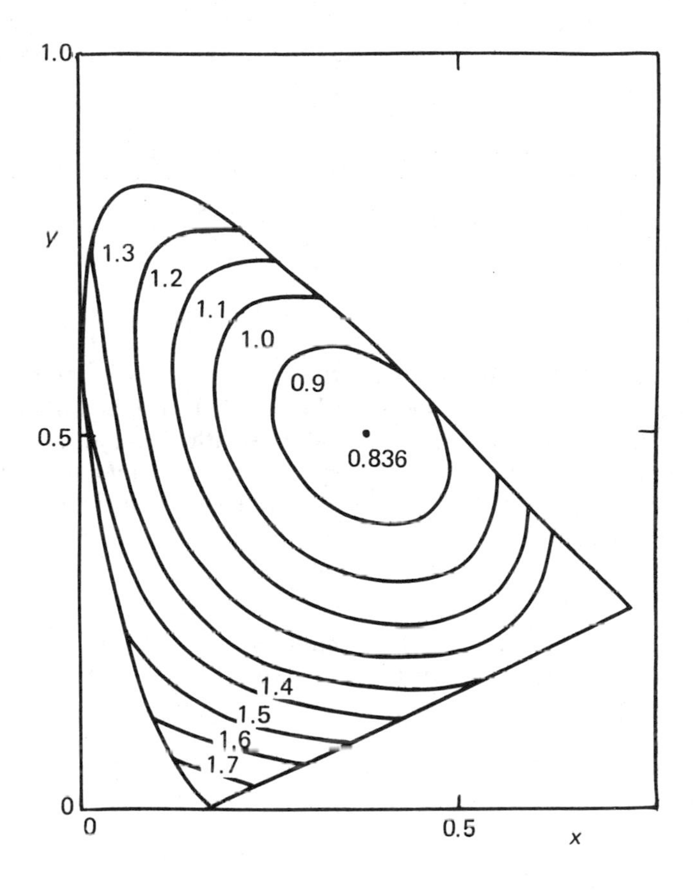

Fig. 11.1 — Contours of constant factor F representing the difference in log luminances of colours that, on average, appear equally bright under the same viewing conditions. The figures on the curves are for 10^F. For example, if $F = 0.3$, then $10^{0.3} = 2$. The figures on the contours therefore show the factors by which a stimulus whose chromaticity lies on the 1.0 contour would have to be increased (or, for yellowish colours, decreased) to appear to have the same brightness. (After Cowan & Ware 1986).

11.4 IDENTIFYING STIMULI OF EQUAL BRIGHTNESS

As mentioned in sections 2.3 and 3.2, colours of equal luminance, even if seen under the same photopic viewing conditions, will not necessarily look equally bright, there being a tendency for brightness to increase with colour saturation. Experimental work on this effect has shown it to vary very considerably from one study to another, and amongst observers in a given study. But some guidance is desirable for applications where the effect is large enough to be important, and this is particularly the case when displays are designed that use saturated self-luminous colours. An empirical formula has therefore been developed (Cowan & Ware 1986; Kaiser 1986) that makes it possible to identify stimuli that, on average, may be expected to look equally bright. A factor F is evaluated from the x,y chromaticity co-ordinates of the colour as follows:

$$F = 0.256 - 0.184y - 2.527xy + 4.656x^3y + 4.657xy^4 .$$

Then, if two stimuli have luminances, L_1,L_2, and factors, F_1,F_2, the two stimuli are equally bright if:

$$\log(L_1) + F_1 = \log(L_2) + F_2 .$$

If these two expressions are not equal, then whichever is greater indicates the stimulus having the greater brightness. In Fig. 11.1, loci of equal values of F are shown. It is clear that, although F tends to increase with saturation, it does so from a minimum that does not occur at the point representing the equi-energy stimulus, but from a point that is displaced from it towards more yellowish colours.

REFERENCES

Cowan, W. B. & Ware, C. unpublished communication (1986).
Kaiser, P. K. *CIE Journal* **5**, 57 (1986).

12

A model of colour vision

12.1 INTRODUCTION

Most of this book has been concerned with internationally accepted procedures for colorimetry, as defined by the CIE. However, the first chapter was different in that it provided a description of the nature of the colour vision provided by the human eye and brain. This last chapter is devoted to describing a model of colour vision that can be used for predicting the appearance of colours under a very wide range of viewing conditions (Hunt 1987, 1991). Measures are derived that are intended to correlate not only with hue, saturation, lightness, and chroma, as in the CIELUV and CIELAB systems, but also with brightness and colourfulness. Parts of the model have already proved useful in some practical applications (Pointer 1986; MacDonald, Luo & Scrivener 1990; Luo, Clarke, Rhodes, Schappo, Scrivener & Tait 1991), but it must be emphasized that the model is speculative in nature, and it does not represent a system on which there is any national or international agreement. The purpose of including it is partly to provide measures that are at present missing from the CIE systems, such as correlates of unique hues, brightness, and colourfulness; and partly to provide a comprehensive scheme for predicting colour appearance that, even if of only limited usefulness in itself, may be helpful in understanding the various effects that are important and how they are related to one another. Similar models have been proposed by other workers (Seim & Valberg 1986; Nayatani, Takahama & Sobagaki 1986; Nayatani, Hashimoto, Takahama & Sobagaki 1987; Nayatani, Takahama, Sobagaki & Hashimoto 1990).

Related colours, those seen in relation to other colours, are considered first; unrelated colours, those seen in isolation from other colours, will be considered subsequently. The input data and steps needed to use the model, together with worked examples, are given in section 12.27 for related colours, and in section 12.28 for unrelated colours.

12.2 VISUAL AREAS IN THE OBSERVING FIELD

For related colours, five different visual fields are recognized in the model.

The colour element considered:
typically a uniform patch of about 2° angular subtense.
The proximal field:
the immediate environment of the colour element considered, extending typically for about 2° from the edge of the colour element considered in all or most directions.
The background:
the environment of the colour element considered, extending typically for about 10° from the edge of the proximal field in all, or most directions. When the proximal field is the same colour as the background, the latter is regarded as extending from the edge of the colour element considered.
The surround:
the field outside the background.
The adapting field:
the total environment of the colour element considered, including the proximal field, the background, and the surround, and extending to the limit of vision in all directions.

The visual patterns of scenes viewed in practice are almost infinitely variable; but the phenomenon of colour constancy (see section 1.8) indicates that the effects of this variety on colour appearance are, to some extent, limited. The regime of fields described above is an attempt to simplify the situation sufficiently to make it feasible for modelling, while making it possible to include the most important factors that affect colour appearance.

12.3 SPECTRAL SENSITIVITIES OF THE CONES

A basic feature of the model is the choice of a set of spectral sensitivity functions for the cones of the retina. The set chosen is shown by the full lines in Fig. 12.1; these curves are a linear combination of the colour-matching functions for the CIE 1931 Standard Colorimetric Observer, $\bar{x}(\lambda)$, $\bar{y}(\lambda)$, $\bar{z}(\lambda)$. If the colour element considered has an angular subtense of more than 4°, the colour-matching functions of the CIE 1964 Supplementary Standard Colorimetric Observer are used instead, and the subscript 10 is attached to all the symbols used. For both Standard Observers, the spectral sensitivities for the cones are obtained from the colour-matching functions by means of the following set of transformation equations:

$$\begin{aligned}\rho &= 0.38971X + 0.68898Y - 0.07868Z\\ \gamma &= -0.22981X + 1.18340Y + 0.04641Z\\ \beta &= 1.00000Z\,.\end{aligned}$$

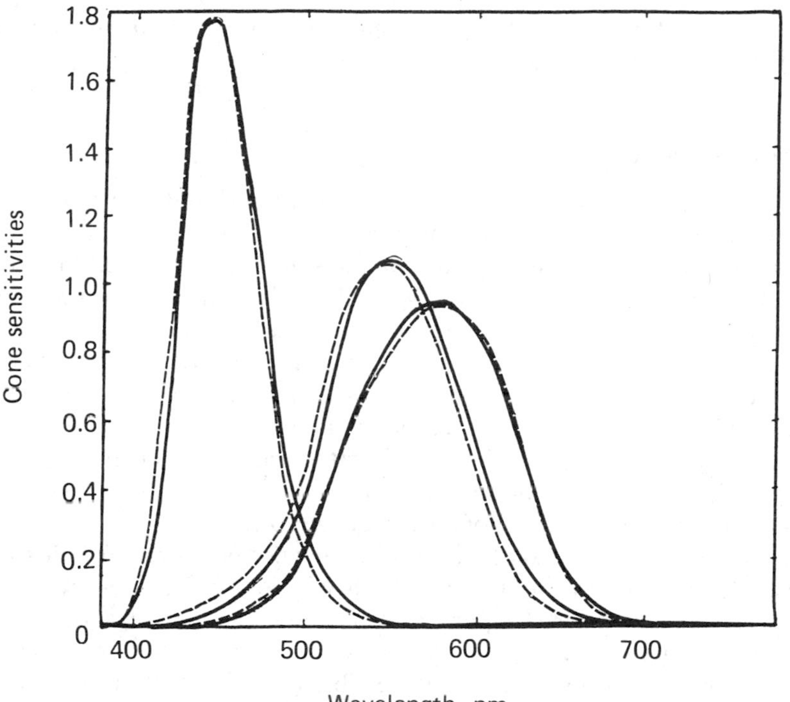

Fig. 12.1 — Spectral sensitivity functions used in the model for cone vision (full lines), compared with those obtained by Estevez (broken lines). These functions are for radiation incident on the cornea of the eye.

The corresponding reverse set of transformation equations is:

$$\begin{aligned} X &= 1.91019\rho - 1.11214\gamma + 0.20195\beta \\ Y &= 0.37095\rho + 0.62905\gamma \\ Z &= \qquad\qquad\qquad\qquad 1.00000\beta \,. \end{aligned}$$

The coefficients used in the above equations are such that the values of ρ, γ, and β are equal to one another for the equi-energy stimulus, S_E. (The similar sets of equations given in section 3.13 are the same except that they are normalized for D_{65} instead of for S_E.)

The broken lines in Fig. 12.1 show the spectral sensitivities derived in a study by Estevez for 2° observations (Estevez 1979). To reproduce these exactly would have required the use of colour-matching functions different from those of the CIE 1931 Standard Colorimetric Observer; this would have been very inconvenient for practical applications, and hence the approximation to these curves provided by the full lines has been used instead.

The above set of transformation equations is used to derive the values of ρ, γ, and β, not only for colour-matching functions, but for any colour. These amounts, ρ, γ, and β, may be considered as the amounts of radiation usefully absorbed per unit area of the retina by the three different types of cone in a given state of adaptation, for light incident on the cornea.

12.4 CONE RESPONSE FUNCTIONS

Under a given set of viewing conditions, there will be a predictable relationship between the responses of the cones and the intensity of the stimulus (intensity denoting here simply the magnitude of the stimulus, not necessarily the flux per unit solid angle). There is much evidence to suggest that this relationship is nonlinear. If the cone responses are taken as being proportional to the square root of the stimulus intensity, the curvatures of lines of constant hue in chromaticity diagrams can be predicted well by using a simple criterion for constant hue (Hunt 1982). A square root relationship would also result in a reduction in the dynamic range of the signals that have to be transmitted from the retina to the brain, and this seems likely on general grounds; thus, for example, a change in stimulus intensity of 1000 to 1 would produce a change in cone response of only about 32 to 1.

A simple square root relationship, however, cannot be correct for all stimulus intensities. When the intensity of the stimulus is very low, noise in the system must prevent extremely small cone responses from being significant; and, when the intensity of the stimulus is very high, the response must eventually reach a maximum level beyond which no further increase is possible (Baylor 1987). These limits are illustrated by our inability to see modulations of colour in extremely dark objects, and by the tendency for very bright colours, such as lamp filaments seen through coloured filters, or coloured flares seen at close quarters, to appear pale or white.

A hyperbolic function is therefore chosen to represent the response for the cones (as suggested by Seim & Valberg 1986, and for which there is physiological evidence as described by Boynton & Whitten 1970, and by Valeton & Van Norren 1983). The responses given by the three different types of cone in a given state of adaptation are then formulated as:

$$f_n(\rho) + 1 = 40[\rho^{0.73}/(\rho^{0.73} + 2)] + 1$$
$$f_n(\gamma) + 1 = 40[\gamma^{0.73}/(\gamma^{0.73} + 2)] + 1$$
$$f_n(\beta) + 1 = 40[\beta^{0.73}/(\beta^{0.73} + 2)] + 1\ .$$

The + 1 terms represent the noise. These responses give maximum values of 41 and minimum values of 1. In Fig. 12.2, $\log [f_n(\rho) + 1]$ is plotted against $\log \rho$. It is clear from the figure that, over the central part of the curve, the response approximates a square root relationship, as shown by the broken line. The responses $f_n(\gamma) + 1$ and $f_n(\beta) + 1$ would be represented by similar graphs.

12.5 ADAPTATION

The actual response produced by the cones is dependent, not only on the intensity of the stimulus, but also on the state of adaptation of the eye. For related colours,

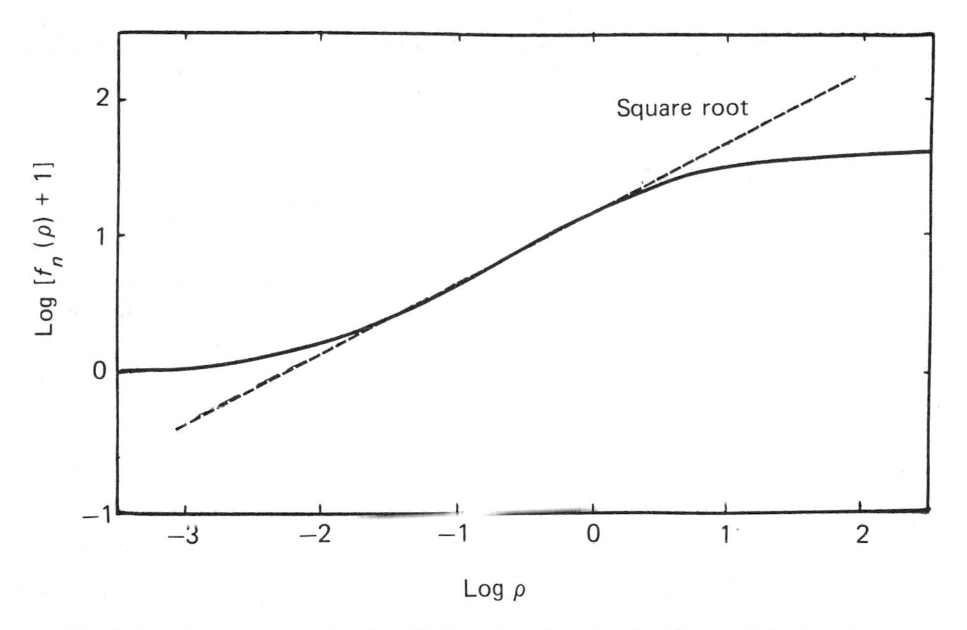

Fig. 12.2 — Cone response function. The log of the function, log $[f_n(\rho) + 1]$, is plotted against the log of the radiation usefully absorbed, log ρ, where $f_n(\rho) = 40\rho^{0.73}/(\rho^{0.73} + 2)$.

adaptation usually provides an approximate compensation for the effects of changes in the level and colour of the illumination, and this results in the phenomenon of colour constancy. For related colours, the cone responses after adaptation are formulated as:

$$\rho_a = B_\rho[f_n(F_L F_\rho \rho/\rho_W) + \rho_D] + 1$$
$$\gamma_a = B_\gamma[f_n(F_L F_\gamma \gamma/\gamma_W) + \gamma_D] + 1$$
$$\beta_a = B_\beta[f_n(F_L F_\beta \beta/\beta_W) + \beta_D] + 1$$

where the function $f_n(I)$ is again of the form:

$$f_n(I) = 40[I^{0.73}/(I^{0.73} + 2)]$$

The factors, ρ_W, γ_W, and β_W, are the ρ, γ, and β values for the reference white. The reference white is normally taken as either the perfect diffuser (or perfect transmitter), or a working standard white (or highly transmitting filter), lit by the illumination prevailing in the area of the colour considered (see section 5.7). The divisions by ρ_W, γ_W, and β_W provide a Von Kries type of allowance for adaptation (see section 3.13), whereby complete compensation would be made for changes in the level and colour of the illumination. The other factors in the above equations make allowance for the fact that such compensation is usually only partial.

12.6 ADAPTATION-REDUCING FACTORS

Compensation for changes in the level and colour of the illumination is usually incomplete, even when the eye is fully adapted to them. Hence, objects look less bright in dim lighting. And a reference white does not generally appear truly achromatic unless the illuminant has a chromaticity close to that of the equi-energy stimulus, S_E; in light of lower correlated colour temperature, such as candlelight, the reference white usually looks yellowish; in light of higher colour temperature, such as that from a blue sky, the reference white usually looks bluish. The factors, F_L, F_ρ, F_γ, F_β, ρ_D, γ_D, and β_D, are introduced to model this reduced compensation.

The factors, F_L, F_ρ, F_γ, F_β, ρ_W, γ_W, and β_W, multiply (or divide) ρ, γ, and β, and may be thought of as occurring at an early stage in the retina, probably as a result of both photochemical and electronic processes. The parameters, ρ_D, γ_D, and β_D, are added to (or subtracted from) the signals, and may be thought of as occurring at a later stage. These latter effects are often characterized by being extremely rapid (apparently instantaneous), whereas the former can take seconds or even minutes to reach equilibrium. The effects of these various factors, and the factors B_ρ, B_γ, and B_β, will be discussed in sections 12.8, 12.9, 12.10, and 12.11. (For another formula for adaptation, see CIE, *CIE Journal*, 1986 and section 3.13.)

12.7 CRITERIA FOR ACHROMACY AND FOR CONSTANT HUE

As mentioned in section 1.6, there is a great deal of evidence that the responses from the three different types of cone are compared by neurons in the retina that result in *colour difference* signals being formed for subsequent transmission along the optic nerve fibres to the brain. We may represent these colour difference signals as:

$$C_1 = \rho_a - \gamma_a$$
$$C_2 = \gamma_a - \beta_a$$
$$C_3 = \beta_a - \rho_a$$

Almost certainly, their complements:

$$C_1' = \gamma_a - \rho_a$$
$$C_2' = \beta_a - \gamma_a$$
$$C_3' = \rho_a - \beta_a$$

also exist, but, for simplicity, we will usually consider only C_1, C_2, and C_3.

Achromatic colours are those that do not exhibit a hue (such as whites, greys, and blacks). As suggested in section 1.7, the criterion adopted for achromacy is:

$$\rho_a = \gamma_a = \beta_a$$

and hence

$$C_1 = C_2 = C_3 = 0$$

and colourfulness increases as C_1, C_2, and C_3, increase.

The criterion for constant hue, as also suggested in section 1.7, is:

C_1 to C_2 to C_3 in constant ratios.

Criteria for unique hues will be considered in section 12.13. (For an overview, omit sections 12.8 to 12.12, which deal with various effects of adaptation).

12.8 CONE BLEACH FACTORS, B_ρ, B_γ, B_β

The factors, B_ρ, B_γ, and B_β, provide reduced cone responses at very high levels of illumination, where appreciable bleaching of the cone pigments occurs. These cone bleach factors are defined as follows:

$$B_\rho = 10^7/[10^7 + 5L_A(\rho_W/100)]$$
$$B_\gamma = 10^7/[10^7 + 5L_A(\gamma_W/100)]$$
$$B_\beta = 10^7/[10^7 + 5L_A(\beta_W/100)]$$

where L_A is the luminance of the adapting field. (For related colours in typical viewing conditions, the luminance of a reference white is often about five times that of the adapting field, and hence $5L_A$ can often be regarded as the luminance of a reference white; this is why $5L_A$ is used in the formula.)

When the cone pigments are bleached, their spectral absorptions can become narrower, and result in narrower sensitivities for the cones; metameric colour matches can then break down, However, the model does not include such changes in cone spectral sensitivity functions.

12.9 LUMINANCE-LEVEL ADAPTATION FACTOR, F_L

The parameter, F_L, in the formulae for ρ_a, γ_a, and β_a, provides allowance for the level of illumination; it is defined as:

$$F_L = 0.2k^4(5L_A) + 0.1(1 - k^4)^2(5L_A)^{1/3}$$

where L_A is the luminance of the adapting field, and

$$k = 1/(5L_A + 1).$$

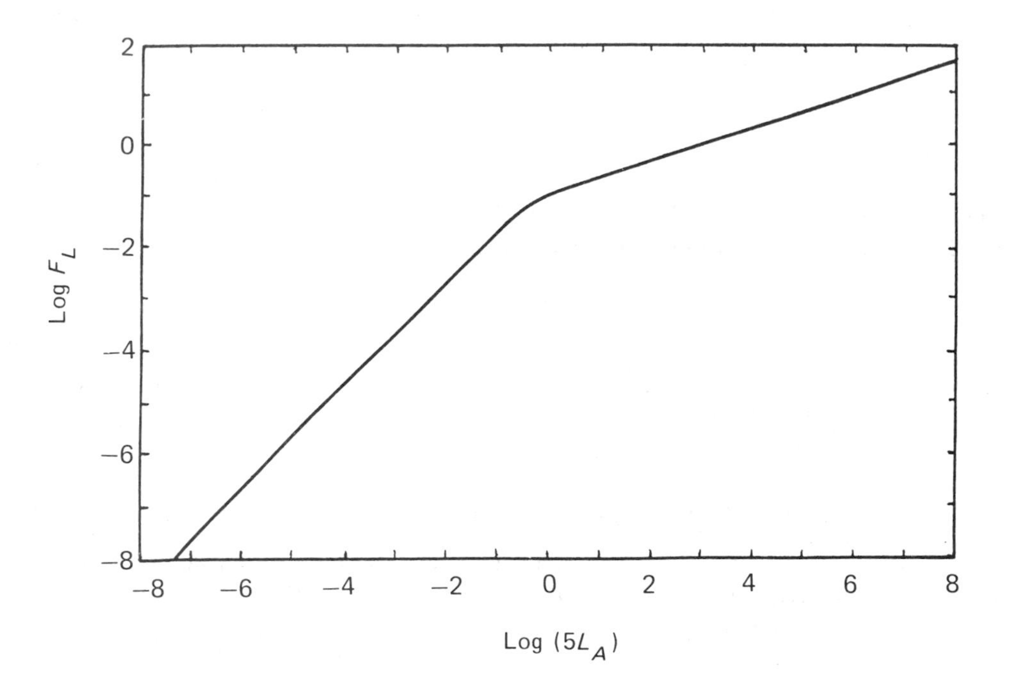

Fig. 12.3 — Luminance level adaptation factor. The log of the factor, log F_L, is plotted against log $(5L_A)$, the log of 5 times the luminance of the adapting field (which is taken to be the luminance of a typical white).

In Fig. 12.3, log F_L is plotted against log $(5L_A)$. The figure shows that, at photopic levels ($5L_A$ greater than 1; log $(5L_A)$ greater than 0), F_L is approximately proportional to the cube root of $5L_A$ (slope of curve equal to 1/3), thus giving partial compensation for changes in adapting luminance; full compensation would occur if F_L were constant. At scotopic levels ($5L_A$ less than about 0.1; log $(5L_A)$ less than − 1) F_L is proportional to $5L_A$ (slope of curve equal to 45°) so that no compensation occurs.

Let us assume, for the moment, that F_ρ, F_γ, and F_β are all equal to unity, and that ρ_D, γ_D, and β_D are all equal to zero (this is true for illuminants having the same chromaticity as that of the equi-energy stimulus, S_E), and that B_ρ, B_γ, and B_β are all equal to unity. Then:

$$\rho_a = f_n(F_L\rho/\rho_W) + 1$$

with similar expressions for γ_a and β_a. In Fig 12.4, ρ_a for these conditions is plotted against log $(5L_A\rho/\rho_W)$ = log I. If $5L_A$ is the luminance of the reference white, and the sample has the same chromaticity as the reference white, then $5L_A\rho/\rho_W$ is equal to the luminance of the sample. The curves shown in Fig. 12.4 are for values of log $(5L_A)$ equal to 8, 7, 6, 5, 4, 3, 2, 1, and 0 log cd/m^2 (full lines), and for dark adaptation (broken line).

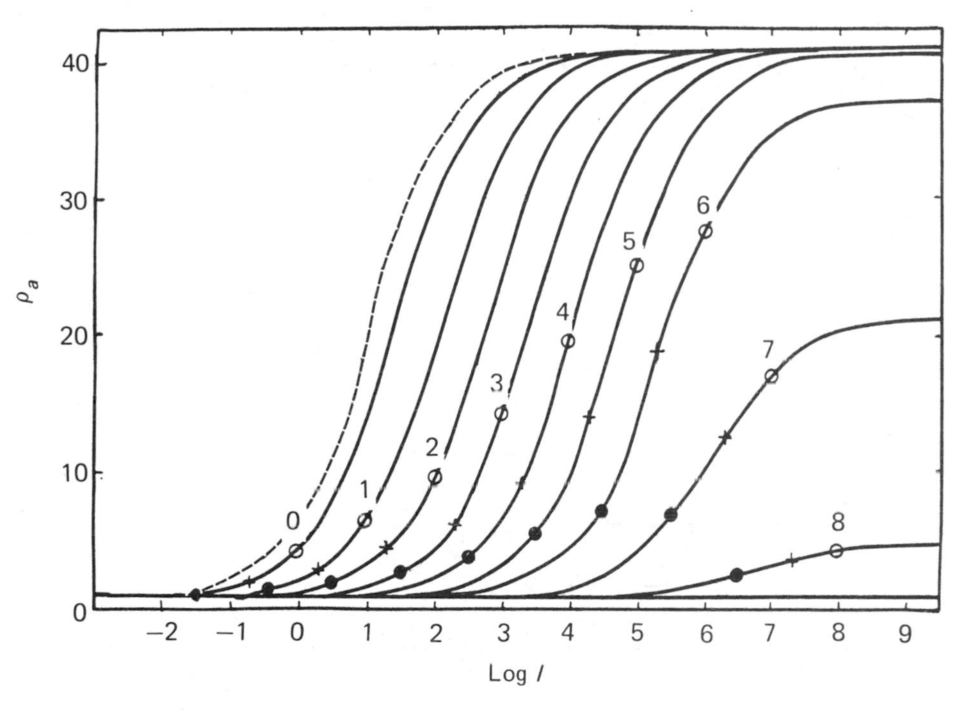

Fig. 12.4 — Response functions for the ρ cones. ρ_a is plotted against log I, where $I = 5L_A\rho/\rho_W$ and L_A is the luminance of the adapting field in cd/m^2, for levels of log $(5L_A)$ equal to 8, 7, 6, 5, 4, 3, 2, 1, and 0 log cd/m^2, full lines, and for dark adaptation, broken line. Similar functions occur for the γ and β responses. (It is assumed that $F_\rho = F_\gamma = F_\beta = 1$, that $\rho_D = \gamma_D = \beta_D = 0$, and that $B_\rho = B_\gamma = B_\beta = 1$.) Open circles: reference white; filled circles: 3.162% black; plus signs: adapting field (luminance one fifth of that of the reference white).

Let us consider the curve labelled 3; this is for log $(5L_A)$ equal to 3, and the open circle on this curve is for a colour having the same value of ρ/ρ_W as for the reference white, and the filled circle for a colour having ρ/ρ_W equal to 0.03162 times that of the reference white (that is, 1.5 less on the log scale). Relationships similar to those shown in the graph in Fig. 12.4 also apply for γ_a and β_a. Hence, when log $(5L_A) = 3$, the reference white would be represented by points at the open-circle positions on curve 3 in all three graphs, and a colour having the same chromaticity as the reference white, but a luminance 0.03162 times (1.5 log units) less, by points at the filled circles on curve 3 in all three graphs. The part of curve 3 between the open and filled points therefore represents the range of colours between white (the reference white) and a black (of luminance 3.162% of that of the reference white), when log $(5L_A)$ is equal to 3 log cd/m^2. The position of the adapting field, L_A (taken as 1/5, that is 20%, of the luminance of the white, or 0.7 less on the log scale) is shown by the plus sign (+) on curve 3.

The other curves of the figure similarly represent the same range of colours for values of log $(5L_A)$ that become progressively smaller as the curve is displaced towards the left, and higher towards the right. Physiological studies show similar families of curves (Valeton & Van Norren 1983).

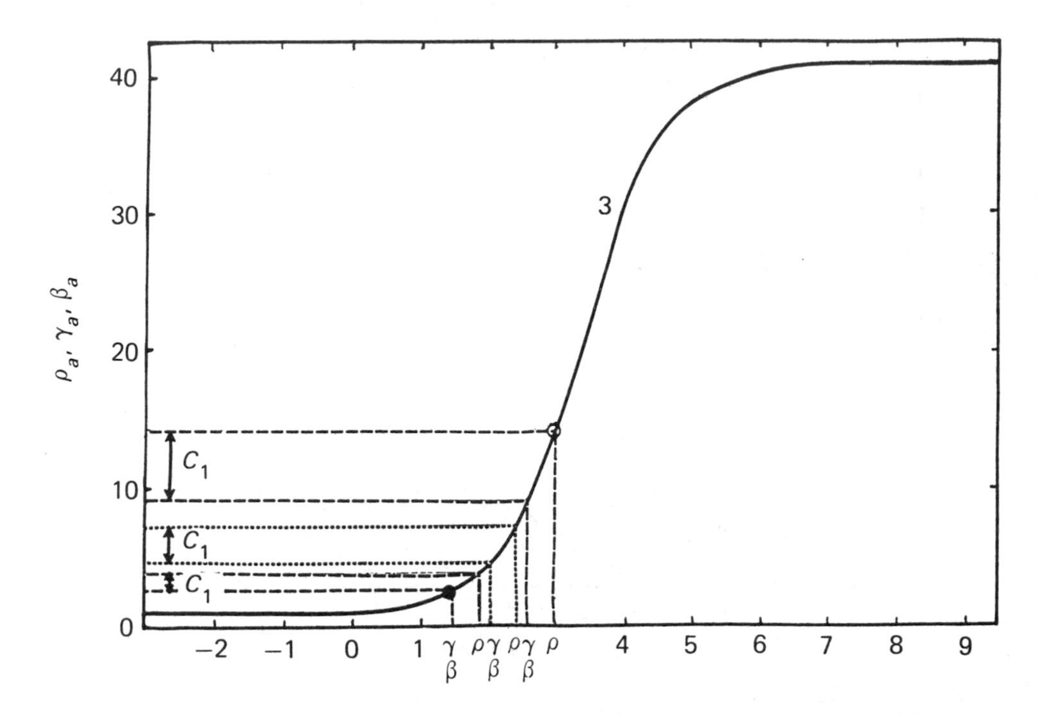

Fig. 12.5 — Curve 3 of Fig. 12.4, together with representations of three stimuli having the same chromaticity but three different luminance factors. As the luminance factor falls, the resulting colour difference signals also fall.

The S-shaped nature of these curves predicts that, for colours of a given chromaticity, as the luminance factor is decreased, the colourfulness will usually decrease. This can be seen as follows. In Fig. 12.5, curve 3 of Fig. 12.4 is shown again, but with vertical lines indicating three red colours of the same chromaticity but different luminance factors. The constancy of chromaticity is represented by these three colours all having the same separation (0.5) on the log I axis between the positions of the radiations usefully absorbed by the ρ, γ, and β cones, ρ being higher than γ and β which are equal. As the luminance factor is decreased, the set of positions on the log I axis for each colour moves to the left; as a result, the responses come from parts of the curve having lower slopes. The difference between the ρ_a and the γ_a responses therefore decreases, and hence C_1 decreases, as shown on the left, indicating reduced colourfulness (C_3 would also decrease similarly). Thus, for a given chromaticity, as the luminance factor decreases, the colourfulness decreases. This is illustrated in Plate 8 (page 113). In this Plate, each vertical column of colours has the same chromaticity, but the luminance factor decreases from top to bottom of each column; this is clearly accompanied by reduced colourfulness.

The lower maxima of curves 6, 7, and 8 in Fig. 12.4 are caused by the cone bleach factors. For the other curves, the following general features can be seen. As the

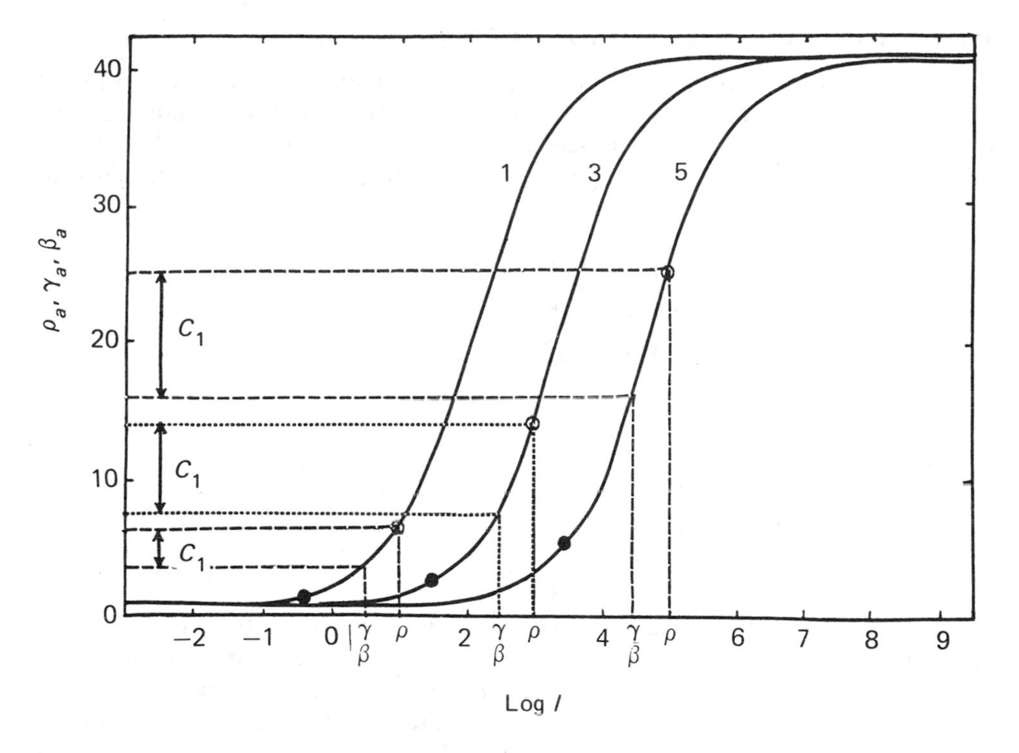

Fig. 12.6—Curves 1, 3, and 5 of Fig. 12.4, together with representations of a colour of the same chromaticity and luminance factor in the three different levels of adapting luminance (assuming the reference white has a luminance of 5 times that of the adapting luminance in each case). As the adapting luminance falls, the resulting colour difference signals also fall.

luminance of the adapting field, L_A, decreases, the curves move to the left, indicating increasing sensitivity. But this movement is insufficient to provide full compensation, and hence the positions of the points representing white (○), the adapting field (+), and black (●) gradually move down each curve to regions of lower slope. This results in reductions in the differences in response between whites, adapting fields, and blacks. For colours, this results in reduced colourfulness. This is illustrated in Fig. 12.6, where curves 1, 3, and 5 of Fig. 12.4 are reproduced. The vertical lines meeting curve 5 indicate a red colour having log I for ρ at the white level, and for γ and β at 0.5 less on the log scale; the corresponding value of C_1 is shown at the left. Similar vertical lines are shown meeting curves 3 and 1, and the corresponding values of C_1 are clearly smaller (as would also be the case for C_3). Hence the model predicts that, as the level of illumination is decreased, colours become of lower colourfulness, as is found in practice.

It is also clear from Fig. 12.6 that the slopes of the curves near the black points (●) become very low for the curves towards the left; and this predicts that dark colours are difficult to distinguish in dim lighting, as is also found in practice.

The curves in Figs 12.4, 12.5, and 12.6 also show that, for stimuli of luminances very much higher than that of the reference white, the responses may approach the maximum level, in which case they will also tend to be reduced in colourfulness.

12.10 CHROMATIC ADAPTATION FACTORS, F_ρ, F_γ, F_β

We must now discuss the parameters F_ρ, F_γ, and F_β. Adaptation to lights of different colours becomes less and less complete as the purity of the colour of the light (relative to the equi-energy stimulus, S_E) increases, and more and more complete as the luminance increases. The following expressions for F_ρ, F_γ, and F_β are designed to represent both these effects:

$$F_\rho = (1 + L_A{}^{1/3} + h_\rho)/(1 + L_A{}^{1/3} + 1/h_\rho)$$
$$F_\gamma = (1 + L_A{}^{1/3} + h_\gamma)/(1 + L_A{}^{1/3} + 1/h_\gamma)$$
$$F_\beta = (1 + L_A{}^{1/3} + h_\beta)/(1 + L_A{}^{1/3} + 1/h_\beta)$$

where

$$h_\rho = 3\rho_W/(\rho_W + \gamma_W + \beta_W)$$
$$h_\gamma = 3\gamma_W/(\rho_W + \gamma_W + \beta_W)$$
$$h_\beta = 3\beta_W/(\rho_W + \gamma_W + \beta_W)$$

In Fig. 12.7 log F_p is plotted against log h_ρ for values of L_A from 10^6 to 10^{-1}. Similar relationships apply between F_γ and h_β, and between F_β and h_β. For the equi-energy stimulus, S_E, $\rho_W = \gamma_W = \beta_W$ and hence $h_\rho = h_\gamma = h_\beta = 1$, and $F_\rho = F_\gamma = F_\beta = 1$.

The effect of these parameters, F_ρ, F_γ, and F_β, can be illustrated by considering, as an example, a yellowish illuminant for which h_ρ and h_γ are greater than unity, and h_β is less than unity. This results in F_ρ and F_γ being greater than unity, and F_β being less than unity. Because F_β is less than unity, $F_\beta\beta/\beta_W$ will be smaller than β/β_W; hence the lower values of β typically produced by a yellowish illuminant are not increased as much as if F_β were equal to unity, and thus the extent of the adaptation is reduced. Furthermore, because F_ρ and F_γ are greater than unity, $F_\rho\rho/\rho_W$ and $F_\gamma\gamma/\gamma_W$ will be larger than ρ/ρ_W and γ/γ_W, respectively, and the higher values of ρ and γ typically produced by a yellowish illuminant are not decreased as much as if F_ρ and F_γ were equal to unity, and thus, again, the extent of the adaptation is reduced.

When observers attempt to identify the colours of surface objects, they can sometimes make perceptual allowance for the colour of the prevailing illumination (Arend & Reeves 1986). For instance, if an observer passes from an environment in which the illuminant is a daylight to one in which the illuminant is a tungsten light, then, although a piece of white paper generally appears to be yellowish in the tungsten light, it may still be correctly identified as a white, not as a yellow, object. This effect is sometimes referred to as *discounting the colour of the illuminant*. This mode of perception can be modelled by setting F_ρ, F_γ, and F_β all equal to unity.

12.11 THE HELSON–JUDD EFFECT FACTORS, ρ_D, γ_D, β_D

When the chromaticity of the illuminant is substantially different from that of the equi-energy stimulus, S_E, white colours tend to appear to be tinged with the hue of

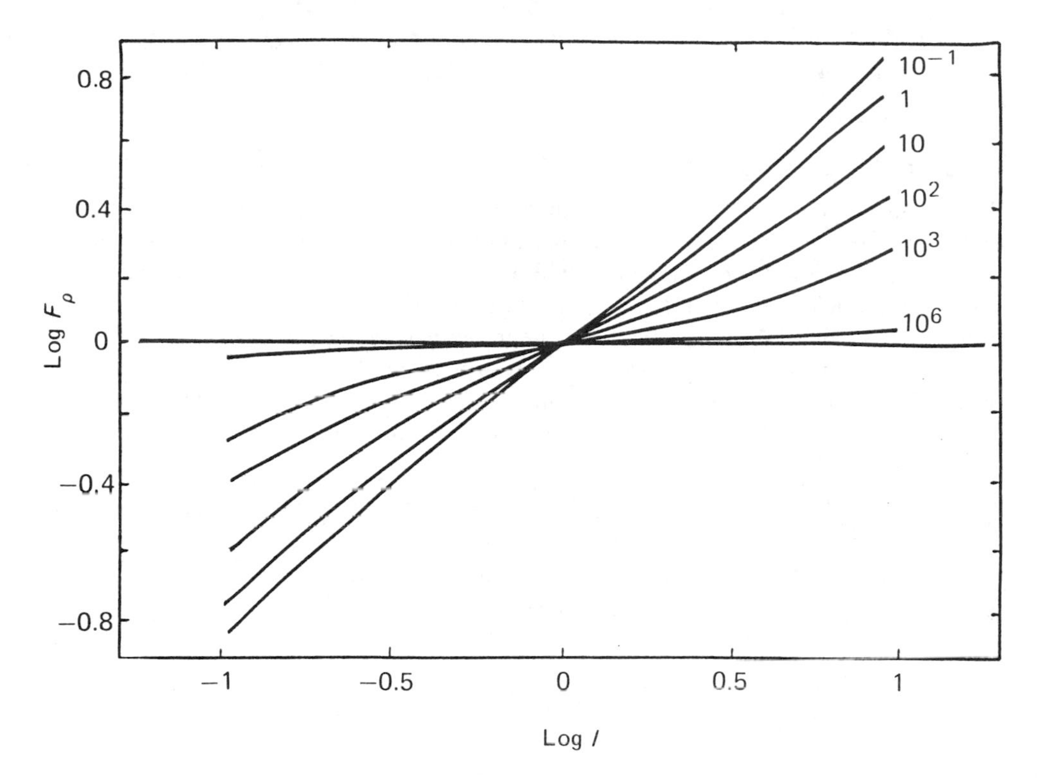

Fig. 12.7 — Chromatic adaptation factor. The logs of the factor, log F_ρ, are shown against the log of h_ρ (a measure of the purity of the colour of the adapting illuminant), for different levels of adapting luminance, 6, 3, 2, 1, 0, and − 1 log cd/m². Similar relationships occur between log F_γ and log h_γ, and between log F_β and log h_β.

the illuminant, and very dark greys with the complementary hue; this is usually referred to as the *Helson–Judd effect*. To allow for this, the parameters ρ_D, γ_D, β_D are set so that

$$\rho_D = f_n[(Y_b/Y_W)F_L F_\gamma] - f_n[(Y_b/Y_W)F_L F_\rho]$$
$$\gamma_D = 0$$
$$\beta_D = f_n[(Y_b/Y_W)F_L F_\gamma] - f_n[(Y_b/Y_W)F_L F_\beta]$$

where Y_b and Y_W are the luminance factors of the background and of the reference white, respectively. This ensures that (except at the very high levels where B_ρ, B_γ, and B_β depart from unity), when $\rho/\rho_W = \gamma/\gamma_W = \beta/\beta_W = Y_b/Y_W$, then $\rho_a = \gamma_a = \beta_a$, and the colour (a grey having the same relative luminance factor as the background) is predicted to appear achromatic. The choice of γ_D as the parameter to be set equal to zero is arbitrary, and, if ρ_D or β_D had been chosen instead, it would not have made any difference to the correlates to be developed in later sections.

If the mode of perception entails discounting the colour of the illuminant, ρ_D, γ_D, and β_D, can all be set to zero.

12.12 USE OF A MODIFIED REFERENCE WHITE

When simultaneous contrast occurs, the proximal field, which provides the immediate environment, causes the appearance of the colour element considered to move towards the colour that is opposite in hue, saturation, and lightness to the colour of the proximal field. But when the angular subtense of the colour element considered becomes less than about a third of a degree, instead of simultaneous contrast, assimilation occurs, the appearance of the colour then tending to move towards that of the proximal field. These effects are modelled by computing a modified reference white, whose cone responses ρ'_W, γ'_W, and β'_W, are given by:

$$\rho'_W = \rho_W[(1-p)r + (1+p)/r]^{1/2}/[(1+p)r + (1-p)/r]^{1/2}$$
$$\gamma'_W = \gamma_W[(1-p)g + (1+p)/g]^{1/2}/[(1+p)g + (1-p)/g]^{1/2}$$
$$\beta'_W = \beta_W[(1-p)b + (1+p)/b]^{1/2}/[(1+p)b + (1-p)/b]^{1/2}$$

where

$$r = (\rho_p/\rho_b)$$
$$g = (\gamma_p/\gamma_b)$$
$$b = (\beta_p/\beta_b)$$

and ρ_p, γ_p, and β_p are the ρ, γ, and β signals for the proximal field, and ρ_b, γ_b, and β_b are those for the background. The value of p depends on the size and shape of the proximal field. It will be between 0 and -1 for simultaneous contrast; it will be between 0 and $+1$ for assimilation. In both cases, it will be nearer to 0, the smaller the area of the proximal field relative to that of the colour considered. When p is negative (simultaneous contrast), the modified reference white becomes more like the proximal field, so that, relative to the modified reference white, colours become less like the proximal field; when p is positive (assimilation), the modified reference white becomes less like the proximal field, so that, relative to the modified reference white, the colours become more like the proximal field.

12.13 CRITERIA FOR UNIQUE HUES

As discussed in section 7.6, there are four unique hues: red, green, yellow, and blue. The model predicts these as occurring at the following ratios of C_1 to C_2 or C_3 (because $C_1 + C_2 + C_3 = 0$, if one of these ratios is constant, the other will also be constant, and need not be specified in addition):

Unique red	$C_1 = C_2$
Unique green	$C_1 = C_3$
Unique yellow	$C_1 = C_2/11$
Unique blue	$C_1 = C_2/4$

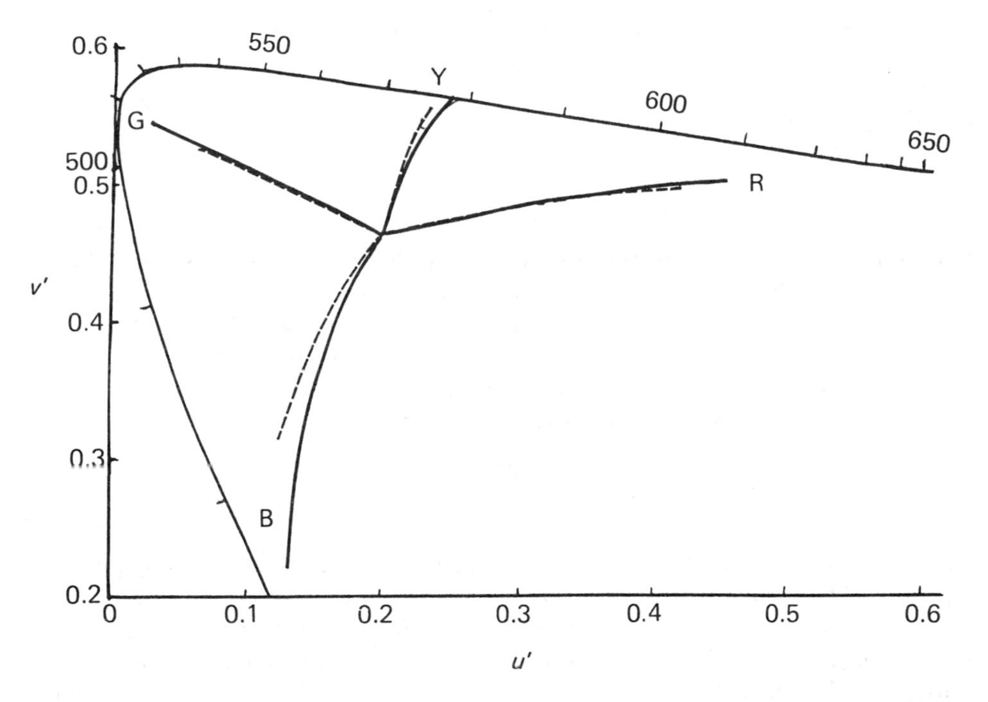

Fig. 12.8 — Unique hue loci predicted by the model (full lines) compared with those of the NCS (broken lines), for Standard Illuminant C. For the model, the hue loci shown are for the following conditions. The adapting luminance was 200 cd/m^2; ρ_D, γ_D, and β_D were all put equal to zero; F_ρ, F_γ, and F_β were all put equal to 1; and the sum of the three cone responses, $\rho_a + \gamma_a + \beta_a$, was put equal to 30, this value for the perfect diffuser being 43.0 (because, as shown in Fig. 12.2, the hyperbolic function approximates a simple power function over its central range, the chromaticities corresponding to loci of constant hue are only slightly dependent on the values of $\rho_a + \gamma_a + \beta_a$ used, provided that ρ_a, γ_a, and β_a are in their central ranges). For the NCS, the luminance factors were the highest available in the system.

The predictions given by these criteria are shown in Fig. 12.8 by the full lines (for colours seen at high levels of adapting illumination, such as when L_A is around 200 cd/m^2, and illuminant S_C is used); the broken lines show the results obtained experimentally in the Natural Colour System (NCS, see section 7.7), which are very similar.

12.14 HUE ANGLE, h_s

In the case of reddish colours, since the criterion for unique red is $C_1 = C_2$, it is to be expected that increasing departures from the unique hue condition, that is, increasing yellowness or blueness, would be indicated by increasing inequality of C_1 and C_2, that is, $C_2 - C_1$ being increasingly different from zero. ($C_2 - C_1$ is used instead of $C_1 - C_2$ so that positive values indicate yellowness.) Similarly, because the criterion for unique green is $C_1 = C_3$, increasing yellowness or blueness of greenish colours would be indicated by $C_1 - C_3$ being increasingly different from zero (positive values indicating yellowness). A measure of the yellowness or blueness of both reddish and greenish colours is therefore taken as the average of these two differences:

$$\tfrac{1}{2}(C_2 - C_1 + C_1 - C_3)$$

and this is equal to

$$\tfrac{1}{2}(C_2 - C_3)\ .$$

(In section 1.7, $C_2 - C_3$ was taken to indicate yellowness or blueness, and the use here of $\tfrac{1}{2}(C_2 - C_3)$ is the same apart from the scaling factor of $\tfrac{1}{2}$.) By similar arguments, redness or greenness of yellowish colours would be indicated by $C_1 - (C_2/11)$, and of bluish colours by $C_1 - (C_2/4)$; but, in this case, because the unique yellow hue is more sharply apparent than the unique blue hue, an average is not taken, and redness or greenness is taken to be indicated by:

$$C_1 - (C_2/11).$$

(In section 1.7, C_1 was used to indicate redness or greenness, and the use here of $C_1 - (C_2/11)$ is similar because $C_2/11$ is usually fairly small compared with C_1.)

It is now necessary to combine these correlates of yellowness-blueness and redness-greenness to obtain a measure of hue. But, because the number of β cones is only about 1/20th that of the ρ or γ cones (Walraven & Bouman 1966), it is to be expected, on signal-to-noise ratio grounds, that the yellowness-blueness signal should have less weight than the redness-greenness signal; a factor of 1/4.5 (which is approximately equal to $1/20^{1/2}$) is used for this purpose, so that yellowness-blueness is taken to be indicated by:

$$\tfrac{1}{2}(C_2 - C_3)/4.5$$

A measure of hue is then obtained as the *hue-angle*, h_s, defined as:

$$\begin{aligned} h_s &= \arctan\{[\tfrac{1}{2}(C_2 - C_3)/4.5]/[C_1 - (C_2/11)]\} \\ &= \arctan(t/t') \end{aligned}$$

where 'arctan' means 'the angle whose tangent is'. h_s lies between 0°and 90° if t and t' are both positive; between 90° and 180° if t is positive and t' is negative; between 180° and 270° if t and t' are both negative; and between 270° and 360° if t is negative and t' is positive.

12.15 CORRELATES OF COLOURFULNESS, *M*, AND SATURATION, *s*

Colourfulness is the extent to which the hue is apparent, and is therefore a combination of yellowness-blueness and redness-greenness. However, before

$\frac{1}{2}(C_2 - C_3)/4.5$ and $C_1 - (C_2/11)$ can be combined to provide a correlate of colourfulness, various factors have to be applied.

First, an eccentricity factor, e_s, is included. This arises because the position of the achromatic point in contours of small constant saturation is eccentric (Hunt 1985). The achromatic point becomes progressively nearer the contour as the hue considered is changed from yellow to red to green to blue; this is regarded as indicating increasing weight of perceptual colorization in the order yellow, red, green, and blue. To reflect this, the following values are assigned to e_s for the unique hues, whose hue-angles, h_s (derived from their ratios of C_1 to C_2 to C_3), are also given here:

	Red	Yellow	Green	Blue
h_s	20.14	90.00	164.25	237.53
e_s	0.8	0.7	1.0	1.2

The values of e_s at intermediate hues are interpolated linearly by the formula:

$$e_s = e_1 + (e_2 - e_1)(h_s - h_1)/(h_2 - h_1)$$

and e_1 and h_1 are the values of e_s and h_s, respectively, for the unique hue having the nearest lower value of h_s; and e_2 and h_2 are the values of e_s and h_s, respectively, for the unique hue having the nearest higher value of h_s. In Fig. 12.9 e_s is shown plotted against h_s.

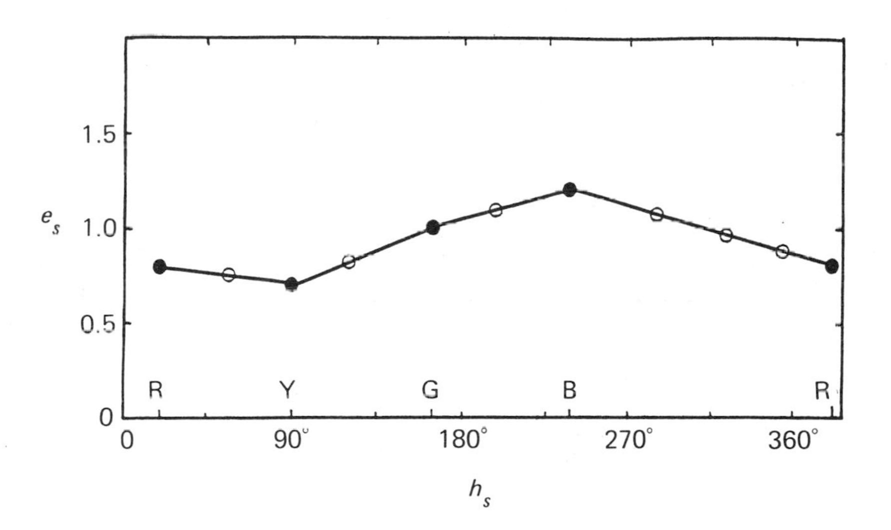

Fig. 12.9 — Eccentricity factor, e_s, plotted against hue angle, h_s.

Second, a factor of 10/13 to allow for cross-channel noise in the system is incorporated.

Third, a chromatic surround induction factor, N_c, is used which makes allowance for the fact that dark or dim surrounds to colours can reduce their colourfulness. N_c is equal to 1.0 for small areas in uniform light backgrounds and surrounds, 1.0 for normal scenes, 0.95 for television and VDU displays in dim surrounds, 0.9 for projected photographs in dark surrounds, and 0.75 for arrays of adjacent colours in dark surrounds.

Fourth, a low-luminance tritanopia factor, F_t, is included in the yellowness-blueness signal, to allow for the fact that, as the illumination level falls, yellowness-blueness discrimination deteriorates earlier than redness-greenness discrimination.

$$F_t = L_A/(L_A + 0.1)$$

In Fig. 12.10, F_t is plotted against $\log L_A$, where L_A is the adapting luminance in cd/m^2; when L_A is above 100 cd/m^2 (2 on the log scale), F_t is nearly equal to unity and thus has no effect.

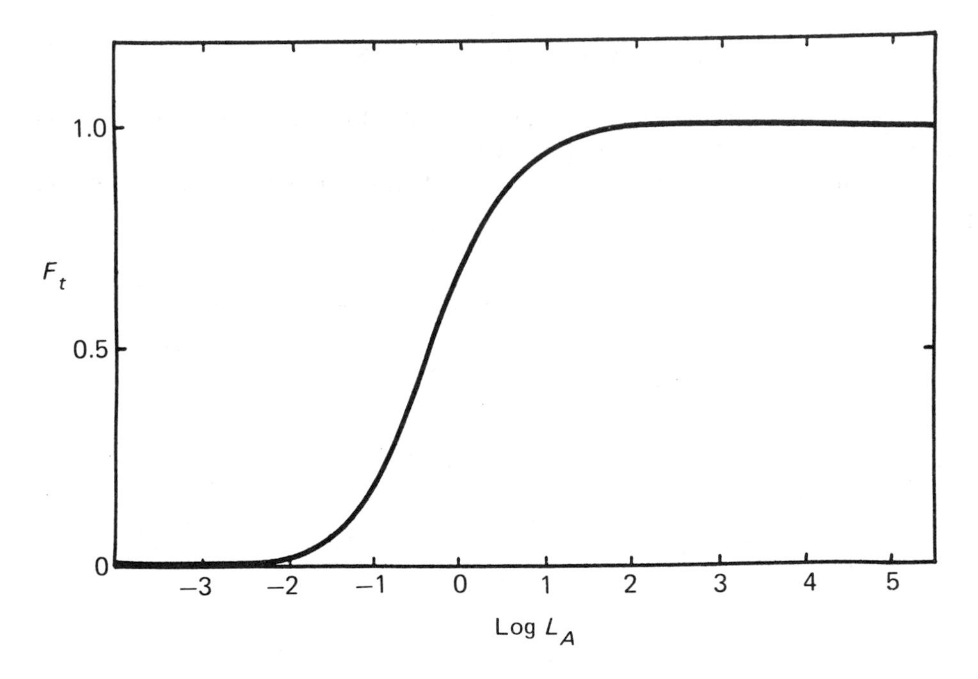

Fig. 12.10 — Low luminance tritanopia factor, F_t, plotted against log L_A, where L_A is the adapting luminance in cd/m^2.

Fifth, to allow for the fact that, compared with their appearance when seen against a grey background, the colourfulness of colours tends to be reduced for light backgrounds, and increased for dark backgrounds (MacDonald, Luo & Scrivener 1990), a chromatic background induction factor, N_{cb}, is introduced where:

$$N_{cb} = 0.725(Y_W/Y_b)^{0.2}$$

Y_b and Y_W being the luminance factors of the background and of the reference white, respectively. If $Y_W/Y_b = 5$ (background luminance 20% of that of the reference white) then $N_{cb} = 1$.

These factors result in correlates of yellowness-blueness, M_{YB}, and redness-greenness, M_{RG}, being formulated as follows:

$$M_{YB} = 100[\tfrac{1}{2}(C_2 - C_3)/4.5][e_s(10/13)N_c N_{cb} F_t]$$
$$M_{RG} = 100[C_1 - (C_2/11)][e_s(10/13)N_c N_{cb}]$$

The constant, 100, is included to give convenient numbers. Correlates of colourfulness, M, and saturation, s, are then given by

$$M = (M_{YB}{}^2 + M_{RG}{}^2)^{1/2}$$
$$s = 50M/(\rho_a + \gamma_a + \beta_a)$$

the constant, 50, being included to give convenient numbers. (The reasons for dividing by $\rho_a + \gamma_a + \beta_a$ will be described in section 12.20.)

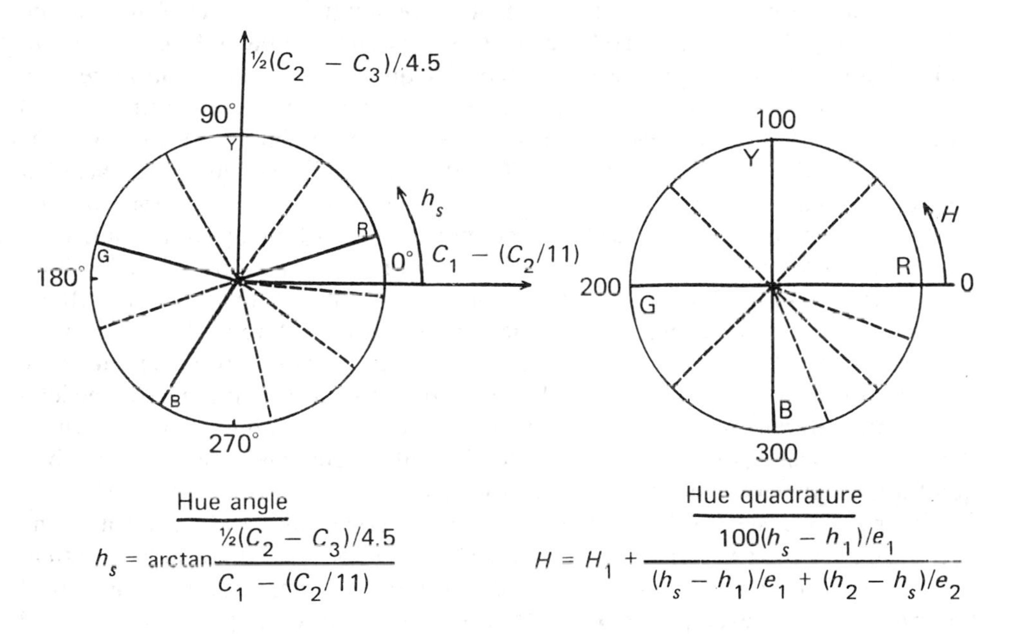

Fig. 12.11 — Left: hue angle, h_s, shown in a plot of $\frac{1}{2}(C_2 - C_3)/4.5$ against $C_1 - (C_2/11)$. Right: hue quadrature, H, shown in a plot where unique red and green are opposite one another, and unique yellow and blue are also opposite one another and at right-angles to the red-green directions.

12.16 CORRELATES OF HUE, H AND H_C

In Fig. 12.11 (left half) is shown a plot of $\frac{1}{2}(C_2 - C_3)/4.5$ against $C_1 - (C_2/11)$. In this figure, the value of h_s is the angle between a horizontal line drawn from the origin towards the right and the line joining the origin to the point representing the colour considered. The positions of the unique hue lines are shown in this diagram by the full lines, R, Y, G, and B.

Hue can also be expressed in terms of the proportions of the unique hues perceived to be present, and the model provides a correlate of hue expressed in this way, *hue quadrature*, H, which is formulated as:

$$H = H_1 + \frac{100[(h_s - h_1)/e_1]}{[(h_s - h_1)/e_1 + (h_2 - h_s)/e_2]}$$

where H_1 is 0, 100, 200, or 300, according to whether red, yellow, green, or blue, respectively, is the hue having the nearest lower value of h_s.

The difference between hue angle, h_s, and hue quadrature, H, is illustrated in Fig. 12.11, where the former is shown on the left, and the latter on the right. The angular positions of the lines representing colours that are perceptually midway between adjacent pairs of unique hues (that is, appearing to contain 50% of each of the two hues) are shown by the broken lines; these broken lines are not at equal angular spacings between the lines representing the unique hues in the figure on the left, because of the effect of the different colorizing weights of the red, yellow, green, and blue unique hues. (In the case of the red-blue quadrant, because of its larger size in the diagram, broken lines are shown that divide it into four perceptually equal parts.) In the case of hue quadrature, H (on the right), the effects of these weights have been included in the derivation of H, and hence the broken lines are spaced at regular intervals. However, because unique red and green are placed opposite one another, with unique yellow and blue also opposite one another and at right angles to the red-green axis, the four quadrants do not represent equal *differences* in hue; while the perceptual difference between the pairs of unique hues red and yellow, yellow and green, and green and blue, are not too different, the perceptual difference between blue and red is about twice as large, and this is represented by the crowding of the three broken lines in this quadrant. However, hue angle (on the left) represents differences in hue more uniformly. The spacing of hue angle (shown on the left) is similar to that of the hues in the Munsell system, while the spacing of hue quadrature (shown on the right) is similar to that of the hues in the NCS system.

H can be expressed both as a number, and as *hue composition*, H_C, in terms of the percentages of the component hues. When the two right-hand digits are more than 50, they indicate the main hue percentage, and the remaining percentage is that of the minor hue; for example, if $H = 262$, the component hues in percentages are 62 Blue and 38 Green, which is abbreviated to 62B 38G. If the two right-hand digits are less than 50, they represent the minor hue percentage, and the remaining percentage is that of the major hue; for example, if $H = 231$, the hue composition is 69 Green 31 Blue, or 69G 31B.

12.17 COMPARISON WITH THE NATURAL COLOUR SYSTEM

In Fig. 12.12, the full lines show the loci of constant hue and saturation predicted by the model for a high level of S_C adapting luminance, such as L_A equal to 200 cd/m^2;

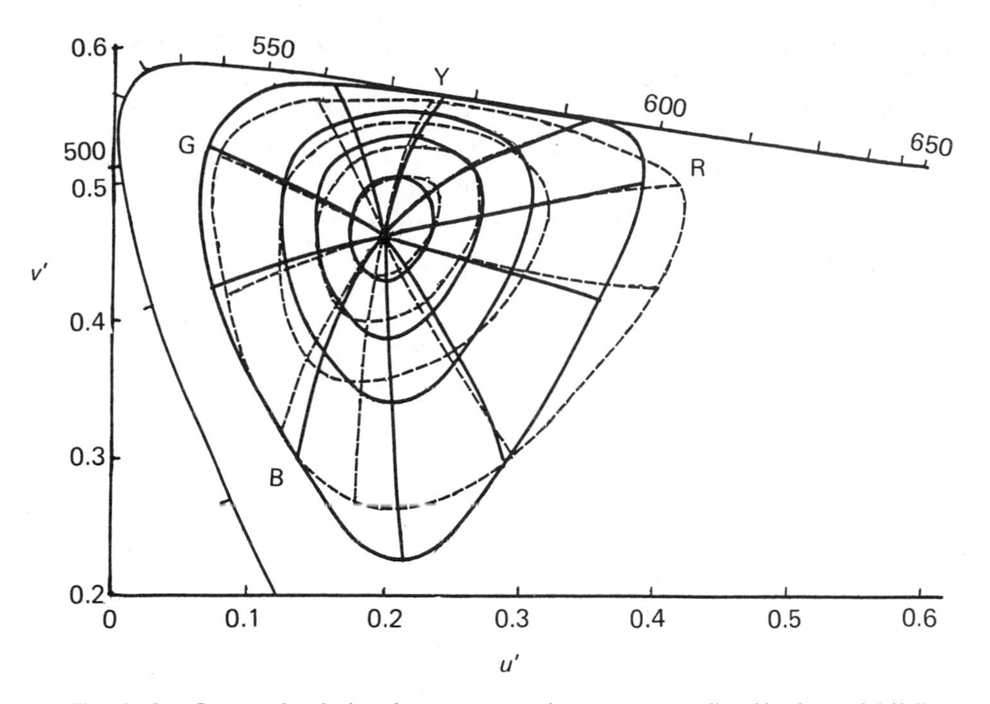

Fig. 12.12 — Constant hue loci, and constant saturation contours, predicted by the model (full lines) compared with those of the NCS (broken lines), for Standard Illuminant C. The unique hue loci are labelled R, Y, G, and B; the intermediate hue loci divide each quadrant into perceptually equal segments of hue. The contours shown are for values of the saturation, s, of the model equal to 60, 120, 180, and 320, and for values of NCS chromaticness, c, equal to 30, 50, 70, and 90. The conditions for the model were as follows. The adapting luminance was 200 cd/m^2; ρ_D, γ_D, and β_D were all put equal to zero; F_ρ, F_γ, and F_β were all put equal to 1; N_c and N_{cb} were put equal to 1; and the sum of the three cone responses, $\rho_a + \gamma_a + \beta_a$, was put equal to 30, this value for the perfect diffuser being 43.0 (because, as shown in Fig. 12.2, the hyperbolic function approximates a simple power function over its central range, the chromaticities corresponding to hue and saturation are only slightly dependent on the values of $\rho_a + \gamma_a + \beta_a$ used, provided that ρ_a, γ_a, and β_a are in their central ranges). For the NCS, the luminance factors were the highest available in the system.

the broken lines show the results obtained experimentally in the Natural Colour System (NCS), and the two sets of lines are seen to be broadly similar. (For an overview, omit section 12.18, which deals with the rod response.)

12.18 THE ROD RESPONSE

The same f_n function is used for the rods as for the cones. However, a scotopic luminance-level adaptation factor, F_{LS}, is used instead of F_L, and is evaluated as follows:

$$F_{LS} = 3800j^2 5L_{AS}/2.26 + 0.2(1 - j^2)^4(5L_{AS}/2.26)^{1/6}$$

where

$$j = 0.00001/(5L_{AS}/2.26 + 0.00001)$$

and L_{AS} is the scotopic luminance of the adapting field (the reason for dividing L_{AS} by 2.26 will be explained shortly). The second of the two terms in the above expression for F_{LS} provides for adaptation, and the first term limits this adaptation at very low levels of illumination. This can be seen in Fig. 12.13 where $\log(F_{LS})$ is

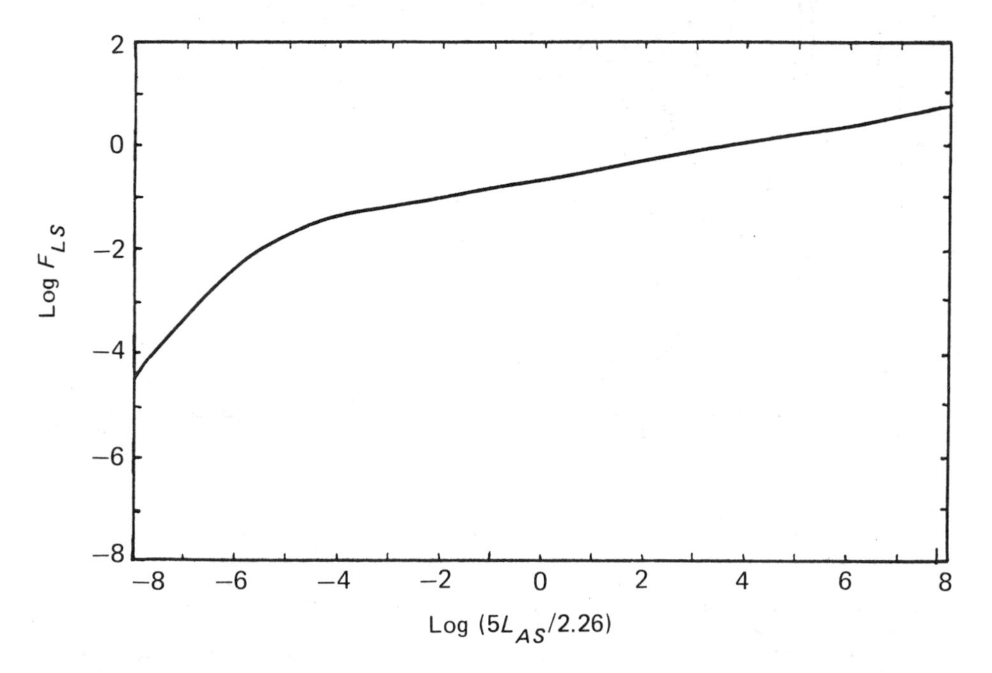

Fig. 12.13 — Scotopic luminance level adaptation factor. The log of the factor, $\log F_{LS}$, is plotted against $\log(5L_{AS}/2.26)$, the log of 5 times the scotopic luminance in scotopic cd/m^2 (divided by 2.26) of the adapting field (which is taken to be typical of a white).

plotted against $\log(5L_{AS}/2.26)$; above $\log(5L_{AS}/2.26)$ equal to -5, the curve has a low slope, thus giving partial compensation for changes in level of adapting luminance (full compensation would correspond to F_{LS} being constant), and below $\log(5L_{AS}/2.26)$ equal to -6, F_{LS} is proportional to L_{AS} so that no compensation occurs.

The scotopic adapting luminance, L_{AS}, may not be known, and it is not usually important to know it accurately. If the adapting field has a chromaticity not too far from that of the Planckian locus, its correlated colour temperature, T, can be used in the following formula to obtain an approximation to L_{AS}:

$$L_{AS}/2.26 = L_A(T/4000 - 0.4)^{1/3} .$$

For other adapting fields, their scotopic luminances, L_{AS}, should be derived from their spectral power distributions by using the $V'(\lambda)$ function. $L_{AS}/2.26$ is used throughout the model (instead of L_{AS}), because, for the equi-energy stimulus, S_E, $L_{AS}/2.26$ is equal to L_A. The value of T used for S_E is 5600 in order to achieve this result.

To provide an upper limit to the rod response at high levels of adaptation and at high stimulus intensities, a rod bleach or saturation factor

$$B_S = 0.5/\{1 + 0.3[(5L_{AS}/2.26)(S/S_W)]^{0.3}\} + 0.5/\{1 + 5[5L_{AS}/2.26]\}$$

is introduced. The rod response after adaptation is then given by

$$A_S = B_S(3.05)[f_n(F_{LS}S/S_W)] + 0.3$$

where 0.3 represents the noise in the signal, and S/S_W are the scotopic luminances relative to that of the reference white. If the true values of S/S_W are not known, the equivalent photopic values, Y/Y_W, can usually be used as an approximation instead. The factor, 3.05, which is equal to $2 + 1 + 1/20$, ensures, for achromatic colours, a smooth transition from cone to rod vision in the total achromatic signal (to be described in the next section).

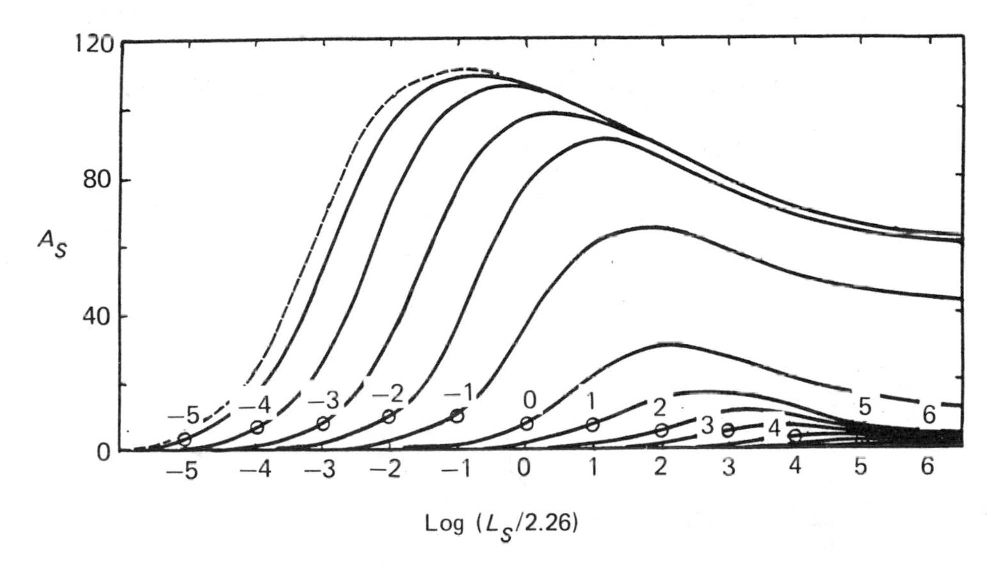

Fig. 12.14 — Response functions for the rods. A_S is plotted against log $(L_S/2.26)$, where L_S is the luminance of the stimulus in scotopic cd/m^2, for levels of log $(5L_{AS}/2.26)$ equal to 6, 5, 4, 3, 2, 1, 0, −1, −2, −3, −4, and −5 , full lines, and for dark adaptation, broken line. Open circles: reference white.

Fig. 12.14 shows rod response curves for different levels of $L_A/2.26$, the adapting scotopic luminance (divided by 2.26), and it can be seen that the rod bleach factor reduces the response for high levels of adapting luminance, and especially so for stimuli of high luminance.

12.19 THE ACHROMATIC RESPONSE, *A*

As mentioned in section 1.6, in addition to the colour difference signals, the retina sends to the brain an achromatic signal. The photopic part of the achromatic signal is given by

$$A_a = 2\rho_a + \gamma_a + (1/20)\beta_a - 3.05 + 1 \ ,$$

assuming the relative ρ, γ, β cone abundances to be in the ratios 2:1:1/20, respectively (Walraven & Bouman 1966). The sum, 3.05, of the separate noises of ρ_a, γ_a, and β_a is replaced by a noise of 1. The total achromatic signal is then given by

$$A = N_{bb}[A_a - 1 + A_S - 0.3 + (1^2 + 0.3^2)^{1/2}] \ ,$$

where A_S is the scotopia contribution from the rods, and the combined noise is taken as the square root of the sum of the squares of the two noise components, 1 and 0.3. N_{bb} is a factor to allow for the brightness induction of the background. $N_{bb} = 1$ if $Y_W/Y_b = 5$ (background luminance 20% of that of the reference white); in general:

$$N_{bb} = 0.725(Y_W/Y_b)^{0.2}$$

In Fig. 12.15, the achromatic signal, A, and its photopic and scotopic components

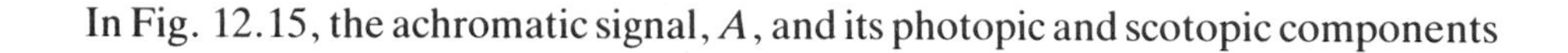

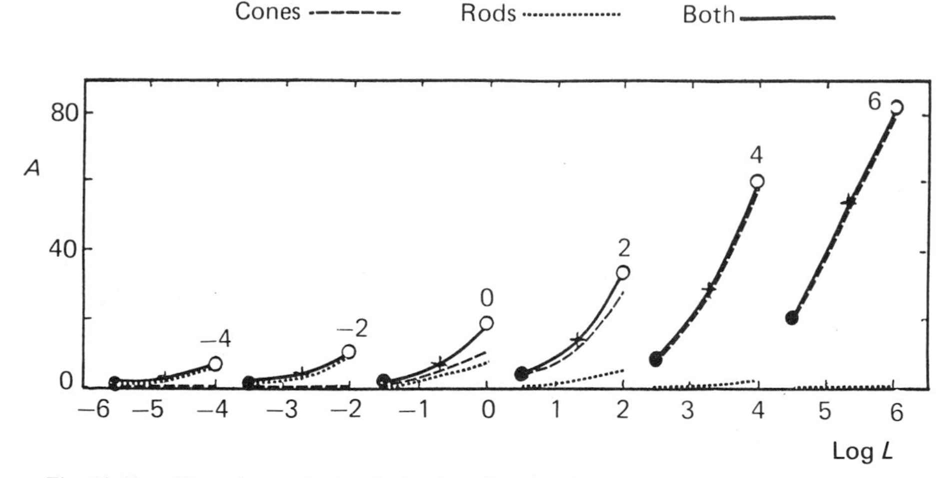

Fig. 12.15 — The achromatic signal, A, plotted against log L, where L is the luminance (or scotopic luminance divided by 2.26) of the stimulus in cd/m^2, for levels of adapting luminance L_A, such that log $(5L_A)$ is equal to 6, 4, 2, 0, −2, and, −4 log cd/m^2. Broken lines: cone contribution; dotted lines: rod contribution; full line: combined cone and rod response. The stimulus is assumed to be such that scotopic (1/2.26) cd/m^2 is equal to photopic cd/m^2. Open circles: reference white; filled circles: 3.162% black; plus signs: adapting field (luminance one fifth of that of the reference white).

are shown over the range from white (○) to black (●) for six different levels of adaptation. (The stimulus is assumed to be such that scotopic (1/2.26) cd/m^2 is equal to photopic cd/m^2.) It is clear that the scotopic contribution to the achromatic signal becomes very small at high levels of adaptation, as is to be expected on the grounds that the rods play very little part in vision at these levels; and the contribution from the cones at very low levels of adaptation is also very small, colour vision then being absent.

12.20 CORRELATES OF RELATIVE YELLOWNESS-BLUENESS, m_{YB}, AND RELATIVE REDNESS-GREENNESS, m_{RG}

Correlates of relative yellowness-blueness, m_{YB} and relative redness-greenness, m_{RG}, are given by

$$m_{YB} = M_{YB}/(\rho_a + \gamma_a + \beta_a)$$
$$m_{RG} = M_{RG}/(\rho_a + \gamma_a + \beta_a) .$$

As shown in section 12.15, a correlate of saturation, s, is given by

$$s = 50M/(\rho_a + \gamma_a + \beta_a)$$

Because saturation is defined (by the CIE) as colourfulness judged in proportion to brightness (see section 1.9), it might have been expected that s would be equal to M divided by a correlate of brightness. However, the saturations of the colours of the spectrum are predicted well (Hunt 1982) by dividing M by $\rho_a + \gamma_a + \beta_a$, and this expression is therefore used as the divisor in preference. (The physiological equivalent of $\rho_a + \gamma_a + \beta_a$ would presumably be some combination of the achromatic signal, A, and the colour difference signals, C_1, C_2, and C_3, that indicated the strength of the retinal signals. For instance, $A + \frac{1}{2}(C_1' + C_2' + C_3)$ is equal to $\rho_a + \gamma_a + (21/20)\beta_a + S_a$, which for photopic levels would be nearly equal to $\rho_a + \gamma_a + \beta_a$, since S_a would be very small. The sum $C_1 + C_2 + C_3$ cannot be used, because it is equal to zero.)

12.21 CORRELATE OF BRIGHTNESS, Q

The brightness response is regarded as being mainly a function of the achromatic signal; but a small contribution from the colour difference signals is added so that allowance can be made for the increase in brightness with increasing purity for colours of constant luminance (the Helmholtz–Kohlrausch effect, see sections 2.3, 3.2, and 11.4). Hence, the brightness response in the model depends on:

$$A + (M/100) .$$

The correlate of brightness, Q, is then evaluated as:

$$Q = \{7[A + (M/100)]\}^{0.6} N_1 - N_2$$

where

$$N_1 = (7A_W)^{0.5}/(5.33 N_b^{0.13})$$
$$N_2 = 7A_W N_b^{0.362}/200 .$$

A_W is the value of A for the reference white, and N_b is the brightness surround induction factor, which has the following values (the values of N_c, introduced in section 12.15, are also listed here for convenience):

	N_b	N_c
Small areas in uniform light backgrounds and surrounds	300	1.0
Normal scenes	75	1.0
Television and VDU displays in dim surrounds	25	0.95
Projected photographs in dark surrounds	10	0.9
Arrays of adjacent colours in dark surrounds	5	0.75

12.22 CORRELATE OF LIGHTNESS, *J*

Lightness is brightness judged relative to that of the reference white. The correlate of lightness, J, is made to be dependent on the luminance factor, Y_b, of the background, and is evaluated as:

$$J = 100(Q/Q_W)^z$$

where

$$z = 1 + (Y_b/Y_W)^{1/2} .$$

In Fig. 12.16 log Q is plotted against log luminance, log L, for the case in which

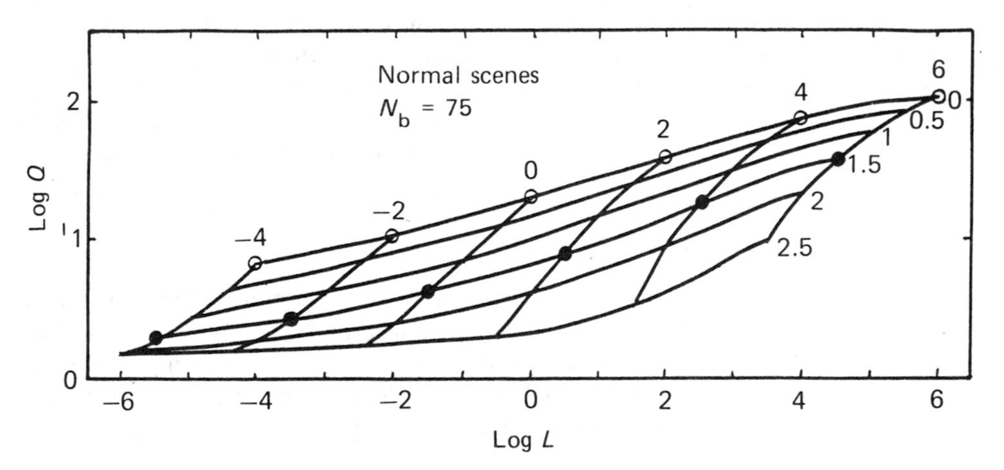

Fig. 12.16 — Brightness-luminance relationships. Log Q is plotted against log L, where Q is the brightness response, and L is the luminance (or scotopic luminance divided by 2.26) in cd/m^2. The relationships are shown for different levels of log adapting luminance in cd/m^2, L_A, such that log $(5L_A)$ is equal to 6, 4, 2, 0, −2, and −4. The curves labelled 0, 0.5, 1, 1.5, 2.0, and 2.5 are for samples having these densities. The brightness induction factor, N_b, is 75. The stimulus is assumed to be such that the luminance in scotopic cd/m^2 (divided by 2.26) is equal to the luminance in photopic cd/m^2.

$N_b = 75$, a value typical for reflecting objects seen in natural surroundings. Results are shown for $M = 0$ for values of $5L_A$ equal to 1000000, 10000, 100, 1, 0.01, 0.0001 cd/m^2, that is, 6, 4, 2, 0, −2, and −4 on the log scale, and for values of L/L_W

(where L_W is the value of L for the reference white) of 1.0, 0.3162, 0.1, 0.03162, 0.01, and 0.003162, that is, densities of 0, 0.5, 1.0, 1.5, 2.0, and 2.5. The following features are evident from this figure. First, brightnesses are all reduced as the level of illumination is reduced. Second, the slopes of the more nearly vertical lines, representing grey scales, are reduced for dark colours at low illumination levels and for light colours at high illumination levels. Third, the lines representing constant L/L_W on the grey scales are approximately parallel, indicating that lightness is approximately constant with illumination level; however, there is a gradual convergence of these lines towards one another as the illumination level falls, and this indicates that the lightnesses of darker colours increase relative to white as the level of illumination is reduced. Results similar to those shown in Fig. 12.16, but over a more limited range, have been obtained from both experimental scaling and modelling (Jameson & Hurvich 1964). In this type of figure, equal differences in log Q represent approximately equal differences in brightness.

12.23 CORRELATE OF CHROMA, C

The correlate of chroma, C, also includes an allowance for the luminance factor, Y_b, of the background, relative to that, Y_W, of the reference white, and is given by:

$$C = 4s^{0.69}(Q/Q_W)^{Y_b/Y_W}(1.31 - 0.31^{Y_b/Y_W})$$

where s is the saturation, Q is the brightness, and Q_W is the brightness of the reference white. In this formula, for a given value of s, the Q/Q_W term usually causes C to diminish as the luminance factor of the sample decreases, and the 0.31 term causes C to diminish as the luminance factor of the background decreases. For white backgrounds for which $Y_b/Y_W = 1$, this formula reduces to $C = 4s^{0.69}(Q/Q_W)$; and for very black backgrounds for which $Y_b/Y_W = 0$, it reduces to $C = 4s^{0.69}(0.31)$.

12.24 CORRELATE OF WHITENESS-BLACKNESS, Q_{WB}

Nayatani and his co-workers have introduced the concept of a whiteness-blackness perception (Nayatani, Hashimoto, Takahama & Sabagaki 1987), and the model provides a correlate of this perception as:

$$Q_{WB} = 20(Q^{0.7} - Q_b^{0.7})$$

where Q_b is the value of Q for the background. In Fig. 12.17, Q_{WB} is plotted against log L, for colours for which $M = 0$, seen against a background having a luminance 0.7 log units less than (one fifth of) that of the reference white, L_W, for values of log $(5L_A)$ equal to 5, 4, 3, 2, 1, 0, − 1, − 2, − 3, − 4, and − 5. Results are shown for values of log (L_W/L) equal to 0 (whites), 0.5, 0.7 (the background), 1, and 1.5 (blacks). It is seen that Q_{WB} has a constant value of zero for the background, but that, as the illumination level increases, the whiteness of the whites increases and the

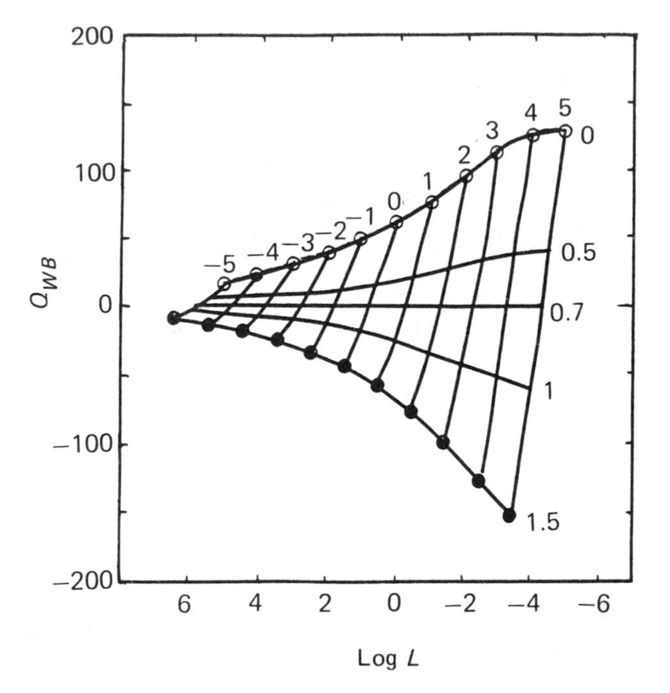

Fig. 12.17 — Whiteness-blackness, Q_{WB}, plotted against log L, where L is the luminance (or scotopic luminance divided by 2.26) in cd/m^2, for colours for which $M = 0$, seen against a background having a luminance 0.7 log units less than (one fifth of) that of the reference white, L_W, for values of log $(5L_A)$ equal to 5, 4, 3, 2, 1, 0, −1, −2, −3, −4, and −5. Results are shown for values of log (L_W/L) equal to 0 (whites), 0.5, 0.7 (the background), 1, and 1.5 (blacks), by the curves labelled with these numbers. The stimulus is assumed to be such that the luminance in scotopic cd/m^2 (divided by 2.26) is equal to the luminance in photopic cd/m^2.

blackness of the blacks decreases. These results are in broad agreement with those found by Stevens (Stevens 1961).

12.25 USING THE MODEL FOR RELATED COLOURS

The input data and steps required for using the model for related colours, together with a worked example, are given in section 12.27. (For an overview, omit sections 12.26 and 12.28, which deal with unrelated colours.)

12.26 UNRELATED COLOURS

Unrelated colours are those that are seen in isolation from other colours: bright light sources, and uniform areas seen against unilluminated backgrounds, are examples. Thus unrelated colours are seen in environments of luminances very much lower than that of the sample, frequently in completely dark fields. But, even in a completely dark field, it is not realistic to take the adapting field luminance, L_A, as zero, because the sample stimulus, and scattered light from it in the eye, will provide an effective adapting luminance above zero. Hence, if L is the luminance of the sample, the luminance of the adapting field is derived as

$$L_A = L^{2/3}/200 \ ,$$

and if $L_S/2.26$ is the scotopic luminance of the sample (divided by 2.26), the scotopic luminance of the adapting field (divided by 2.26) is derived as

$$L_{AS}/2.26 = (L_S/2.26)^{2/3}/200 \ .$$

The above expressions mean that the effective adapting luminance is 1/2000th of the sample luminance when the latter is equal to 1000 cd/m^2, 1/200th at 1 cd/m^2, 1/20th at 0.001 cd/m^2, and equal to the sample luminance when the latter has a value of 0.000 000 125 cd/m^2.

The chromaticity of the adapting field for unrelated colours is taken as that of the equi-energy stimulus, S_E, because this is similar to the stimulus that appears most neutral to the dark-adapted eye (Hurvich & Jameson 1951); it is not modified as a function of the chromaticity of the sample, because the light taken as the adapting luminance has less than 1/100th of the luminance of the sample, for stimuli whose luminances are in the photopic range.

Sometimes, unrelated colours are seen immediately after another field has been viewed. For instance, a pilot, flying at night, may look first at his flight deck displays, and then out of the aircraft at signal lights. To allow for this situation, the concept of a conditioning field is introduced. The conditioning field is regarded as a field that is seen just prior to viewing the unrelated colour; its chromaticity is denoted as x_C, y_C, and its luminance as L_C and scotopic luminance (divided by 2.26) as $L_{CS}/2.26$. If there is no such conditioning field, the values of x_C, y_C, L_C and $L_{CS}/2.26$ are taken to be the same as those of the adapting field. The factor $(L_A/L_C)^c$ is used to reduce the cone response when L_C is greater than L_A, and $[(L_{AS}/2.26)/(L_{CS}/2.26)]^c$ to reduce the rod response when $L_{CS}/2.26$ is greater than $L_{AS}/2.26$ (see section 12.28.2, Step 6; and Step 15, where $[(L_{AS}/2.26)/(L_{CS}/2.26)]^c$ is simplified to $(L_{AS}/L_{CS})^c$); c might be about 0.2.

If the scotopic luminances, L_S, of the stimuli are not known, and their chromaticities are not too far from the Planckian locus, their correlated colour temperatures, T, can be used in the same formula as was used for related colours to obtain $L_S/2.26$ from the photopic luminance, L, as follows:

$$L_S/2.26 = L(T/4000 - 0.4)^{1/3} \ .$$

For other stimuli, their scotopic luminances (divided by 2.26), $L_S/2.26$, should be derived from their spectral power distributions using the $V'(\lambda)$ function.

For unrelated colours the concept of a reference white does not apply. However, to provide a partial adjustment of sensitivity analogous to that provided in the case of related colours (obtained by dividing ρ by ρ_W, γ by γ_W, and β by β_W), ρ, γ, and β are now divided by:

$$W = [(1/3)(\rho + \gamma + \beta)]^{1/2}$$

(as shown in section 12.28.2, Step 6). This allows for some effect of the stimulus intensity on the sensitivity of the cone system; part of this will be caused by changes in the pupil diameter. A similar adjustment of the sensitivity of the rod system is allowed for by dividing the scotopic luminance (divided by 2.26), $L_S/2.26$, by its square root (as shown in section 12.28.2, Step 15, by the term $(L_S/2.26)^{1/2}$ which is equal to $(L_S/2.26)/(L_S/2.26)^{1/2}$). In the cone bleach factors for unrelated colours ($B_{\rho u}$, $B_{\gamma u}$, $B_{\beta u}$), $\rho_W/100$, $\gamma_W/100$, and $\beta_W/100$ are replaced by $3\rho_C/(\rho_C + \gamma_C + \beta_C)$, $3\gamma_C/(\rho_C + \gamma_C + \beta_C)$, and $3\beta_C/(\rho_C + \gamma_C + \beta_C)$, where ρ_C, γ_C, and β_C are the values of ρ, γ, and β for the conditioning field (as shown in section 12.28.2, Step 6). In the rod bleach or saturation factor for unrelated colours (B_{Su}), S/S_W is replaced by $(L_S/2.26)/(L_S/2.26)^{1/2}$ (included as $(L_S/2.26)^{1/2}$ in section 12.28.2, Step 15).

The low-luminance tritanopia factor for unrelated colours, F_{tu} is formulated by using L instead of L_A:

$$F_{tu} = L/(L + 0.1)$$

The luminance level of a stimulus has an effect on its apparent hue if it is an unrelated colour, but not if it is a related colour (Hunt 1989). This phenomenon, known as the *Bezold–Brücke effect*, is allowed for by making the eccentricity factor, e_s, for unrelated colours, depend on the luminance, L, of the stimulus in the case of its values for the unique yellow and blue hues, as follows:

Yellow $e_s = 0.7[L/(L + 10)] + 0.3[10/(L + 10)]$

Blue $e_s = 1.2[L/(L + 10)] + 0.2[10/(L + 10)]$.

For unrelated colours, the background and surround usually have luminances very much lower than that of the sample; hence the chromatic induction surround factor, N_c, is put equal to 0.5. The formulae for the correlates of hue, H and H_C, colourfulness, M, and saturation, s, for unrelated colours are given in section 12.28.2, steps 9, 10, and 13.

The correlate of brightness, Q, for unrelated colours is then given by

$$Q = \{[1.1][A + (M/100)]\}^{0.9}$$

In Fig. 12.18, log Q is plotted against log stimulus luminance, log L , for unrelated colours (full line), and compared to experimental results (circles) obtained by Bartleson (Bartleson 1980); the agreement is seen to be quite good.

Fig. 12.18 — Brightness–luminance relationship for unrelated colours. Log Q is plotted against log L, where Q is the brightness response, and L is the luminance (or scotopic luminance divided by 2.26) in cd/m^2. The stimulus is assumed to be such that $M = 0$, and the luminance in scotopic cd/m^2 (divided by 2.26) is equal to the luminance in photopic cd/m^2.

The input data and steps required for using the model for unrelated colours, together with a worked example, are given in section 12.28.

12.27 STEPS IN USING THE MODEL FOR RELATED COLOURS

12.27.1 Input data required for the model

The following input data are required:

Chromaticity co-ordinates, x, y, *and luminance factors*, Y, *in the illuminant considered*:

Illuminant	x_I, y_I
Adapting field	x_A, y_A
Background	x_b, y_b, Y_b
Proximal field	x_p, y_p, Y_p
Reference white	x_W, y_W, Y_W
Samples:	x, y, Y

It is often possible to take some of these chromaticities as the same:

if the adapting field is a typical average scene, or generally white, grey, or black, it is usually acceptable to take:

$x_A = x_I$ and $y_A = y_I$

if the background is a typical average scene, or generally white, grey, or black, it is usually acceptable to take:

$x_b = x_I$ and $y_b = y_I$

if the background is a typical average scene, or generally white, grey, or black, and if the proximal field is the same as the background, it is usually acceptable to take a non-selective neutral as the reference white and hence:

$x_W = x_I$ and $y_W = y_I$

It is also often possible to take the following values of Y:

if the reference white is the perfect diffuser:

$Y_W = 100$

if the background is a typical average scene, it is usually acceptable to take:

$Y_b = 20$

If the proximal field is different from the background, then a modified reference white can be derived (as described in Step 2).

Luminances of stimuli, and scotopic data:

Photopic luminance of reference white in cd/m^2: $L_W = L_P Y_W/100$

where L_P is the luminance of the perfect diffuser in cd/m^2; $L_P = E/\pi$ where E is the illuminance in lux. (The photopic luminances, L, of the samples, in cd/m^2, are given by $L = L_P Y/100$.)

Photopic luminance of adapting field in cd/m^2: L_A

If the value, L_A, is not available, $L_W/5$ can be used as an approximation.

Scotopic luminance of adapting field in scotopic cd/m^2: L_{AS}

If the value, L_{AS}, is not available, an approximation to it can be derived from L_A as:

$$L_{AS}/2.26 = L_A(T/4000 - 0.4)^{1/3}$$

where T is the correlated colour temperature of the illuminant, if its chromaticity is not too far from the Planckian locus; for other illuminants, L_{AS} should be derived from their spectral power distributions, using the $V'(\lambda)$ function.

Scotopic luminances relative to reference white: S/S_W

If the scotopic values, S/S_W, are not available, the equivalent photopic values, Y/Y_W, can be used instead as an approximation.

Values of the chromatic and brightness surround induction factors:

If optimised values of N_c, the chromatic surround induction factor, and N_b, the brightness surround induction factor, are not available, the following values can be used:

	N_c	N_b
Small areas in uniform light backgrounds and surrounds	1.0	300
Normal scenes	1.0	75
Television and VDU displays in dim surrounds	0.95	25
Projected photographs in dark surrounds	0.9	10
Arrays of adjacent colours in dark surrounds	0.75	5

Values of the chromatic and brightness background induction factors:

Chromatic background induction factor: N_{cb}

Brightness background induction factor: N_{bb}

If optimized values of N_{cb} and N_{bb} are not available, the following values can be used:

$$N_{cb} = 0.725(Y_W/Y_b)^{0.2} \qquad N_{bb} = 0.725(Y_W/Y_b)^{0.2} .$$

12.27.2 Steps in using the model

Step 1 Calculate X, Y, Z for the reference white, for the background, for the proximal field (if different from the background), and for the samples.

$$X = xY/y \qquad Y = Y \qquad Z = (1 - x - y)Y/y$$

Step 2 Calculate ρ, γ, β for the reference white, for the background, for the proximal field (if different from the background), and for the samples.

$$\begin{aligned} \rho &= \quad 0.38971X + 0.68898Y - 0.07868Z \\ \gamma &= -0.22981X + 1.18340Y + 0.04641Z \\ \beta &= \qquad\qquad\qquad\qquad\qquad 1.00000Z \end{aligned}$$

Note: when account is being taken of simultaneous contrast or assimilation, the values for the reference white, ρ_W, γ_W, β_W, for the background, ρ_b, γ_b, β_b, and for the proximal field, ρ_p, γ_p, β_p, are used to derive values for a modified reference white, ρ'_W, γ'_W, β'_W:

$$\begin{aligned} \rho'_W &= \rho_W[(1-p)r + (1+p)/r]^{1/2}/[(1+p)r + (1-p)/r]^{1/2} \\ \gamma'_W &= \gamma_W[(1-p)g + (1+p)/g]^{1/2}/[(1+p)g + (1-p)/g]^{1/2} \\ \beta'_W &= \beta_W[(1-p)b + (1+p)/b]^{1/2}/[(1+p)b + (1-p)/b]^{1/2} \end{aligned}$$

where

$$\begin{aligned} r &= (\rho_p/\rho_b) \\ g &= (\gamma_p/\gamma_b) \\ b &= (\beta_p/\beta_b) . \end{aligned}$$

The values of p are between 0 and -1 for simultaneous contrast, and between 0 and $+1$ for assimilation.

Step 3 Calculate ρ/ρ_W, γ/γ_W, β/β_W for the samples.

ρ_W, γ_W, β_W are the values of ρ, γ, β for the reference white, or for the modified reference white, as appropriate.

Step 4 Calculate F_L

$$F_L = 0.2k^4(5L_A) + 0.1(1-k^4)^2(5L_A)^{1/3}$$

where

$$k = 1/(5L_A + 1)$$

Step 5 Calculate F_ρ, F_γ, F_β

$$h_\rho = 3\rho_W/(\rho_W + \gamma_W + \beta_W)$$
$$h_\gamma = 3\gamma_W/(\rho_W + \gamma_W + \beta_W)$$
$$h_\beta = 3\beta_W/(\rho_W + \gamma_W + \beta_W)$$

$$F_\rho = (1 + L_A{}^{1/3} + h_\rho)/(1 + L_A{}^{1/3} + 1/h_\rho)$$
$$F_\gamma = (1 + L_A{}^{1/3} + h_\gamma)/(1 + L_A{}^{1/3} + 1/h_\gamma)$$
$$F_\beta = (1 + L_A{}^{1/3} + h_\beta)/(1 + L_A{}^{1/3} + 1/h_\beta)$$

If the colour of the illuminant is discounted, $F_\rho = F_\gamma = F_\beta = 1$.

Step 6 Calculate ρ_D, γ_D, β_D

$$\rho_D = f_n[(Y_b/Y_W)F_LF_\gamma] - f_n[(Y_b/Y_W)F_LF_\rho]$$
$$\gamma_D = 0$$
$$\beta_D = f_n[(Y_b/Y_W)F_LF_\gamma] - f_n[(Y_b/Y_W)F_LF_\beta]$$

where

$$f_n[I] = 40[I^{0.73}/(I^{0.73} + 2)]$$

If the colour of the illuminant is discounted, $\rho_D = \gamma_D = \beta_D = 0$.

Step 7 Calculate ρ_a, γ_a, β_a

$$\rho_a = B_\rho[f_n(F_LF_\rho\rho/\rho_W) + \rho_D] + 1$$
$$\gamma_a = B_\gamma[f_n(F_LF_\gamma\gamma/\gamma_W) + \gamma_D] + 1$$
$$\beta_a = B_\beta[f_n(F_LF_\beta\beta/\beta_W) + \beta_D] + 1$$

where

$$B_\rho = 10^7/[10^7 + 5L_A(\rho_W/100)]$$
$$B_\gamma = 10^7/[10^7 + 5L_A(\gamma_W/100)]$$
$$B_\beta = 10^7/[10^7 + 5L_A(\beta_W/100)]$$

and

$$f_n[I] = 40[I^{0.73}/(I^{0.73} + 2)]$$

Step 8 Calculate A_a, C_1, C_2, C_3

$$A_a = 2\rho_a + \gamma_a + (1/20)\beta_a - 3.05 + 1$$
$$C_1 = \rho_a - \gamma_a$$
$$C_2 = \gamma_a - \beta_a$$
$$C_3 = \beta_a - \rho_a$$

Step 9 Calculate h_s

$$h_s = \arctan\{[\tfrac{1}{2}(C_2 - C_3)/4.5]/[C_1 - (C_2/11)]\}$$
$$= \arctan(t/t')$$

where 'arctan' means 'the angle whose tangent is'. h_s lies between 0° and 90° if t and t' are both positive; between 90° and 180° if t is positive and t' is negative; between 180° and 270° if t and t' are both negative; and between 270° and 360° if t is negative and t' is positive.

Step 10 Calculate the hue quadrature H

$$H = H_1 + \frac{100[(h_s - h_1)/e_1]}{[(h_s - h_1)/e_1 + (h_2 - h_s)/e_2]}$$

where H_1 is 0, 100, 200, or 300, according to whether red, yellow, green, or blue, respectively, is the hue having the nearest lower value of h_s. The values of h_s and e_s for the four unique hues are:

	Red	Yellow	Green	Blue
h_s	20.14	90.00	164.25	237.53
e_s	0.8	0.7	1.0	1.2

e_1 and h_1 are the values of e_s and h_s, respectively, for the unique hue having the nearest lower value of h_s; and e_2 and h_2 are these values for the unique hue having the nearest higher value of h_s.

Step 11 Calculate the hue composition, H_C
Where H_P is the part of H after its hundreds digit, if:

$H = H_P$, the hue composition is H_P Yellow, $100 - H_P$ Red
$H = 100 + H_P$, the hue composition is H_P Green, $100 - H_P$ Yellow
$H = 200 + H_P$, the hue composition is H_P Blue, $100 - H_P$ Green
$H = 300 + H_P$, the hue composition is H_P Red, $100 - H_P$ Blue

Step 12 Calculate e_s

$$e_s = e_1 + (e_2 - e_1)(h_s - h_1)/(h_2 - h_1)$$

where e_1 and h_1 are the values of e_s and h_s, respectively, for the unique hue having the nearest lower value of h_s; and e_2 and h_2 are these values for the unique hue having the nearest higher value of h_s.

Step 13 Calculate F_t

$$F_t = L_A/(L_A + 0.1)$$

Step 14 Calculate the yellowness-blueness, M_{YB}, the redness-greenness, M_{RG}, the colourfulness, M, the relative yellowness-blueness, m_{YB}, the relative redness-greenness, m_{RG}, and the saturation, s

$$M_{YB} = 100[\tfrac{1}{2}(C_2 - C_3)/4.5][e_s(10/13)N_cN_{cb}F_t]$$
$$M_{RG} = 100[C_1 - (C_2/11)][e_s(10/13)N_cN_{cb}]$$
$$M = (M_{YB}{}^2 + M_{RG}{}^2)^{1/2}$$
$$m_{YB} = M_{YB}/(\rho_a + \gamma_a + \beta_a)$$
$$m_{RG} = M_{RG}/(\rho_a + \gamma_a + \beta_a)$$
$$s = 50M/(\rho_a + \gamma_a + \beta_a)$$

Step 15 Calculate F_{LS}

$$F_{LS} = 3800j^2 5L_{AS}/2.26 + 0.2(1 - j^2)^4(5L_{AS}/2.26)^{1/6}$$

where

$$j = 0.00001/(5L_{AS}/2.26 + 0.00001)$$

Step 16 Calculate A_S

$$A_S = B_S(3.05)[f_n(F_{LS}S/S_W)] + 0.3$$

where

$$B_S = 0.5/\{1 + 0.3[(5L_{AS}/2.26)(S/S_W)]^{0.3}\} + 0.5/\{1 + 5[5L_{AS}/2.26]\}$$

and

$$f_n[I] = 40[I^{0.73}/(I^{0.73} + 2)]$$

Step 17 Calculate A

$$A = N_{bb}[A_a - 1 + A_S - 0.3 + (1^2 + 0.3^2)^{1/2}]$$

Step 18 Calculate $A + (M/100)$

$$A + (M/100)$$

Step 19 Calculate the brightness, Q, and the brightness of the reference white, Q_W

$$Q = \{7[A + (M/100)]\}^{0.6} N_1 - N_2$$

where

$$N_1 = (7A_W)^{0.5}/(5.33 N_b^{0.13})$$
$$N_2 = 7A_W N_b^{0.362}/200$$

and A_W is the value of A for the reference white.

Step 20 Calculate the lightness, J

$$J = 100(Q/Q_W)^z$$

where

$$z = 1 + (Y_b/Y_W)^{1/2}$$

Step 21 Calculate the chroma, C

$$C = 4s^{0.69}(Q/Q_W)^{Y_b/Y_W}(1.31 - 0.31^{Y_b/Y_W})$$

Step 22 Calculate the whiteness-blackness, Q_{WB}

$$Q_{WB} = 20(Q^{0.7} - Q_b^{0.7})$$

where Q_b is the value of Q for the background.

Step 23
Tabulate the values of H, H_C, M, s, Q, J, C, and Q_{WB}.

12.27.3 A sample colour taken as a worked example

To illustrate the way in which the model operates, a reflecting surface colour whose chromaticity coordinates, x,y, and luminance factors, Y, are known in illuminants D_{65} and S_A, is taken as an example. Its appearance is predicted for illumination levels corresponding to $5L_A$ being equal to 10000, 100, 1, and 0.01 cd/m^2.

The input data are as follows, where the sample is identified as S1 and the reference white as RW:

	x	y	Y
RW in D_{65}	0.3127	0.3290	100.00
S1 in D_{65}	0.2525	0.3571	27.04
RW in S_A	0.4476	0.4074	100.00
S1 in S_A	0.3618	0.4483	23.93

Luminance levels of adapting fields, L_A:

	$5L_A$	L_A	$5L_{AS}/2.26$	$L_{AS}/2.26$
D_{65}	10000	2000	10710	2142
	100	20	107.10	21.42
	1	0.2	1.0710	0.2142
	0.01	0.002	0.010710	0.002142
S_A	10000	2000	6630	1326
	100	20	66.30	13.26
	1	0.2	0.6630	0.1326
	0.01	0.002	0.006630	0.001326

The scotopic values were obtained by using the formula given earlier, with $T = 6504$ for D_{65}, and $T = 2856$ for S_A.

The photopic ratio, Y/Y_W, was used as an approximation for the scotopic luminance relative to that of the reference white, S/S_W.

As the colours are reflecting surface samples, it is appropriate to set: $N_c = 1$ and $N_b = 75$.

The adapting field is assumed to be a 20% reflecting neutral grey.
The background is assumed to be a 20% reflecting neutral grey.
The proximal field is assumed to be the same as the background.
The reference white is taken as the perfect diffuser.

Predictions

The predictions for this sample in the two illuminants at the four different illumination levels are given in Table 12.1, from which the following inferences can be drawn.

Table 12.1 — Predictions for the reference white (RW) and for the sample (S1) in illuminants D_{65} and S_A at four different illumination levels

		$5L_A$	L_A	H	H_C	M	s	Q	J	C	Q_{WB}
D_{65}	RW	10000	2000	266	66B 34G	1.1	1	74	100	2	126
		100	20	266	66B 34G	4.3	7	38	100	8	95
		1	0.2	267	67B 33G	3.7	14	20	100	13	64
		0.01	0.002	267	67B 33G	0.3	4	11	100	5	41
S_A	RW	10000	2000	87	87Y 13R	4.4	4	75	100	5	128
		100	20	79	79Y 21R	16.1	27	41	100	20	100
		1	0.2	68	68Y 32R	10.3	38	22	100	26	70
		0.01	0.002	61	61Y 39R	0.5	7	11	100	8	40
D_{65}	S1	10000	2000	228	72G 28B	71	108	49	55	48	24
		100	20	229	71G 29B	34	114	23	47	49	16
		1	0.2	229	71G 29B	12.7	90	11	44	41	10
		0.01	0.002	230	70G 30B	0.8	12	6.2	44	10	6
S_A	S1	10000	2000	243	57G 43B	80	124	48	51	53	15
		100	20	237	63G 37B	38	129	23	43	53	10
		1	0.2	231	69G 31B	14.6	102	12	39	44	6
		0.01	0.002	228	72G 28B	0.9	14	5.7	41	11	4

The reference white in D_{65} is greenish-blue, but the colourfulness, M, is very low, so that the hue will not be very apparent. The reference white in S_A is reddish-yellow, with lower colourfulness, M, at high illumination (because of the more complete adaptation) and at low illumination (because the response is mainly scotopic), with higher colourfulness at the intermediate illumination levels. The sample is a bluish-green in both illuminations, but is slightly more bluish in S_A; it is also slightly darker, lower J, and has a slightly greater colourfulness, higher M, in S_A. Its saturation, s, chroma, C, and lightness, J, remain approximately the same at the three higher illumination levels, and this indicates approximate *illuminance constancy*; but, at the lowest level, the saturation and chroma are greatly reduced because the level is at the bottom of the photopic range, and the colourfulness is therefore very small. As is to be expected, the brightness, Q, of both the reference white and the sample, and the colourfulness, M, of the sample, decrease considerably as the illumination level falls, but the lightness, J, remains nearly constant. The whiteness-blackness, Q_{WB}, falls with illumination level for both the reference white and the sample; a dark sample would have shown negative values of decreasing negative magnitude with falling illuminance.

12.28 STEPS IN USING THE MODEL FOR UNRELATED COLOURS

12.28.1 Input data required for the model

The following input data are required:

Chromaticity co-ordinates, x,y, and photopic and scotopic luminances, L and L_S:

Sample	x,	y,	L,	L_S
Adapting field	x_A,	y_A,	L_A,	L_{AS}
Conditioning field	x_C,	y_C,	L_C,	L_{CS}

The photopic luminance of the adapting field, L_A, is taken as:

$$L^{2/3}/200 .$$

The scotopic luminance (divided by 2.26) of the adapting field, L_{AS}, is taken as:

$$L_{AS}/2.26 = (L_S/2.26)^{2/3}/200 .$$

The chromaticity of the adapting field is taken as that of S_E, so that $x_A = 1/3$, $y_A = 1/3$. The conditioning field is the field seen just prior to viewing the unrelated colour. If there is no conditioning field, the values of x_C, y_C, L_C, L_{CS} are taken to be the same as those of the adapting field.

Scotopic luminances

If the scotopic luminances of the stimuli, L_S, are not known, they may be derived from the photopic luminances, L, as:

$$L_S/2.26 = L(T/4000 - 0.4)^{1/3}$$

where T is the correlated colour temperature, if the samples have chromaticities not too far from the Planckian locus; for other samples, the scotopic luminances, L_S, should be derived from their spectral power distributions, using the $V'(\lambda)$ function.

Chromatic surround induction factor
For unrelated colours: $N_c = 0.5$

12.28.2 Steps in using the model for unrelated colours

Step 1 Calculate X, Y, Z for the sample, and for the conditioning field.

$$X = xL/y \quad Y = L \quad Z = (1 - x - y)L/y$$

Step 2 Calculate ρ, γ, β for the sample, and for the conditioning field.

$$\rho = 0.38971X + 0.68898Y - 0.07868Z$$
$$\gamma = -0.22981X + 1.18340Y + 0.04641Z$$
$$\beta = 1.00000Z$$

Step 3 Calculate W

$$W = [(1/3)(\rho + \gamma + \beta)]^{1/2}$$

Step 4 Calculate F_L

$$F_L = 0.2k^4(5L_A) + 0.1(1 - k^4)^2(5L_A)^{1/3}$$

where

$$k = 1/(5L_A + 1)$$

Step 5 Calculate F_ρ, F_γ, F_β

$$h_\rho = (3\rho_C/(\rho_C + \gamma_C + \beta_C)$$
$$h_\gamma = (3\gamma_C/(\rho_C + \gamma_C + \beta_C)$$
$$h_\beta = (3\beta_C/(\rho_C + \gamma_C + \beta_C)$$
$$F_\rho = (1 + L_A{}^{1/3} + h_\rho)/(1 + L_A{}^{1/3} + 1/h_\rho)$$
$$F_\gamma = (1 + L_A{}^{1/3} + h_\gamma)/(1 + L_A{}^{1/3} + 1/h_\gamma)$$
$$F_\beta = (1 + L_A{}^{1/3} + h_\beta)/(1 + L_A{}^{1/3} + 1/h_\beta)$$

If there is no conditioning field, $h_\rho = h_\gamma = h_\beta = 1$, and $F_\rho = F_\gamma = F_\beta = 1$.

Step 6 Calculate ρ_a, γ_a, β_a

$$\rho_a = B_{\rho u}\{f_n[F_L F_\rho (L_A/L_C)^c \rho/W]\} + 1$$
$$\gamma_a = B_{\gamma u}\{f_n[F_L F_\gamma (L_A/L_C)^c \gamma/W]\} + 1$$
$$\beta_a = B_{\beta u}\{f_n[F_L F_\beta (L_A/L_C)^c \beta/W]\} + 1$$

where

$$B_{\rho u} = 10^7/[10^7 + (5L_A)3\rho_C/(\rho_C + \gamma_C + \beta_C)]$$
$$B_{\gamma u} = 10^7/[10^7 + (5L_A)3\gamma_C/(\rho_C + \gamma_C + \beta_C)]$$
$$B_{\beta u} = 10^7/[10^7 + (5L_A)3\beta_C/(\rho_C + \gamma_C + \beta_C)]$$

and ρ_C, γ_C, β_C are the values of ρ, γ, β for the conditioning field, and

$$f_n[I] = 40[I^{0.73}/(I^{0.73} + 2)]$$

A typical value for c is 0.2. If there is no conditioning field, $\rho_C = \rho_a$, $\gamma_C = \gamma_a$, and $\beta_C = \beta_a$ (and, since $\rho_a = \gamma_a = \beta_a$, the ratios that follow $5L_A$ in the equations for $B_{\rho u}$, $B_{\gamma u}$, and $B_{\beta u}$ reduce to unity).

Step 7 Calculate A_a, C_1, C_2, C_3

$$A_a = 2\rho_a + \gamma_a + (1/20)\beta_a - 3.05 + 1$$
$$C_1 = \rho_a - \gamma_a$$
$$C_2 = \gamma_a - \beta_a$$
$$C_3 = \beta_a - \rho_a$$

Step 8 Calculate h_s

$$h_s = \arctan\{[\tfrac{1}{2}(C_2 - C_3)/4.5]/[C_1 - (C_2/11)]\}$$
$$= \arctan(t/t')$$

where 'arctan' means 'the angle whose tangent is'. h_s lies between 0° and 90° if t and t' are both positive; between 90° and 180° if t is positive and t' is negative; between 180° and 270° if t and t' are both negative; and between 270° and 360° if t is negative and t' is positive.

Step 9 Calculate the hue quadrature H

$$H = H_1 + \frac{100[(h_s - h_1)/e_1]}{[(h_s - h_1)/e_1 + (h_2 - h_s)/e_2]}$$

where H_1 is either 0, 100, 200, or 300, according to whether red, yellow, green, or blue, respectively, is the hue having the nearest lower value of h_s. The values of h_s and e_s for the four unique hues are:

	h_s	e_s
Red	20.14	0.8
Yellow	90.00	$0.7[L/(L+10)]+0.3[10/(L+10)]$
Green	164.25	1.0
Blue	237.53	$1.2[L/(L+10)]+0.2[10/(L+10)]$

e_1 and h_1 are the values of e_s and h_s, respectively, for the unique hue having the nearest lower value of h_s; and e_2 and h_2 are these values for the unique hue having the nearest higher value of h_s.

Step 10 Calculate the hue composition, H_C

Where H_P is the part of H after its hundreds digit, if:

$H = H_P$, the hue composition is H_P yellow, $100 - H_P$ red
$H = 100 + H_P$, the hue composition is H_P green, $100 - H_P$ yellow
$H = 200 + H_P$, the hue composition is H_P blue, $100 - H_P$ green
$H = 300 + H_P$, the hue composition is H_P red, $100 - H_P$ Blue

Step 11 Calculate e_s

$$e_s = e_1 + (e_2 - e_1)(h_s - h_1)/(h_2 - h_1)$$

where e_1 and h_1 are the values of e_s and h_s, respectively, for the unique hue having the nearest lower value of h_s; and e_2 and h_2 are these values for the unique hue having the nearest higher value of h_s.

Step 12 Calculate F_{tu}

$$F_{tu} = L/(L+0.1)$$

Step 13 Calculate the yellowness-blueness, M_{YB}, the redness-greenness, M_{RG}, the colourfulness, M, the relative yellowness-blueness, m_{YB}, the relative redness-greenness, m_{RG}, and the saturation, s

$$M_{YB} = 100[\tfrac{1}{2}(C_2 - C_3)/4.5][e_s(10/13)N_cF_{tu}]$$
$$M_{RG} = 100[C_1 - (C_2/11)][e_s(10/13)N_c]$$
$$M = (M_{YB}^2 + M_{RG}^2)^{1/2}$$
$$m_{YB} = M_{YB}/(\rho_a + \gamma_a + \beta_a)$$
$$m_{RG} = M_{RG}/(\rho_a + \gamma_a + \beta_a)$$
$$s = 50M/(\rho_a + \gamma_a + \beta_a)$$

Step 14 Calculate F_{LS}

$$F_{LS} = 3800j^2 5L_{AS}/2.26 + 0.2(1 - j^2)^4 (5L_{AS}/2.26)^{1/6}$$

where

$$j = 0.00001/(5L_{AS}/2.26 + 0.00001)$$

Step 15 Calculate A_S

$$A_S = B_{Su}(3.05)\{f_n[F_{LS}(L_{AS}/L_{CS})^c (L_S/2.26)^{1/2}]\} + 0.3$$

where

$$B_{Su} = 0.5/\{1 + 0.3[(5L_{AS}/2.26)(L_S/2.26)^{1/2}]^{0.3}\} + 0.5/\{1 + 5[5L_{AS}/2.26]\}$$

and

$$f_n[I] = 40[I^{0.73}/(I^{0.73} + 2)]$$

A typical value for c is 0.2.

Step 16 Calculate A

$$A = A_a - 1 + A_S - 0.3 + (1^2 + 0.3^2)^{1/2}$$

Step 17 Calculate $A + (M/100)$

$$A + (M/100)$$

Step 18 Calculate the brightness, Q

$$Q = \{[1.1][A + (M/100)]\}^{0.9}$$

Step 19 Tabulate the values of H, H_C, M, s, and Q.

12.28.3 A sample colour taken as a worked example

To illustrate the way in which the model operates, the following worked example is given:

	x	y	L
Sample	0.3580	0.3900	L
Adapting field	1/3	1/3	$L^{2/3}/200$
Conditioning field	1/3	1/3	$L^{2/3}/200$

The scotopic luminances divided by 2.26, are assumed to be equal to the photopic luminances: $L_S/2.26 = L$.

The chromatic induction factor, $N_c = 0.5$.

The appearance of the sample is predicted for sample luminances from 100 000 to 0.000 001 cd/m^2, and the results given in Table 12.2.

Table 12.2 — Predictions for the unrelated sample at various luminances, and for a related sample of the same chromaticity

THE UNRELATED SAMPLE

L	H	H_C	M	s	Q
100000	142	58Y 42G	5.8	2	84.2
10000	143	57Y 43G	14.7	7	80.4
1000	143	57Y 43G	23.7	16	67.3
100	145	55Y 45G	18.3	25	44.8
10	152	48Y 52G	4.5	22	22.1
1	162	38Y 62G	0.6	7	10.3
0.1	164	36Y 64G	0.1	1	4.8
0.01	—	—	0.0	0	2.6
0.001	—	—	0.0	0	1.7
0.0001	—	—	0.0	0	1.3
0.00001	—	—	0.0	0	1.2
0.000001	—	—	0.0	0	1.1

THE RELATED SAMPLE

L	H	H_C	M	s	Q
140	143	57Y 43G	26.2	39	33.8

Data for the related sample:
Illuminant: equi-energy stimulus, S_E: $x = 0.3333$, $y = 0.3333$
Luminance of reference white: $L_W = 200$
Brightness surround induction factor: $N_b = 75$
Chromatic surround induction factor: $N_c = 1$
Luminance factor of background: $Y_b = 100$
Luminance factor of sample: $Y = 70$ (hence $L = 140$)

Also included in Table 12.2 are the results for a related colour having the same chromaticity, a luminance factor of 70%, and a luminance of 140 cd/m^2.

Predictions

The following trends can be seen from the results in Table 12.2. The colourfulness, M, increases from zero at scotopic levels to a maximum at 1000 cd/m^2, and then falls off again because of the maximum response of the visual system being approached. The saturation, s, varies rather less in the main part of the photopic range. The brightness, Q, increases from very low values at scotopic levels to about 80 at 10000 cd/m^2, after which, although further increases do occur, the rate of increase slows down because of approaching the maximum response. The hue composition, H_C, shows a shift from about 57Y43G at high levels, to about 37Y63G at low photopic levels, because of the Bezold–Brucke effect.

The related colour at a luminance of 140 cd/m^2 has a brightness, Q, of 33.8, and a colourfulness, M, of 26.2; it is clear from the upper part of the table that the unrelated colour at a similar brightness would have a colourfulness of only about 12. This decrease in colourfulness occurs as a result of the unrelated colour being viewed in a dark background and surround.

12.29 REVERSING THE MODEL

In some applications it is useful to be able to operate the model in reverse. For instance, if it is known that a certain colour appearance is required in a defined set of viewing conditions, it may be necessary to compute the corresponding tristimulus values.

Starting with the hue quadrature, H, the colourfulness, M, and the brightness, Q, the procedure is as follows:

(1) From H, calculate h_s, which is a function of C_1, C_2, and C_3.
(2) Express M in terms of C_1, C_2, and C_3.
(3) Remember that $C_1 + C_2 + C_3 = 0$.
(4) From (1), (2), and (3) obtain the values of C_1, C_2, and C_3.
(5) From Q and M, calculate the value of A.
(6) Using A_S and A obtain the value of A_a.
(7) From the values of C_1, C_2, C_3, and A_a, calculate the values of ρ_a, γ_a, and β_a, remembering that $A_a = 2\rho_a + \gamma_a + (1/20)\beta_a - 2.05$.
(8) From ρ_a, γ_a, *and* β_a calculate the values of ρ, γ, and β.
(9) From ρ, γ, and β calculate the values of X, Y, and Z.

For related colours, the value of A_S depends on S/S_W; this not usually known, and Y/Y_W is usually used in the model instead as an approximation. But, when using the model in reverse, if Y/Y_W is not known, some extra steps are necessary. There are three alternative methods of dealing with this situation. The first method is to set S/S_W equal to zero for all colours. This results in a model that deals with photopic vision only; this is useful for levels of illumination for which $5L_A$ is above about 10 cd/m^2. Using S/S_W equal to zero instead of equal to Y/Y_W results in M and s being unchanged, C being changed very slightly, and J being changed slightly; for many applications these changes are negligible, and if Q, which is changed significantly, is not required, then this form of the model can be used for both forward and reverse calculations. The second method is to use J to calculate $(S/S_W)_J$, an approximate value for S/S_W. The following formula for $(S/S_W)_J$ can be used for photopic conditions for colours of luminance factor not less than 3%.

$$(S/S_W)_J = \{J - [\log(5L_A)]^2/4\}^{1.8}/\{100 - [\log(5L_A)]^2/4\}^{1.8},$$

$(S/S_W)_J$ can then be used in the reverse model to calculate X, Y, and Z. Using $(S/S_W)_J$ instead of Y/Y_W in the forward model results in M and s being unchanged, C being changed very slightly, and Q and J being changed slightly. For many applications these changes are negligible. If more precise results are required, the resulting value of Y/Y_W can be used as S/S_W to derive a new set of values of X, Y, and Z; this procedure can then be iterated until stable values of Y/Y_W are obtained. The third method is to require that S/S_W be equal to Y/Y_W from the outset, and to use methods of successive numerical approximation to complete the calculation.

For related colours, the starting data may include J instead of Q, but Q can be calculated from J and Q_W, the latter being obtained by using the model in the forward direction with the reference white stimulus data as starting point.

If the starting data include saturation, s, or chroma, C, instead of M, then it is required to obtain $\rho_a + \gamma_a + \beta_a$ in order to derive M. One way of doing this is to use $A + (M/100) - 2.05 = d$ as a first approximation to $\rho_a + \gamma_a + \beta_a$, and thus obtain approximate values of X, Y, and Z. The model can then be used forwards to obtain Q', a new value of Q, from which a new value d', of d is obtained where $d' = d(Q/Q')^{1/0.6}$; the model is then used in reverse to obtain new X, Y, and Z; this procedure can then be iterated until stable values of $\rho_a + \gamma_a + \beta_a$ are obtained. Because of the complexity of this procedure, the values of M obtained when deriving s and C should not be discarded.

REFERENCES

Arend, L. & Reeves, A. *J. Opt. Soc. Amer. A.* **3**, 1743 (1986).

Bartleson, C. J. *Die Farbe* **28**, 132 (1980).

Baylor, D. A. *Investig. Ophthalmol.* **28**, 34 (1987).

Boynton, R. M. & Whitten, D. N. *Science* **170**, 1423 (1970).

CIE, *CIE Journal* **5**, 16 (1986).

Estevez, O. *On the fundamental data-base of normal and dichromatic colour vision.* PhD thesis, University of Amsterdam (1979).

Hunt, R. W. G. *Color Res. Appl.* **7**, 95 (1982).

Hunt, R. W. G. *Color Res. Appl.* **10**, 12 (1985).

Hunt, R. W. G. *Color Res. Appl.* **12**, 297 (1987).

Hunt, R. W. G. *Color Res. Appl.* **14**, 238 (1989).

Hunt, R. W. G. *Color Res. Appl.* **16**, 146 (1991)

Hurvich, L. M. & Jameson, D. *J. Opt. Soc. Amer.* **41**, 787 (1951).

Jameson, D. & Hurvich, L. M. *Vision Res.* **4**, 135 (1964).

Luo, M. R., Clarke, A. A., Rhodes, P. A., Schappo, A., Scrivener, S. A. R. & Tait, C.J. *Color Res. Appl.* **16**, 166, 181 (1991)

MacDonald, L. W., Luo, M. R. & Scrivener, S. A. R. *J. Phot. Sci.* **38**, 177 (1990).

McCann, J. J. & Houston, K. L. In *Colour Vision*, eds. J. D. Mollon & L. T. Sharpe, pp. 535–544, Academic Press, London (1983).

Nayatani, Y., Hashimoto, K., Takahama, K. & Sobagaki, H. *Color Res. Appl.* **12**, 121 and 231 (1987).

Nayatani, Y., Takahama, K. & Sobagaki, H. *Color Res. Appl.* **11**, 62 (1986).

Nayatani, Y., Takahama, K., Sobagaki, H. & Hashimoto, K. *Color Res. Appl.* **15**, 210 (1990).

Pointer, M. R. *J. Phot. Sci.* **34**, 81 (1986).

Seim, T. & Valberg, A. *Color Res. Appl.* **11**, 11 (1986).

Stevens, S. S. *Science* **133**, 80 (1961).

Valeton, J. M. & Van Norren, D. *Vision Res.* **23**, 1539 (1983)

Walraven, P. L. & Bouman, M. A. *Vision Res.* **6**, 567 (1966).

Appendix 1
Radiometric and photometric terms and units

A1.1 Introduction

Photometry is normally based on the *photopic spectral luminous efficiency* function $V(\lambda)$, tabulated in Appendix 2, which was standardized by the CIE in 1924. The $V(\lambda)$ function is used as a weighting function for evaluating the total amount of light in a mixture of radiations of different wavelengths. If the radiant flux at wavelength λ_1 is P_1, that at wavelength λ_2 is P_2, etc., and the values of the $V(\lambda)$ function at these wavelengths are V_1, V_2, etc., then the luminous flux is given by:

$$L = K_m(P_1V_1 + P_2V_2 + \ldots\ldots\ldots\ldots) \ .$$

The values, P, represent the radiant fluxes (powers) per small constant-width wavelength intervals centred on the wavelengths, λ, the intervals together covering all of the visible spectrum. The value of the constant, K_m, will be discussed later. (Although the values of the $V(\lambda)$ function at wavelengths below about 450 nm are now known to be too low, the resulting errors are negligible in most practical cases. More accurate values, are given in Appendix 2.)

When the level of illumination is so low that vision is mediated by the rods instead of by the cones, photometry is based on the *scotopic spectral luminance efficiency* function, $V'(\lambda)$, also tabulated in Appendix 2, which was standardized by the CIE in 1951. Unless otherwise stated, it is always assumed that the photopic function, $V(\lambda)$, is being used.

A1.2 Physical detectors

Practical photometry is now almost always carried out using some kind of photoelectric physical detector, because this provides more precise and quicker measurements than are possible with visual photometry. But these physical detectors do not have spectral sensitivities that correspond to the $V(\lambda)$ or $V'(\lambda)$ functions of the eye. They therefore have to be used with filters that modify their spectral sensitivities to match these functions as closely as possible. Such filtered detectors then measure *luminous* flux (or *scotopic luminous* flux in the case of the $V'(\lambda)$ function).

Detectors used for radiometry should be equally sensitive to radiant flux of all parts of the spectrum; this is usually only the case for thermopiles and bolometers, and other detectors require correcting filters to obtain equal sensitivity throughout the spectrum as closely as possible. Such filtered detectors then measure *radiant* flux.

Photochemical and photobiological processes often depend on the number of quanta, irrespective of their energies, above a threshold. For these applications, measurement of the rate of flow of quanta is required, and physical detectors then need yet another type of correcting filter. Such filtered detectors then measure *quantum* flux.

For physical detectors to produce measurements that are true functions of these fluxes, it is necessary for them to have responses that are linearly proportional to the radiation incident upon them, and for the effects of radiation of different parts of the spectrum to add together linearly.

An alternative to using filtered detectors, is to measure the radiant power at each small constant-width wavelength interval throughout the spectrum, and then to use the appropriate weighting function to convert the spectro-radiometric results obtained to luminous or quantum measures as required.

A1.3 Photometric units and terms

Lumen, *lm*.
Unit of luminous flux (radiant flux weighted by the $V(\lambda)$ function). The luminous flux of a beam of monochromatic radiation whose frequency is 540×10^{12} hertz, and whose radiant flux is 1/683 watt. The frequency of 540×10^{12} hertz is closely equal to a wavelength of 555 nm, the wavelength for which the $V(\lambda)$ function has its maximum value of 1.0.

> The constant, K_m, in the equation given in section A1.1 for evaluating the amount of light, L, is put equal to 683 to obtain the luminous flux in lumens when the radiant flux is expressed in watts. (If the scotopic function $V'(\lambda)$ is being used, K'_m is used instead of K_m, where K'_m is equal to 1700.)

Candela, cd.
Unit of luminous intensity (luminous flux per unit solid angle). The luminous intensity, in a given direction, of a point source emitting 1 lumen per steradian.

Lux, lx.
Unit of illuminance (luminous flux per unit area incident on a surface). The illuminance produced by a luminous flux of 1 lumen uniformly distributed over a surface of area 1 square metre.

Foot-candle or *Lumen per square foot*.
Unit of illuminance. The illuminance produced by a luminous flux of 1 lumen uniformly distributed over a surface of area of 1 square foot. 1 lumen per square foot = 10.76 lux.

Candela per square metre, cd.m^{-2}.
Unit of luminance (luminous flux per unit solid angle and per unit projected area, in a given direction, at a point on a surface). The luminance produced by a luminous intensity of 1 candela uniformly distributed over a surface area of 1 square metre. (Obsolete name: *nit*).

Other units of luminance are the lambert, the millilambert, the foot-lambert (sometimes called the equivalent foot-candle), and the apostilb.

1 lambert = $10^4/\pi$ cd.m^{-2}.
1 millilambert = $10/\pi$ cd.m^{-2}.
1 foot-lambert = $10.76/\pi$ = 3.426 cd.m^{-2}.
1 apostilb = $1/\pi$ = 0.3183 cd.m^{-2} .

Luminance factor, β.
Ratio of the luminance of an area to that of the perfect diffuser (an ideal isotropic diffuser with a reflectance equal to unity) identically illuminated.

Calculation of luminance from illuminance.
An isotropic diffuser of luminance factor, β, under an illuminance, E lux, has a luminance, L, given by:

$$L = E\beta/\pi \text{ cd.m}^{-2}.$$

Lumen per square metre, lm.m^{-2}.
Unit of luminous exitance (luminous flux per unit area leaving a surface).

Radiant efficiency.
Ratio of the radiant flux emitted to the power consumed by a source.

Luminous efficacy
(a) Of a source. Ratio of luminous flux emitted to power consumed. Unit: lm.W^{-1}.
(b) Of a radiation. Ratio of luminous flux to radiant flux. Unit lm.W^{-1}.

The maximum spectral luminous efficacy for photopic vision occurs at a wavelength of 555 nm and is equal to 683 lm.W^{-1}; and for scotopic vision occurs at a wavelength of 510 nm and is equal to 1700 lm.W^{-1}.

Luminous efficiency.
Ratio of radiant flux weighted according to the $V(\lambda)$ function, to the radiant flux.

Point brilliance.
The illuminance produced by a source on a plane at the observer's eye, when the apparent diameter of the source is inappreciable. Unit: lx.

Troland

Unit used to express a quantity proportional to retinal illuminance produced by a light stimulus. When the eye views a surface of uniform luminance, the number of trolands is equal to the product of the area in square millimetres of the limiting pupil, natural or artificial, times the luminance of the surface in candelas per square metre.

A1.4 Radiant and quantum units and terms

Most of the above terms and units have radiant and quantum equivalents, in which radiant flux and quantum flux, respectively, are used instead of luminous flux. All three types of term and unit are listed in Table A1.1.

Table A1.1 — Luminous, radiant, and quantum terms and units

Luminous	*Radiant*	*Quantum*
luminous flux, F lm	radiant flux, F_e W	quantum flux, F_q s^{-1}
luminous intensity, I cd ($lm.sr^{-1}$)	radiant intensity, I_e $W.sr^{-1}$	quantum intensity, I_q $s^{-1}.sr^{-1}$
luminance, L $cd.m^{-2}$	radiance, L_e $W.sr^{-1}.m^{-2}$	quantum radiance, L_q $s^{-1}.sr^{-1}.m^{-2}$
illuminance, E 1x ($lm.m^{-2}$)	irradiance, E_e $W.m^{-2}$	quantum irradiance, E_q $s^{-1}.m^{-2}$
light exposure, H lx.s	radiant exposure, H_e $J.m^{-2}$	quantum exposure, H_q m^{-2}
luminous exitance, M $1m.m^{-2}$	radiant exitance, M_e $W.m^{-2}$	quantum exitance, M_q $s^{-1}.m^{-2}$

Abbreviations:

lm: lumen
cd: candela
lx: lux
W: watt
J: joule (watt second)
sr: steradian (unit solid angle)
s: second
m: metre

A1.5 Radiation sources

Incandescent sources.

Incandescent sources emit radiation in accordance with their temperature and emissive properties.

Planckian (black-body) sources.

Planckian sources emit radiation in accordance with Planck's law: this gives the spectral concentration of radiant exitance, M_e, in watts per square metre per wavelength interval, as a function of wavelength, λ, in metres, and temperature, T, in kelvins, by the formula:

$$M_e = c_1\lambda^{-5}(e^{c_2/\lambda T} - 1)^{-1}$$

in $W.m^{-3}$, where $c_1 = 3.74183 \times 10^{-16}$ $W.m^2$, $c_2 = 1.4388 \times 10^{-2}$ m.K, and $e =$ 2.718282.

A1.6 Terms for measures of reflection and transmission

The terms *reflection* and *transmission* refer to the processes of radiation being returned by a medium, or passed on through a medium, respectively. The amount of radiation reflected or transmitted by the medium relative to that by the perfect diffuser is expressed by means of different terms according to the geometrical arrangement for collecting the radiation. *Reflectance factor* and *transmittance factor* are used when the radiations reflected or transmitted by the sample and by the perfect diffuser lie within a defined cone. If this cone is a hemisphere, the terms *reflectance* and *transmittance* are used. If the cone is very small, the term *luminance factor* or *radiance factor* is used. *Reflectometer value* is used for measurement by means of a particular instrument for measuring reflected radiation. The following adjectives are used to denote the type of radiation involved:

Spectral, when the radiation is monochromatic. In this case the symbol is followed by (λ), thus $\rho(\lambda)$, or $\tau(\lambda)$, for example.

Radiant, when the radiation is evaluated in terms of its total power or energy. In this case the symbol has a subscript e, thus ρ_e, or τ_e, for example.

Luminous, when the radiation is evaluated by using the $V(\lambda)$ function as a weighting function.

The terms themselves are defined as follows:

Reflectance, ρ [*transmittance*, τ]
Ratio of the reflected [transmitted] radiant or luminous flux to the incident flux under specified conditions of irradiation.

Note 1. Regular reflectance, ρ_r [*regular transmittance,* τ_r] is the ratio of the part of the regularly reflected [transmitted] flux to the incident flux.
Note 2. Diffuse reflectance, ρ_d [*diffuse transmittance,* τ_d] is the ratio of the part of the flux reflected [transmitted] by diffusion to the incident flux
Note 3. $\rho = \rho_r + \rho_d$; $\tau = \tau_r + \tau_d$
Note 4. In the case of transmitting samples, if the irradiation and the collection are both diffuse, the ratio is referred to as *doubly diffuse transmittance,* τ_{dd}.

The perfect reflecting [transmitting] diffuser
An ideal isotropic diffuser with a reflectance [transmittance] equal to unity.
In this definition by *isotropic* is meant that the radiation is reflected [transmitted] equally strongly in all directions, so that the perfect diffuser has a luminance that

is independent of the direction of viewing; by a reflectance [transmittance] *equal to unity*, is meant that the perfect diffuser reflects [transmits] all the light that is incident on it at every wavelength throughout the visible spectrum.

Reflectance factor, R [*transmittance factor*, T]

(at a representative surface element, for the part of the radiation contained in a given cone with apex at the representative surface element, and for incident radiation of given spectral composition and geometrical distribution)

Ratio of the radiant or luminous flux reflected [transmitted] in the directions limited by the cone to that reflected [transmitted] in the same directions by the perfect diffuser identically irradiated.

Note 1. For regularly reflecting surfaces that are irradiated by a source of small solid angle, the reflectance factor may be much larger than 1 if the cone contains the mirror image of the source.
Note 2. If the solid angle of the cone approaches a hemisphere, the reflectance [transmittance] factor approaches the reflectance [transmittance]. In this case, a regular component (if present) must be included; instruments employing integrating spheres may approximate this condition. A regular component can be excluded if it is sufficiently defined and a suitable trap is used; then the reflectance [transmittance] factor approaches the diffuse reflectance [transmittance]; instruments employing integrating spheres with gloss traps may approximate this condition.
Note 3. If the solid angle of the cone approaches zero, the reflectance [transmittance] factor approaches the radiance or luminance factor.

Radiance factor, β [*luminance factor*, β]

(at a representative surface element of a non-self-radiating medium, in a given direction, under specified conditions of irradiation)

Ratio of the radiance [luminance] of the medium to that of the perfect diffuser identically irradiated.

Note. In the case of photoluminescent media, the radiance [luminance] factor is the sum, β_T, of two portions, the *reflected radiance* [*luminance*] *factor*, β_S, and the *luminescent radiance* [*luminance*] *factor*, β_L; $\beta_T = \beta_S + \beta_L$.

Density, D

Logarithm to the base 10 of the reciprocal of the reflectance factor or the transmittance factor, or other similar measure.

Appendix 2
Spectral luminous efficiency functions

Wave-length, nm,	$V(\lambda)$	$V_M(\lambda)$	$V'(\lambda)$	log $V_{b,2}(\lambda)$	log $V_{b,10}(\lambda)$
380	0.000 039	0.000 200	0.000 589		
390	0.000 120	0.000 800	0.002 209		
400	0.000 396	0.002 800	0.009 29	−2.06	−2.06
410	0.001 21	0.007 40	0.034 84	−1.64	−1.57
420	0.004 00	0.017 50	0.096 6	−1.35	−1.14
430	0.011 6	0.027 30	0.199 8	−1.19	−0.94
440	0.023 0	0.037 90	0.328 1	−1.07	−0.77
450	0.038 0	0.046 80	0.455	−0.99	−0.67
460	0.060 0	0.060 00	0.567	−0.89	−0.56
470	0.091 0	0.090 98	0.676	−0.75	−0.44
480	0.139	0.139 02	0.793	−0.63	−0.33
490	0.208	0.208 02	0.904	−0.53	−0.25
500	0.323	0.323 00	0.982	−0.36	−0.14
510	0.503	0.503 00	0.997	−0.15	−0.03
520	0.710	0.710 00	0.935	−0.01	0.06
530	0.862	0.862 00	0.811	0.08	0.08
540	0.954	0.954 00	0.650	0.10	0.10
550	0.995	0.994 95	0.481	0.11	0.11
560	0.995	0.995 00	0.328 8	0.06	0.06
570	0.952	0.952 00	0.207 6	0.00	0.00
580	0.870	0.870 00	0.121 2	−0.03	−0.03
590	0.757	0.757 00	0.065 5	−0.05	−0.05
600	0.631	0.631 00	0.033 15	−0.09	−0.09
610	0.503	0.503 00	0.015 93	−0.14	−0.14
620	0.381	0.381 00	0.007 37	−0.21	−0.21
630	0.265	0.265 00	0.003 335	−0.32	−0.32
640	0.175	0.175 00	0.001 497	−0.48	−0.48
650	0.107	0.107 00	0.000 677	−0.68	−0.68
660	0.061	0.061 00	0.000 312 9	−0.89	−0.89
670	0.032	0.032 00	0.000 148 0	−1.15	−1.15
680	0.017	0.017 00	0.000 071 5	−1.44	−1.44
690	0.008 2	0.008 21	0.000 035 33	−1.74	−1.74
700	0.004 1	0.004 102	0.000 017 80	−2.03	−2.03
710	0.002 1	0.002 091	0.000 009 14	−2.35	−2.35
720	0.001 05	0.001 047	0.000 004 78	−2.65	−2.65
730	0.000 52	0.000 520	0.000 002 546	−2.97	−2.97
740	0.000 25	0.000 249 2	0.000 001 379		
750	0.000 12	0.000 120 0	0.000 000 760		
760	0.000 06	0.000 060 0	0.000 000 425		
770	0.000 03	0.000 030 0	0.000 000 241		
780	0.000 015	0.000 014 99	0.000 000 139		

Appendix 3

CIE colour-matching functions

Wave-length, nm	$\bar{x}(\lambda)$	$\bar{y}(\lambda)$	$\bar{z}(\lambda)$	$\bar{x}_{10}(\lambda)$	$\bar{y}_{10}(\lambda)$	$\bar{z}_{10}(\lambda)$
380	0.0014	0.0000	0.0065	0.0002	0.0000	0.0007
385	0.0022	0.0001	0.0105	0.0007	0.0001	0.0029
390	0.0042	0.0001	0.0201	0.0024	0.0003	0.0105
395	0.0076	0.0002	0.0362	0.0072	0.0008	0.0323
400	0.0143	0.0004	0.0679	0.0191	0.0020	0.0860
405	0.0232	0.0006	0.1102	0.0434	0.0045	0.1971
410	0.0435	0.0012	0.2074	0.0847	0.0088	0.3894
415	0.0776	0.0022	0.3713	0.1406	0.0145	0.6568
420	0.1344	0.0040	0.6456	0.2045	0.0214	0.9725
425	0.2148	0.0073	1.0391	0.2647	0.0295	1.2825
430	0.2839	0.0116	1.3856	0.3147	0.0387	1.5535
435	0.3285	0.0168	1.6230	0.3577	0.0496	1.7985
440	0.3483	0.0230	1.7471	0.3837	0.0621	1.9673
445	0.3481	0.0298	1.7826	0.3867	0.0747	2.0273
450	0.3362	0.0380	1.7721	0.3707	0.0895	1.9948
455	0.3187	0.0480	1.7441	0.3430	0.1063	1.9007
460	0.2908	0.0600	1.6692	0.3023	0.1282	1.7454
465	0.2511	0.0739	1.5281	0.2541	0.1528	1.5549
470	0.1954	0.0910	1.2876	0.1956	0.1852	1.3176
475	0.1421	0.1126	1.0419	0.1323	0.2199	1.0302
480	0.0956	0.1390	0.8130	0.0805	0.2536	0.7721
485	0.0580	0.1693	0.6162	0.0411	0.2977	0.5701
490	0.0320	0.2080	0.4652	0.0162	0.3391	0.4153
495	0.0147	0.2586	0.3533	0.0051	0.3954	0.3024
500	0.0049	0.3230	0.2720	0.0038	0.4608	0.2185
505	0.0024	0.4073	0.2123	0.0154	0.5314	0.1592
510	0.0093	0.5030	0.1582	0.0375	0.6067	0.1120
515	0.0291	0.6082	0.1117	0.0714	0.6857	0.0822
520	0.0633	0.7100	0.0782	0.1177	0.7618	0.0607
525	0.1096	0.7932	0.0573	0.1730	0.8233	0.0431
530	0.1655	0.8620	0.0422	0.2365	0.8752	0.0305
535	0.2257	0.9149	0.0298	0.3042	0.9238	0.0206
540	0.2904	0.9540	0.0203	0.3768	0.9620	0.0137
545	0.3597	0.9803	0.0134	0.4516	0.9822	0.0079
550	0.4334	0.9950	0.0087	0.5298	0.9918	0.0040
555	0.5121	1.0000	0.0057	0.6161	0.9991	0.0011
560	0.5945	0.9950	0.0039	0.7052	0.9973	0.0000
565	0.6784	0.9786	0.0027	0.7938	0.9824	0.0000
570	0.7621	0.9520	0.0021	0.8787	0.9556	0.0000

Wave-length, nm	$\bar{x}(\lambda)$	$\bar{y}(\lambda)$	$\bar{z}(\lambda)$	$\bar{x}_{10}(\lambda)$	$\bar{y}_{10}(\lambda)$	$\bar{z}_{10}(\lambda)$
575	0.8425	0.9154	0.0018	0.9512	0.9152	0.0000
580	0.9163	0.8700	0.0017	1.0142	0.8689	0.0000
585	0.9786	0.8163	0.0014	1.0743	0.8256	0.0000
590	1.0263	0.7570	0.0011	1.1185	0.7774	0.0000
595	1.0567	0.6949	0.0010	1.1343	0.7204	0.0000
600	1.0622	0.6310	0.0008	1.1240	0.6583	0.0000
605	1.0456	0.5668	0.0006	1.0891	0.5939	0.0000
610	1.0026	0.5030	0.0003	1.0305	0.5280	0.0000
615	0.9384	0.4412	0.0002	0.9507	0.4618	0.0000
620	0.8544	0.3810	0.0002	0.8563	0.3981	0.0000
625	0.7514	0.3210	0.0001	0.7549	0.3396	0.0000
630	0.6424	0.2650	0.0000	0.6475	0.2835	0.0000
635	0.5419	0.2170	0.0000	0.5351	0.2283	0.0000
640	0.4479	0.1750	0.0000	0.4316	0.1798	0.0000
645	0.3608	0.1382	0.0000	0.3437	0.1402	0.0000
650	0.2835	0.1070	0.0000	0.2683	0.1076	0.0000
655	0.2187	0.0816	0.0000	0.2043	0.0812	0.0000
660	0.1649	0.0610	0.0000	0.1526	0.0603	0.0000
665	0.1212	0.0446	0.0000	0.1122	0.0441	0.0000
670	0.0874	0.0320	0.0000	0.0813	0.0318	0.0000
675	0.0636	0.0232	0.0000	0.0579	0.0226	0.0000
680	0.0468	0.0170	0.0000	0.0409	0.0159	0.0000
685	0.0329	0.0119	0.0000	0.0286	0.0111	0.0000
690	0.0227	0.0082	0.0000	0.0199	0.0077	0.0000
695	0.0158	0.0057	0.0000	0.0138	0.0054	0.0000
700	0.0114	0.0041	0.0000	0.0096	0.0037	0.0000
705	0.0081	0.0029	0.0000	0.0066	0.0026	0.0000
710	0.0058	0.0021	0.0000	0.0046	0.0018	0.0000
715	0.0041	0.0015	0.0000	0.0031	0.0012	0.0000
720	0.0029	0.0010	0.0000	0.0022	0.0008	0.0000
725	0.0020	0.0007	0.0000	0.0015	0.0006	0.0000
730	0.0014	0.0005	0.0000	0.0010	0.0004	0.0000
735	0.0010	0.0004	0.0000	0.0007	0.0003	0.0000
740	0.0007	0.0002	0.0000	0.0005	0.0002	0.0000
745	0.0005	0.0002	0.0000	0.0004	0.0001	0.0000
750	0.0003	0.0001	0.0000	0.0003	0.0001	0.0000
755	0.0002	0.0001	0.0000	0.0002	0.0001	0.0000
760	0.0002	0.0001	0.0000	0.0001	0.0000	0.0000
765	0.0001	0.0000	0.0000	0.0001	0.0000	0.0000
770	0.0001	0.0000	0.0000	0.0001	0.0000	0.0000
775	0.0001	0.0000	0.0000	0.0000	0.0000	0.0000
780	0.0000	0.0000	0.0000	0.0000	0.0000	0.0000

Appendix 4

CIE spectral chromaticity co-ordinates

Wave-length nm	$x(\lambda)$	$y(\lambda)$	$u'(\lambda)$	$v'(\lambda)$	$x_{10}(\lambda)$	$y_{10}(\lambda)$	$u'_{10}(\lambda)$	$v'_{10}(\lambda)$
380	0.1741	0.0050	0.2569	0.0165	0.1813	0.0197	0.2524	0.0617
385	0.1740	0.0050	0.2567	0.0165	0.1809	0.0195	0.2519	0.0612
390	0.1738	0.0049	0.2564	0.0163	0.1803	0.0194	0.2512	0.0606
395	0.1736	0.0049	0.2560	0.0163	0.1795	0.0190	0.2502	0.0597
400	0.1733	0.0048	0.2558	0.0159	0.1784	0.0187	0.2488	0.0587
405	0.1730	0.0048	0.2553	0.0159	0.1771	0.0184	0.2472	0.0578
410	0.1726	0.0048	0.2545	0.0159	0.1755	0.0181	0.2449	0.0569
415	0.1721	0.0048	0.2537	0.0160	0.1732	0.0178	0.2417	0.0599
420	0.1714	0.0051	0.2522	0.0169	0.1706	0.0179	0.2376	0.0559
425	0.1703	0.0058	0.2496	0.0191	0.1679	0.0187	0.2325	0.0583
430	0.1689	0.0069	0.2461	0.0226	0.1650	0.0203	0.2266	0.0627
435	0.1669	0.0086	0.2411	0.0278	0.1622	0.0225	0.2202	0.0687
440	0.1644	0.0109	0.2347	0.0349	0.1590	0.0257	0.2127	0.0774
445	0.1611	0.0138	0.2267	0.0437	0.1554	0.0300	0.2038	0.0886
450	0.1566	0.0177	0.2161	0.0550	0.1510	0.0364	0.1926	0.1046
455	0.1510	0.0227	0.2033	0.0689	0.1459	0.0452	0.1796	0.1252
460	0.1440	0.0297	0.1877	0.0871	0.1389	0.0589	0.1620	0.1546
465	0.1355	0.0399	0.1690	0.1119	0.1295	0.0779	0.1410	0.1907
470	0.1241	0.0578	0.1441	0.1510	0.1152	0.1090	0.1130	0.2406
475	0.1096	0.0868	0.1147	0.2044	0.0957	0.1591	0.0812	0.3035
480	0.0913	0.1327	0.0828	0.2708	0.0728	0.2292	0.0519	0.3618
485	0.0687	0.2007	0.0521	0.3427	0.0452	0.3275	0.0264	0.4310
490	0.0454	0.2950	0.0282	0.4117	0.0210	0.4401	0.0102	0.4807
495	0.0235	0.4127	0.0119	0.4698	0.0073	0.5625	0.0030	0.5200
500	0.0082	0.5384	0.0035	0.5131	0.0056	0.6745	0.0020	0.5477
505	0.0039	0.6548	0.0014	0.5432	0.0219	0.7526	0.0073	0.5650
510	0.0139	0.7502	0.0046	0.5638	0.0495	0.8023	0.0158	0.5763
515	0.0389	0.8120	0.0123	0.5770	0.0850	0.8170	0.0269	0.5820
520	0.0743	0.8338	0.0231	0.5837	0.1252	0.8102	0.0402	0.5847
525	0.1142	0.8262	0.0360	0.5861	0.1664	0.7922	0.0547	0.5857
530	0.1547	0.8059	0.0501	0.5868	0.2071	0.7663	0.0703	0.5854
535	0.1929	0.7816	0.0643	0.5865	0.2436	0.7399	0.0856	0.5846
540	0.2296	0.7543	0.0792	0.5856	0.2786	0.7113	0.1015	0.5831
545	0.2658	0.7243	0.0953	0.5841	0.3132	0.6813	0.1188	0.5812

Wave-length nm	$x(\lambda)$	$y(\lambda)$	$u'(\lambda)$	$v'(\lambda)$	$x_{10}(\lambda)$	$y_{10}(\lambda)$	$u'_{10}(\lambda)$	$v'_{10}(\lambda)$
550	0.3016	0.6923	0.1127	0.5821	0.3473	0.6501	0.1375	0.5789
555	0.3373	0.6589	0.1319	0.5796	0.3812	0.6182	0.1579	0.5762
560	0.3731	0.6245	0.1531	0.5766	0.4142	0.5858	0.1801	0.5730
565	0.4087	0.5896	0.1766	0.5732	0.4469	0.5531	0.2045	0.5693
570	0.4441	0.5547	0.2026	0.5694	0.4790	0.5210	0.2310	0.5653
575	0.4788	0.5202	0.2312	0.5651	0.5096	0.4904	0.2592	0.5611
580	0.5125	0.4866	0.2623	0.5604	0.5386	0.4614	0.2888	0.5567
585	0.5448	0.4544	0.2959	0.5554	0.5654	0.4346	0.3193	0.5521
590	0.5752	0.4242	0.3315	0.5501	0.5900	0.4100	0.3501	0.5457
595	0.6029	0.3965	0.3681	0.5446	0.6116	0.3884	0.3800	0.5430
600	0.6270	0.3725	0.4035	0.5393	0.6306	0.3694	0.4088	0.5387
605	0.6482	0.3514	0.4380	0.5342	0.6471	0.3529	0.4358	0.5346
610	0.6658	0.3340	0.4691	0.5296	0.6612	0.3388	0.4605	0.5309
615	0.6801	0.3197	0.4967	0.5254	0.6731	0.3269	0.4827	0.5276
620	0.6915	0.3083	0.5202	0.5219	0.6827	0.3173	0.5017	0.5247
625	0.7006	0.2993	0.5399	0.5190	0.6898	0.3120	0.5163	0.5225
630	0.7079	0.2920	0.5565	0.5165	0.6955	0.3045	0.5286	0.5207
635	0.7140	0.2859	0.5709	0.5144	0.7010	0.2990	0.5407	0.5189
640	0.7190	0.2809	0.5830	0.5125	0.7059	0.2941	0.5517	0.5172
645	0.7230	0.2770	0.5930	0.5110	0.7103	0.2898	0.5619	0.5157
650	0.7260	0.2740	0.6005	0.5099	0.7137	0.2863	0.5700	0.5145
655	0.7283	0.2717	0.6064	0.5090	0.7156	0.2844	0.5746	0.5138
660	0.7300	0.2700	0.6108	0.5084	0.7168	0.2832	0.5775	0.5134
665	0.7311	0.2689	0.6138	0.5079	0.7179	0.2821	0.5802	0.5130
670	0.7320	0.2680	0.6161	0.5076	0.7187	0.2813	0.5822	0.5127
675	0.7327	0.2673	0.6181	0.5073	0.7193	0.2807	0.5837	0.5124
680	0.7334	0.2666	0.6200	0.5070	0.7198	0.2802	0.5848	0.5123
685	0.7340	0.2660	0.6216	0.5068	0.7200	0.2800	0.5854	0.5122
690	0.7344	0.2656	0.6226	0.5066	0.7202	0.2798	0.5858	0.5121
695	0.7346	0.2654	0.6231	0.5065	0.7203	0.2797	0.5861	0.5121
700	0.7347	0.2653	0.6234	0.5065	0.7204	0.2796	0.5863	0.5121
705	0.7347	0.2653	0.6234	0.5065	0.7203	0.2797	0.5862	0.5121
710	0.7347	0.2653	0.6234	0.5065	0.7202	0.2798	0.5859	0.5121
715	0.7347	0.2653	0.6234	0.5065	0.7201	0.2799	0.5856	0.5122
720	0.7347	0.2653	0.6234	0.5065	0.7199	0.2801	0.5851	0.5122
725	0.7347	0.2653	0.6234	0.5065	0.7197	0.2803	0.5846	0.5123
730	0.7347	0.2653	0.6234	0.5065	0.7195	0.2806	0.5840	0.5124
735	0.7347	0.2653	0.6234	0.5065	0.7192	0.2808	0.5834	0.5125
740	0.7347	0.2653	0.6234	0.5065	0.7189	0.2811	0.5827	0.5126
745	0.7347	0.2653	0.6234	0.5065	0.7186	0.2814	0.5819	0.5127
750	0.7347	0.2653	0.6234	0.5065	0.7183	0.2817	0.5811	0.5128
755	0.7347	0.2653	0.6234	0.5065	0.7180	0.2820	0.5803	0.5129
760	0.7347	0.2653	0.6234	0.5065	0.7176	0.2824	0.5795	0.5131
765	0.7347	0.2653	0.6234	0.5065	0.7172	0.2828	0.5786	0.5132
770	0.7347	0.2653	0.6234	0.5065	0.7169	0.2831	0.5777	0.5134
775	0.7347	0.2653	0.6234	0.5065	0.7165	0.2835	0.5767	0.5135
780	0.7347	0.2653	0.6234	0.5065	0.7161	0.2839	0.5757	0.5136

Appendix 5

Relative spectral power distributions of illuminants

A5.1 Introduction

In section A5.2 relative spectral power distributions are given for CIE Standard Illuminants A, B, C, D_{50}, D_{55}, D_{65}, and D_{75}. In section A5.3 relative spectral power distributions are given for representative fluorescent lamps. In section A5.4, the method is given for deriving the relative spectral power distributions for CIE Standard D illuminants of other correlated colour temperatures. In section A5.5 relative spectral power distributions are given for Planckian radiators. In section A5.6 relative spectral power distributions are given for gas-discharge lamps.

A5.2 CIE Standard Illuminants

Relative spectral power distributions of standard illuminants

Wave-length nm	A	B	C	D_{50}	D_{55}	D_{65}	D_{75}
300	0.93			0.02	0.02	0.03	0.04
305	1.13			1.03	1.05	1.66	2.59
310	1.36			2.05	2.07	3.29	5.13
315	1.62			4.91	6.65	11.77	17.47
320	1.93	0.02	0.01	7.78	11.22	20.24	29.81
325	2.27	0.26	0.20	11.26	15.94	28.64	42.37
330	2.66	0.50	0.40	14.75	20.65	37.05	54.93
335	3.10	1.45	1.55	16.35	22.27	38.50	56.09
340	3.59	2.40	2.70	17.95	23.88	39.95	57.26
345	4.14	4.00	4.85	19.48	25.85	42.43	60.00
350	4.74	5.60	7.00	21.01	27.82	44.91	62.74
355	5.41	7.60	9.95	22.48	29.22	45.78	62.86
360	6.14	9.60	12.90	23.94	30.62	46.64	62.98
365	6.95	12.40	17.20	25.45	32.46	49.36	66.65
370	7.82	15.20	21.40	26.96	34.31	52.09	70.31
375	8.77	88.80	27.50	25.72	33.45	51.03	68.51
380	9.80	22.40	33.00	24.49	32.58	49.98	66.70
385	10.90	26.85	39.92	27.18	35.34	52.31	68.33
390	12.09	31.30	47.40	29.87	38.09	54.65	69.96
395	13.35	36.18	55.17	39.59	49.52	68.70	85.95
400	14.71	41.30	63.30	49.31	60.95	82.75	101.93
405	16.15	46.62	71.81	52.91	64.75	87.12	106.91
410	17.68	52.10	80.60	56.51	68.55	91.49	111.89
415	19.29	57.70	89.53	58.27	70.07	92.46	112.35
420	20.99	63.20	98.10	60.03	71.58	93.43	112.80

Wave-length nm	A	B	C	D_{50}	D_{55}	D_{65}	D_{75}
425	22.79	68.37	105.80	58.93	69.75	90.06	107.94
430	24.67	73.10	112.40	57.82	67.91	86.68	103.09
435	26.64	77.31	117.75	66.32	76.76	95.77	112.14
440	28.70	80.80	121.50	74.82	85.61	104.86	121.20
445	30.85	83.44	123.45	81.04	91.80	110.94	127.10
450	33.09	85.40	124.00	87.25	97.99	117.01	133.01
455	35.41	86.88	123.60	88.93	99.23	117.41	132.68
460	37.81	88.30	123.10	90.61	100.46	117.81	132.36
465	40.30	90.08	123.30	90.99	100.19	116.34	129.84
470	42.87	92.00	123.80	91.37	99.91	114.86	127.32
475	45.52	93.75	124.09	93.24	101.33	115.39	127.06
480	48.24	95.20	123.90	95.11	102.74	115.92	126.80
485	51.04	96.23	122.92	93.54	100.41	112.37	122.29
490	53.91	96.50	120.70	91.96	98.08	108.81	117.78
495	56.85	95.71	116.90	93.84	99.38	109.08	117.19
500	59.86	94.20	112.10	95.72	100.68	109.35	116.59
505	62.93	92.37	106.98	96.17	100.69	108.58	115.15
510	66.06	90.70	102.30	96.61	100.70	107.80	113.70
515	69.25	89.65	98.81	96.87	100.34	106.30	111.18
520	72.50	89.50	96.90	97.13	99.99	104.79	108.66
525	75.79	90.43	96.78	99.61	102.10	106.24	109.55
530	79.13	92.20	98.00	102.10	104.21	107.69	110.44
535	82.52	94.46	99.94	101.43	103.16	106.05	108.37
540	85.95	96.90	102.10	100.75	102.10	104.41	106.29
545	89.41	99.16	103.95	101.54	102.53	104.23	105.60
550	92.91	101.00	105.20	102.32	102.97	104.05	104.90
555	96.44	102.20	105.67	101.16	101.48	102.02	102.45
560	100.00	102.80	105.30	100.00	100.00	100.00	100.00
565	103.58	102.92	104.11	98.87	98.61	98.17	97.81
570	107.18	102.60	102.30	97.74	97.22	96.33	95.62
575	110.80	101.90	100.15	98.33	97.48	96.06	94.91
580	114.44	101.00	97.80	98.92	97.75	95.79	94.21
585	118.08	100.07	95.43	96.21	94.59	92.24	90.60
590	121.73	99.20	93.20	93.50	91.43	88.69	87.00
595	125.39	98.44	91.22	95.59	92.93	89.35	87.11
600	129.04	98.00	89.70	97.69	94.42	90.01	87.23
605	132.70	98.08	88.83	98.48	94.78	89.80	86.68
610	136.35	98.50	88.40	99.27	95.14	89.60	86.14
615	139.99	99.06	88.19	99.16	94.68	88.65	84.86
620	143.62	99.70	88.10	99.04	94.22	87.70	83.58
625	147.24	100.36	88.06	97.38	92.33	85.49	81.16
630	150.84	101.00	88.00	95.72	90.45	83.29	78.75
635	154.42	101.56	87.86	97.29	91.39	83.49	78.59
640	157.98	102.20	87.80	98.86	92.33	83.70	78.43
645	161.52	103.05	87.99	97.26	90.59	81.86	76.61

Wavelength nm	A	B	C	D_{50}	D_{55}	D_{65}	D_{75}
650	165.03	103.90	88.20	95.67	88.85	80.03	74.80
655	168.51	104.59	88.20	96.93	89.59	80.12	74.56
660	171.96	105.00	87.90	98.19	90.32	80.21	74.32
665	175.38	105.08	87.22	100.60	92.13	81.25	74.87
670	178.77	104.90	86.30	103.00	93.95	82.28	75.42
675	182.12	104.55	85.30	101.07	91.95	80.28	73.50
680	185.43	103.90	84.00	99.13	89.96	78.28	71.58
685	188.70	102.84	82.21	93.26	84.82	74.00	67.71
690	191.93	101.60	80.20	87.38	79.68	69.72	63.85
695	195.12	100.38	78.24	89.49	81.26	70.67	64.46
700	198.26	99.10	76.30	91.60	82.84	71.61	65.08
705	201.36	97.70	74.36	92.25	83.84	72.98	66.57
710	204.41	96.20	72.40	92.89	84.84	74.35	68.07
715	207.41	94.60	70.40	84.87	77.54	67.98	62.26
720	210.36	92.90	68.30	76.85	70.24	61.60	56.44
725	213.27	91.10	66.30	81.68	74.77	65.74	60.34
730	216.12	89.40	64.40	86.51	79.30	69.89	64.24
735	218.92	88.00	62.80	89.55	82.15	72.49	66.70
740	221.67	86.90	61.50	92.58	84.99	75.09	69.15
745	224.36	85.90	60.20	85.40	78.44	69.34	63.89
750	227.00	85.20	59.20	78.23	71.88	63.59	58.63
755	229.59	84.80	58.50	67.96	62.34	55.01	50.62
760	232.12	84.70	58.10	57.69	52.79	46.42	42.62
765	234.59	84.90	58.00	70.31	64.36	56.61	51.98
770	237.01	85.40	58.20	82.92	75.93	66.81	61.35
775	239.37	86.10	58.50	80.60	73.87	65.09	59.84
780	241.68	87.00	59.10	78.27	71.82	63.38	58.32
785	243.92			78.91	72.38	63.84	58.73
790	246.12			79.55	72.94	64.30	59.14
795	248.25			76.48	70.14	61.88	56.94
800	250.33			73.40	67.35	59.45	54.73
805	252.35			68.66	63.04	55.71	51.32
810	254.31			63.92	58.73	51.96	47.92
815	256.22			67.35	61.86	54.70	50.42
820	258.07			70.78	64.99	57.44	52.92
825	259.86			72.61	66.65	58.88	54.23
830	261.60			74.44	68.31	60.31	55.54

Chromaticity co-ordinates (based on 5 nm intervals from 380 to 780 nm)

	X	Y	Z	x	y	u'	v'
A	109.85	100.00	35.58	0.4476	0.4074	0.2560	0.5243
C	98.07	100.00	118.23	0.3101	0.3162	0.2009	0.4609
D_{50}	96.42	100.00	82.49	0.3457	0.3585	0.2092	0.4881
D_{55}	95.68	100.00	92.14	0.3324	0.3474	0.2044	0.4807
D_{65}	95.04	100.00	108.89	0.3127	0.3290	0.1978	0.4683
D_{75}	94.96	100.00	122.61	0.2990	0.3149	0.1935	0.4585

	X_{10}	Y_{10}	Z_{10}	x_{10}	y_{10}	u'_{10}	v'_{10}
A	111.15	100.00	35.20	0.4512	0.4059	0.2590	0.5242
C	97.28	100.00	116.14	0.3104	0.3191	0.2000	0.4626
D_{50}	96.72	100.00	81.41	0.3478	0.3595	0.2102	0.4889
D_{55}	95.79	100.00	90.93	0.3341	0.3488	0.2051	0.4816
D_{65}	94.81	100.00	107.33	0.3138	0.3310	0.1979	0.4695
D_{75}	94.41	100.00	120.63	0.2997	0.3174	0.1930	0.4601

A5.3. Representative fluorescent lamps

Spectral power distributions are given in this section for twelve types of fluorescent lamps, F1 to F12. These distributions do not constitute CIE Standard Illuminants, but they have been compiled by the CIE for use as representative distributions for practical purposes. The twelve lamps represented are in three different groups, 'normal, 'broad-band', and 'three-band'. The distributions F2, F7, and F11, which are asterisked, are intended for use in preference to the others when the choice within each group is not critical. Details of the lamps are as follows:

Group	Lamp	Chromaticity Co-ordinates (based on 5 nm intervals from 380 to 780 nm)				Correlated colour temperature	Colour rendering index
		x	y	u'	v'	K	R_a
Normal	F1	0.3131	0.3371	0.1951	0.4726	6430	76
	F2*	0.3721	0.3751	0.2203	0.4996	4230	64
	F3	0.4091	0.3941	0.2368	0.5132	3450	57
	F4	0.4402	0.4031	0.2531	0.5215	2940	51
	F5	0.3138	0.3452	0.1927	0.4769	6350	72
	F6	0.3779	0.3882	0.2190	0.5062	4150	59
Broad-band	F7*	0.3129	0.3292	0.1979	0.4685	6500	90
	F8	0.3458	0.3586	0.2092	0.4881	5000	95
	F9	0.3741	0.3727	0.2225	0.4988	4150	90
Three-band	F10	0.3458	0.3588	0.2091	0.4882	5000	81
	F11*	0.3805	0.3769	0.2251	0.5017	4000	83
	F12	0.4370	0.4042	0.2506	0.5214	3000	83

Each of the distributions in the normal group consists of two semi-broad band emissions of antimony and manganese activations in calcium halophosphate phosphor. Those in the broad-band group are more or less enhanced in colour rendering properties as compared with those in the normal group, usually using multiple phosphors; this results in the distributions being flatter and having a wider range in the visible spectrum. The distributions in the three-band group consist mostly of three narrow-band emissions in red, green, and blue wavelength regions; in most cases the narrow band emissions are caused by ternary compositions of rare-earth phosphors. The CIE Colour Rendering Index, R_a, is described in section 4.17. The unit of these spectral power distributions is $W.nm^{-1}.lm^{-1}$. Below 380 nm and above 780 nm, these spectral power distributions should be taken as zero.

Spectral power distributions of representative fluorescennt lamps

Wave-length nm	F1	F2*	F3	F4	F5	F6	F7*	F8	F9	F10	F11*	F12
380	1.87	1.18	0.82	0.57	1.87	1.05	2.56	1.21	0.90	1.11	0.91	0.96
385	2.36	1.48	1.02	0.70	2.35	1.31	3.18	1.50	1.12	0.80	0.63	0.64
390	2.94	1.84	1.26	0.87	2.92	1.63	3.84	1.81	1.36	0.62	0.46	0.45
395	3.47	2.15	1.44	0.98	3.45	1.90	4.53	2.13	1.60	0.57	0.37	0.33
400	5.17	3.44	2.57	2.01	5.10	3.11	6.15	3.17	2.59	1.48	1.29	1.19
405	19.49	15.69	14.36	13.75	18.91	14.80	19.37	13.08	12.80	12.16	12.68	12.48
410	6.13	3.85	2.70	1.95	6.00	3.43	7.37	3.83	3.05	2.12	1.59	1.12
415	6.24	3.74	2.45	1.59	6.11	3.30	7.05	3.45	2.56	2.70	1.79	0.94
420	7.01	4.19	2.73	1.76	6.85	3.68	7.71	3.86	2.86	3.74	2.46	1.08
425	7.79	4.62	3.00	1.93	7.58	4.07	8.41	4.42	3.30	5.14	3.38	1.37
430	8.56	5.06	3.28	2.10	8.31	4.45	9.15	5.09	3.82	6.75	4.49	1.78
435	43.67	34.98	31.85	30.28	40.76	32.61	44.14	34.10	32.62	34.39	33.94	29.05
440	16.94	11.81	9.47	8.03	16.06	10.74	17.52	12.42	10.77	14.86	12.13	7.90
445	10.72	6.27	4.02	2.55	10.32	5.48	11.35	7.68	5.84	10.40	6.95	2.65
450	11.35	6.63	4.25	2.70	10.91	5.78	12.00	8.60	6.57	10.76	7.19	2.71
455	11.89	6.93	4.44	2.82	11.40	6.03	12.58	9.46	7.25	10.67	7.12	2.65
460	12.37	7.19	4.59	2.91	11.83	6.25	13.08	10.24	7.86	10.11	6.72	2.49
465	12.75	7.40	4.72	2.99	12.17	6.41	13.45	10.84	8.35	9.27	6.13	2.33
470	13.00	7.54	4.80	3.04	12.40	6.52	13.71	11.33	8.75	8.29	5.46	2.10
475	13.15	7.62	4.86	3.08	12.54	6.58	13.88	11.71	9.06	7.29	4.79	1.91
480	13.23	7.65	4.87	3.09	12.58	6.59	13.95	11.98	9.31	7.91	5.66	3.01
485	13.17	7.62	4.85	3.09	12.52	6.56	13.93	12.17	9.48	16.64	14.29	10.83
490	13.13	7.62	4.88	3.14	12.47	6.56	13.82	12.28	9.61	16.73	14.96	11.88
495	12.85	7.45	4.77	3.06	12.20	6.42	13.64	12.32	9.68	10.44	8.97	6.88
500	12.52	7.28	4.67	3.00	11.89	6.28	13.43	12.35	9.74	5.94	4.72	3.43
505	12.20	7.15	4.62	2.98	11.61	6.20	13.25	12.44	9.88	3.34	2.33	1.49
510	11.83	7.05	4.62	3.01	11.33	6.19	13.08	12.55	10.04	2.35	1.47	0.92
515	11.50	7.04	4.73	3.14	11.10	6.30	12.93	12.68	10.26	1.88	1.10	0.71
520	11.22	7.16	4.99	3.41	10.96	6.60	12.78	12.77	10.48	1.59	0.89	0.60
525	11.05	7.47	5.48	3.90	10.97	7.12	12.60	12.72	10.63	1.47	0.83	0.63
530	11.03	8.04	6.25	4.69	11.16	7.94	12.44	12.60	10.78	1.80	1.18	1.10
535	11.18	8.88	7.34	5.81	11.54	9.07	12.33	12.43	10.96	5.71	4.90	4.56
540	11.53	10.01	8.78	7.32	12.12	10.49	12.26	12.22	11.18	40.98	39.59	34.40
545	27.74	24.88	23.82	22.59	27.78	25.22	29.52	28.96	27.71	73.69	72.84	65.40
550	17.05	16.64	16.14	15.11	17.73	17.46	17.05	16.51	16.29	33.61	32.61	29.48
555	13.55	14.59	14.59	13.88	14.47	15.63	12.44	11.79	12.28	8.24	7.52	7.16
560	14.33	16.16	16.63	16.33	15.20	17.22	12.58	11.76	12.74	3.38	2.83	3.08
565	15.01	17.56	18.49	18.68	15.77	18.53	12.72	11.77	13.21	2.47	1.96	2.47
570	15.52	18.62	19.95	20.64	16.10	19.43	12.83	11.84	13.65	2.14	1.67	2.27
575	18.29	21.47	23.11	24.28	18.54	21.97	15.46	14.61	16.57	4.86	4.43	5.09
580	19.55	22.79	24.69	26.26	19.50	23.01	16.75	16.11	18.14	11.45	11.28	11.96
585	15.48	19.29	21.41	23.28	15.39	19.41	12.83	12.34	14.55	14.79	14.76	15.32
590	14.91	18.66	20.85	22.94	14.64	18.56	12.67	12.53	14.65	12.16	12.73	14.27
595	14.15	17.73	19.93	22.14	13.72	17.42	12.45	12.72	14.66	8.97	9.74	11.86

Wave-length nm	F1	F2*	F3	F4	F5	F6	F7*	F8	F9	F10	F11*	F12
600	13.22	16.54	18.67	20.91	12.69	16.09	12.19	12.92	14.61	6.52	7.33	9.28
605	12.19	15.21	17.22	19.43	11.57	14.64	11.89	13.12	14.50	8.31	9.72	12.31
610	11.12	13.80	15.65	17.74	10.45	13.15	11.60	13.34	14.39	44.12	55.27	68.53
615	10.03	12.36	14.04	16.00	9.35	11.68	11.35	13.61	14.40	34.55	42.58	53.02
620	8.95	10.95	12.45	14.42	8.29	10.25	11.12	13.87	14.47	12.09	13.18	14.67
625	7.96	9.65	10.95	12.56	7.32	8.95	10.95	14.07	14.62	12.15	13.16	14.38
630	7.02	8.40	9.51	10.93	6.41	7.74	10.76	14.20	14.72	10.52	12.26	14.71
635	6.20	7.32	8.27	9.52	5.63	6.69	10.42	14.16	14.55	4.43	5.11	6.46
640	5.42	6.31	7.11	8.18	4.90	5.71	10.11	14.13	14.40	1.95	2.07	2.57
645	4.73	5.43	6.09	7.01	4.26	4.87	10.04	14.34	14.58	2.19	2.34	2.75
650	4.15	4.68	5.22	6.00	3.72	4.16	10.02	14.50	14.88	3.19	3.58	4.18
655	3.64	4.02	4.45	5.11	3.25	3.55	10.11	14.46	15.51	2.77	3.01	3.44
660	3.20	3.45	3.80	4.36	2.83	3.02	9.87	14.00	15.47	2.29	2.48	2.81
665	2.81	2.96	3.23	3.69	2.49	2.57	8.65	12.58	13.20	2.00	2.14	2.42
670	2.47	2.55	2.75	3.13	2.19	2.20	7.27	10.99	10.57	1.52	1.54	1.64
675	2.18	2.19	2.33	2.64	1.93	1.87	6.44	9.98	9.18	1.35	1.33	1.36
680	1.93	1.89	1.99	2.24	1.71	1.60	5.83	9.22	8.25	1.47	1.46	1.49
685	1.72	1.64	1.70	1.91	1.52	1.37	5.41	8.62	7.57	1.79	1.94	2.14
690	1.67	1.53	1.55	1.70	1.48	1.29	5.04	8.07	7.03	1.74	2.00	2.34
695	1.43	1.27	1.27	1.39	1.26	1.05	4.57	7.39	6.35	1.02	1.20	1.42
700	1.29	1.10	1.09	1.18	1.13	0.91	4.12	6.71	5.72	1.14	1.35	1.61
705	1.19	0.99	0.96	1.03	1.05	0.81	3.77	6.16	5.25	3.32	4.10	5.04
710	1.08	0.88	0.83	0.88	0.96	0.71	3.46	5.63	4.80	4.49	5.58	6.98
715	0.96	0.76	0.71	0.74	0.85	0.61	3.08	5.03	4.29	2.05	2.51	3.19
720	0.88	0.68	0.62	0.64	0.78	0.54	2.73	4.46	3.80	0.49	0.57	0.71
725	0.81	0.61	0.54	0.54	0.72	0.48	2.47	4.02	3.43	0.24	0.27	0.30
730	0.77	0.56	0.49	0.49	0.68	0.44	2.25	3.66	3.12	0.21	0.23	0.26
735	0.75	0.54	0.46	0.46	0.67	0.43	2.06	3.36	2.86	0.21	0.21	0.23
740	0.73	0.51	0.43	0.42	0.65	0.40	1.90	3.09	2.64	0.24	0.24	0.28
745	0.68	0.47	0.39	0.37	0.61	0.37	1.75	2.85	2.43	0.24	0.24	0.28
750	0.69	0.47	0.39	0.37	0.62	0.38	1.62	2.65	2.26	0.21	0.20	0.21
755	0.64	0.43	0.35	0.33	0.59	0.35	1.54	2.51	2.14	0.17	0.24	0.17
760	0.68	0.46	0.38	0.35	0.62	0.39	1.45	2.37	2.02	0.21	0.32	0.21
765	0.69	0.47	0.39	0.36	0.64	0.41	1.32	2.15	1.83	0.22	0.26	0.19
770	0.61	0.40	0.33	0.31	0.55	0.33	1.17	1.89	1.61	0.17	0.16	0.15
775	0.52	0.33	0.28	0.26	0.47	0.26	0.99	1.61	1.38	0.12	0.12	0.10
780	0.43	0.27	0.21	0.19	0.40	0.21	0.81	1.32	1.12	0.09	0.09	0.05

A5.4 Method of calculating D Illuminant distributions

For daylight illuminants, CIE Standard Illuminant D_{65} should be used whenever possible; if D_{65} is not appropriate, then either D_{50}, D_{55}, or D_{75}, should be used if possible. If none of these daylight illuminants is appropriate, then one of the other CIE D illuminants, defined below, should be used.

Chromaticity. The chromaticity co-ordinates must be such that:

$$y_D = -3.000x_D^2 + 2.870x_D - 0.275$$

with x_D being within the range of 0.250 to 0.380. The correlated colour temperature T_c (calculated with c_2 of Planck's Law being equal to 1.4388×10^{-2}mK) of daylight D is related to x_D by the following formulae based on normals to the Planckian locus on a chromaticity diagram in which v is plotted against u (not v' against u'):

(a) for correlated colour temperatures from 4000 K to 7000 K:

$$x_D = -4.6070(10^9/T_c^3) + 2.9678(10^6/T_c^2) + 0.09911(10^3/T_c) + 0.244063 \ .$$

(b) for correlated colour temperatures from 7000 K to 25000 K

$$x_D = -2.0064(10^9/T_c^3) + 1.9018(10^6/T_c^2) + 0.24748(10^3/T_c) + 0.237040 \ .$$

The relative spectral power distributions, $S(\lambda)$, of the D illuminants are given by:

$$S(\lambda) = S_0(\lambda) + M_1 S_1(\lambda) + M_2 S_2(\lambda) \ ,$$

where $S_0(\lambda)$, $S_1(\lambda)$, $S_2(\lambda)$ are functions of wavelength, λ, as given in the Table below, and M_1, M_2 are factors whose values are related to the chromaticity co-ordinates x_D, y_D as follows:

$$M_1 = \frac{-1.3515 - 1.7703x_D + 5.9114y_D}{0.0241 + 0.2562x_D - 0.7341y_D}$$

$$M_2 = \frac{0.0300 - 31.4424x_D + 30.0717y_D}{0.0241 + 0.2562x_D - 0.7341y_D} \ .$$

Values of x_D, y_D, M_1, and M_2 for correlated colour temperatures in the range 4000 K to 25000 K are given below.

Wave-length nm	$S_0(\lambda)$	$S_1(\lambda)$	$S_2(\lambda$	Wave-length nm	$S_0(\lambda)$	$S_1(\lambda)$	$S_2(\lambda)$
300	0.04	0.02	0.00	400	94.80	43.40	−1.10
305	3.02	2.26	1.00	405	99.80	44.85	−0.80
310	6.00	4.50	2.00	410	104.80	46.30	−0.50
315	17.80	13.45	3.00	415	105.35	45.10	−0.60
320	29.60	22.40	4.00	420	105.90	43.90	−0.70
325	42.45	32.20	6.25	425	101.35	40.50	−0.95
330	55.30	42.00	8.50	430	96.80	37.10	−1.20
335	56.30	41.30	8.15	435	105.35	36.90	−1.90
340	57.30	40.60	7.80	440	113.90	36.70	−2.60
345	59.55	41.10	7.25	445	119.75	36.30	−2.75
350	61.80	41.60	6.70	450	125.60	35.90	−2.90
355	61.65	39.80	6.00	455	125.55	34.25	−2.85
360	61.50	38.00	5.30	460	125.50	32.60	−2.80
365	65.15	40.20	5.70	465	123.40	30.25	−2.70
370	68.80	42.40	6.10	470	121.30	27.90	−2.60
375	66.10	40.45	4.55	475	121.30	26.10	−2.60
380	63.40	38.50	3.00	480	121.30	24.30	−2.60
385	64.60	36.75	2.10	485	117.40	22.20	−2.20
390	65.80	35.00	1.20	490	113.50	20.10	−1.80
395	80.30	39.20	0.05	495	113.30	18.15	−1.65

Wave-length nm	$S_0(\lambda)$	$S_1(\lambda)$	$S_2(\lambda$	Wave-length nm	$S_0(\lambda)$	$S_1(\lambda)$	$S_2(\lambda)$
500	113.10	16.20	−1.50	675	83.10	−13.80	10.00
505	111.95	14.70	−1.40	680	81.30	−13.60	10.20
510	110.80	13.20	−1.30	685	76.60	−12.80	9.25
515	108.65	10.90	−1.25	690	71.90	−12.00	8.30
520	106.50	8.60	−1.20	695	73.10	−12.65	8.95
525	107.65	7.35	−1.10	700	74.30	−13.30	9.60
530	108.80	6.10	−1.00	705	75.35	−13.10	9.05
535	107.05	5.15	−0.75	710	76.40	−12.90	8.50
540	105.30	4.20	−0.50	715	69.85	−11.75	7.75
545	104.85	3.05	−0.40	720	63.30	−10.60	7.00
550	104.40	1.90	−0.30	725	67.50	−11.10	7.30
555	102.20	0.95	−0.15	730	71.70	−11.60	7.60
560	100.00	0.00	0.00	735	74.35	−11.90	7.80
565	98.00	−0.80	0.10	740	77.00	−12.20	8.00
570	96.00	−1.60	0.20	745	71.10	−11.20	7.35
575	95.55	−2.55	0.35	750	65.20	−10.20	6.70
580	95.10	−3.50	0.50	755	56.45	−9.00	5.95
585	92.10	−3.50	1.30	760	47.70	−7.80	5.20
590	89.10	−3.50	2.10	765	58.15	−9.50	6.30
595	89.80	−4.65	2.65	770	68.60	−11.20	7.40
600	90.50	−5.80	3.20	775	66.80	−10.80	7.10
605	90.40	−6.50	3.65	780	65.00	−10.40	6.80
610	90.30	−7.20	4.10	785	65.50	−10.50	6.90
615	89.35	−7.90	4.40	790	66.00	−10.60	7.00
620	88.40	−8.60	4.70	795	63.50	−10.15	6.70
625	86.20	−9.05	4.90	800	61.00	−9.70	6.40
630	84.00	−9.50	5.10	805	57.15	−9.00	5.95
635	84.55	−10.20	5.90	810	53.30	−8.30	5.50
640	85.10	−10.90	6.70	815	56.10	−8.80	5.80
645	83.50	−10.80	7.00	820	58.90	−9.30	6.10
650	81.90	−10.70	7.30	825	60.40	−9.55	6.30
655	82.25	−11.35	7.95	830	61.90	−9.80	6.50
660	82.60	−12.00	8.60				
665	83.75	−13.00	9.20				
670	84.90	−14.00	9.80				

Chromaticity co-ordinates x_D,y_D and factors M_1,M_2 used in the calculation of the relative spectral power distributions of CIE D Illuminants. The corresponding chromaticity co-ordinates u'_D,v'_D are also given.

T_c	x_D	y_D	u'_D	v'_D	M_1	M_2
4000	0.3823	0.3838	0.2236	0.5049	−1.505	2.827
4100	0.3779	0.3812	0.2217	0.5031	−1.464	2.460
4200	0.3737	0.3786	0.2200	0.5014	−1.422	2.127
4300	0.3697	0.3760	0.2183	0.4997	−1.378	1.825
4400	0.3658	0.3734	0.2168	0.4979	−1.333	1.550
4500	0.3621	0.3709	0.2153	0.4962	−1.286	1.302
4600	0.3585	0.3684	0.2139	0.4946	−1.238	1.076
4700	0.3551	0.3659	0.2126	0.4929	−1.190	0.871
4800	0.3519	0.3634	0.2114	0.4913	−1.140	0.686
4900	0.3487	0.3610	0.2102	0.4897	−1.090	0.518

T_c	x_D	y_D	u'_D	v'_D	M_1	M_2
5000	0.3457	0.3587	0.2091	0.4882	−1.040	0.367
5100	0.3429	0.3564	0.2081	0.4866	−0.989	0.230
5200	0.3401	0.3541	0.2071	0.4851	−0.939	0.106
5300	0.3375	0.3619	0.2062	0.4837	−0.888	−0.005
5400	0.3349	0.3497	0.2053	0.4822	−0.837	−0.105
5500	0.3325	0.3476	0.2044	0.4808	−0.786	−0.195
5600	0.3302	0.3455	0.2036	0.4795	−0.736	−0.276
5700	0.3279	0.3435	0.2028	0.4781	−0.685	−0.348
5800	0.3258	0.3416	0.2021	0.4768	−0.635	−0.412
5900	0.3237	0.3397	0.2014	0.4755	−0.586	−0.469
6000	0.3217	0.3378	0.2007	0.4743	−0.536	−0.519
6100	0.3198	0.3360	0.2001	0.4730	−0.487	−0.563
6200	0.3179	0.3342	0.1995	0.4719	−0.439	−0.602
6300	0.3161	0.3325	0.1989	0.4707	−0.391	−0.635
6400	0.3144	0.3308	0.1983	0.4695	−0.343	−0.664
6500	0.3128	0.3292	0.1978	0.4684	−0.296	−0.688
6600	0.3112	0.3276	0.1973	0.4673	−0.250	−0.709
6700	0.3097	0.3260	0.1968	0.4663	−0.204	−0.726
6800	0.3082	0.3245	0.1963	0.4652	−0.159	−0.739
6900	0.3067	0.3231	0.1959	0.4642	−0.114	−0.749
7000	0.3054	0.3216	0.1955	0.4632	−0.070	−0.757
7100	0.3040	0.3202	0.1950	0.4623	−0.026	−0.762
7200	0.3027	0.3189	0.1946	0.4613	0.017	−0.765
7300	0.3015	0.3176	0.1943	0.4604	0.060	−0.765
7400	0.3003	0.3163	0.1939	0.4595	0.102	−0.763
7500	0.2991	0.3150	0.1935	0.4586	0.144	−0.760
7600	0.2980	0.3138	0.1932	0.4578	0.184	−0.755
7700	0.2969	0.3126	0.1928	0.4569	0.225	−0.748
7800	0.2958	0.3115	0.1925	0.4561	0.264	−0.740
7900	0.2948	0.3103	0.1922	0.4553	0.303	−0.730
8000	0.2938	0.3092	0.1919	0.4545	0.342	−0.720
8100	0.2928	0.3081	0.1916	0.4537	0.380	−0.708
8200	0.2919	0.3071	0.1913	0.4530	0.417	−0.695
8300	0.2910	0.3061	0.1911	0.4523	0.454	−0.682
8400	0.2901	0.3051	0.1908	0.4515	0.490	−0.667
8500	0.2892	0.3041	0.1906	0.4508	0.526	−0.652
9000	0.2853	0.2996	0.1894	0.4475	0.697	−0.566
9500	0.2818	0.2956	0.1884	0.4446	0.856	−0.471
10000	0.2788	0.2920	0.1876	0.4419	1.003	−0.369
10500	0.2761	0.2887	0.1868	0.4395	1.139	−0.265
11000	0.2737	0.2858	0.1861	0.4373	1.266	−0.160
12000	0.2697	0.2808	0.1850	0.4335	1.495	0.045
13000	0.2664	0.2767	0.1841	0.4303	1.693	0.239
14000	0.2637	0.2732	0.1834	0.4275	1.868	0.419
15000	0.2614	0.2702	0.1828	0.4252	2.021	0.586
17000	0.2578	0.2655	0.1818	0.4214	2.278	0.878
20000	0.2539	0.2603	0,1809	0.4172	2.571	1.231
25000	0.2499	0.2548	0.1798	0.4126	2.907	1.655
5003	0.3457	0.3586	0.2091	0.4881	−1.039	0.363
5503	0.3324	0.3475	0.2044	0.4808	−0.785	−0.198
6504	0.3127	0.3291	0.1978	0.4684	−0.295	−0.689
7504	0.2990	0.3150	0.1935	0.4586	0.145	−0.760

A5.5 Planckian radiators

Relative spectral power distributions (c_2=1.4388×10^{-2} m.K)

Wave-length, nm	1000 K	2000 K	2500 K	3000 K	4000 K	5000 K	6000 K	7000 K	8000 K	10000 K
380	0.004	1.582	5.343	12.03	33.11	60.62	90.33	119.6	147.0	194.5
385	0.006	1.895	6.093	13.27	35.07	62.65	91.86	120.2	146.5	191.6
390	0.008	2.258	6.919	14.60	37.07	64.65	93.30	120.7	145.9	188.6
395	0.012	2.676	7.826	16.00	39.09	66.60	94.64	121.1	145.2	185.7
400	0.019	3.155	8.817	17.49	41.13	68.51	95.90	121.4	144.5	182.7
405	0.027	3.702	9.896	19.06	43.19	70.37	97.06	121.7	143.6	179.7
410	0.039	4.324	11.07	20.71	45.27	72.18	98.14	121.8	142.7	176.8
415	0.056	5.028	12.34	22.44	47.36	73.93	99.13	121.8	141.7	173.8
420	0.080	5.821	13.71	24.26	49.46	75.63	100.0	121.7	140.6	170.8
425	0.113	6.711	15.18	26.15	51.56	77.28	100.9	121.6	139.5	167.8
430	0.159	7.707	16.76	28.12	53.66	78.87	101.6	121.4	138.3	164.9
435	0.220	8.816	18.45	30.17	55.76	80.40	102.3	121.1	137.0	162.0
440	0.302	10.05	20.25	32.30	57.85	81.87	102.9	120.7	135.8	159.1
445	0.413	11.41	22.17	34.50	59.93	83.28	103.4	120.3	134.4	156.2
450	0.559	12.91	24.20	36.78	62.00	84.63	103.8	119.8	133.1	153.3
455	0.751	14.57	26.36	39.13	64.06	85.92	104.2	119.3	131.7	150.5
460	1.003	16.38	28.63	41.54	66.10	87.15	104.5	118.7	130.3	147.7
465	1.331	18.36	31.03	44.03	68.12	88.33	104.8	118.0	128.8	145.0
470	1.753	20.52	33.56	46.58	70.11	89.44	104.9	117.4	127.4	142.2
475	2.295	22.86	36.20	49.18	72.08	90.49	105.1	116.6	125.9	139.5
480	2.986	25.40	38.98	51.85	74.02	91.48	105.1	115.9	124.4	136.9
485	3.861	28.15	41.88	54.58	75.93	92.41	105.1	115.0	122.9	134.3
490	4.965	31.11	44.91	57.35	77.81	93.29	105.1	114.2	121.4	131.7
495	6.349	34.30	48.06	60.18	79.65	94.11	105.0	113.3	119.8	129.2
500	8.074	37.72	51.34	63.05	81.46	94.87	104.8	112.4	118.3	126.7
505	10.21	41.39	54.75	65.97	83.24	95.57	104.6	111.5	116.7	124.2
510	12.86	45.30	58.28	68.93	84.97	96.22	104.4	110.5	115.2	121.8
515	16.10	49.48	61.93	71.92	86.67	96.92	104.1	109.5	113.7	119.5
520	20.07	53.92	65.70	74.95	88.32	97.37	103.8	108.5	112.1	117.1
525	24.90	58.64	69.60	78.01	89.93	97.86	103.4	107.5	110.6	114.8
530	30.76	63.64	73.61	81.10	91.51	98.31	103.0	106.5	109.0	112.6
535	37.82	68.94	77.73	84.21	93.03	98.71	102.6	105.4	107.5	110.4
540	46.31	74.53	81.97	87.34	94.52	99.05	102.1	104.4	106.0	108.2
545	56.47	80.43	86.32	90.48	95.95	99.36	101.6	103.3	104.5	106.1
550	68.59	86.63	90.78	93.65	97.35	99.61	101.1	102.2	103.0	104.0
555	82.98	93.16	95.34	96.82	98.70	99.83	100.6	101.1	101.5	102.0
560	100.0	100.0	100.0	100.0	100.0	100.0	100.0	100.0	100.0	100.0
565	120.1	107.2	104.8	103.2	101.3	100.1	99.40	98.90	98.53	98.04
570	143.7	114.7	109.6	106.4	102.5	100.2	98.78	97.79	97.07	96.12
575	171.3	122.5	114.6	109.6	103.6	100.3	98.14	96.68	95.63	94.24
580	203.5	130.7	119.6	112.7	104.8	100.3	97.48	95.57	94.20	92.39
585	241.0	139.2	124.7	115.9	105.8	100.3	96.80	94.45	92.78	90.58
590	284.5	148.0	129.9	119.1	106.9	100.2	96.10	93.34	91.38	88.81
595	334.8	157.2	135.2	122.2	107.8	100.1	95.39	92.23	89.88	87.08

Wave-length, nm	1000 K	2000 K	2500 K	3000 K	4000 K	5000 K	6000 K	7000 K	8000 K	10000 K
600	392.7	166.8	140.5	125.4	108.8	100.0	94.66	91.12	88.62	85.38
605	459.3	176.7	145.9	128.5	109.7	99.85	93.92	90.01	87.27	83.72
610	535.7	186.9	151.4	131.6	110.5	99.67	93.17	88.91	85.93	82.09
615	623.0	197.5	156.9	134.7	111.3	99.46	92.41	87.81	84.61	80.49
620	722.5	208.4	162.5	137.7	112.1	99.22	91.64	86.72	83.30	78.93
625	835.6	219.7	168.2	140.8	112.8	98.96	90.86	85.63	82.01	77.40
630	963.9	231.3	173.9	143.8	113.5	98.67	90.07	84.55	80.74	75.91
635	1109	243.2	179.6	146.7	114.1	98.36	89.28	83.47	79.49	74.44
640	1273	255.5	185.4	149.7	114.7	98.02	88.48	82.40	78.25	73.01
645	1458	268.2	191.1	152.6	115.3	97.66	87.67	81.34	77.03	71.61
650	1665	281.1	197.0	155.4	115.8	97.28	86.86	80.29	75.83	70.23
655	1897	294.4	202.8	158.3	116.3	96.88	86.05	79.25	74.64	68.89
660	2157	308.0	208.7	161.1	116.7	96.47	85.23	78.21	73.47	67.57
665	2447	321.9	214.6	163.8	117.1	96.03	84.41	77.18	72.32	66.29
670	2770	336.2	220.5	166.5	117.5	95.58	83.59	76.17	71.19	65.03
675	3129	350.7	226.4	169.2	117.8	95.11	82.76	75.16	70.07	63.79
680	3528	365.6	232.3	171.8	118.1	94.63	81.94	74.16	68.97	62.59
685	3969	380.7	238.3	174.4	118.3	94.13	81.12	73.17	67.89	61.41
690	4456	396.1	244.2	176.9	118.6	93.62	80.29	72.19	66.82	60.25
695	4994	411.9	250.1	179.4	118.7	93.09	79.47	71.22	65.77	59.12
700	5586	427.8	256.0	181.8	118.9	92.56	78.65	70.26	64.74	58.01
705	6237	444.1	261.9	184.2	119.0	92.01	77.83	69.32	63.72	56.93
710	6951	460.6	267.7	186.6	119.1	91.45	77.02	68.38	62.72	55.87
715	7733	477.4	273.6	188.9	119.2	90.88	76.20	67.45	61.73	54.83
720	8588	494.4	279.4	191.1	119.2	90.30	75.39	66.54	60.76	53.81
725	9522	511.7	285.2	193.3	119.3	89.72	74.58	65.63	59.81	52.82
730	10540	529.2	291.0	195.4	119.2	89.12	73.78	64.74	58.87	51.85
735	11648	546.9	296.7	197.5	119.2	88.52	72.98	63.85	57.95	50.90
740	12852	564.8	302.4	199.6	119.1	87.91	72.18	62.98	57.04	49.96
745	14158	582.9	308.1	201.6	119.0	87.30	71.39	62.12	56.15	49.05
750	15574	601.2	313.7	203.5	118.9	86.68	70.61	61.27	55.27	48.16
755	17106	619.7	319.3	205.4	118.8	86.05	69.82	60.43	54.41	47.28
760	18761	638.4	324.8	207.2	118.6	85.42	69.05	59.60	53.56	46.43
765	20547	657.2	330.3	209.0	118.4	84.78	68.28	58.78	52.73	45.59
770	22472	676.2	335.7	210.7	118.2	84.15	67.51	57.97	51.90	44.77
775	24544	695.4	341.1	212.4	118.0	83.50	66.75	57.17	51.10	43.97
780	26771	714.7	346.4	214.0	117.7	82.86	66.00	56.39	50.30	43.18

Chromaticity co-ordinates (based on 5 nm intervals from 380 to 780 nm):

	1000 K	2000 K	2500 K	3000 K	4000 K	5000 K	6000 K	7000 K	8000 K	10000 K
x	0.6526	0.5266	0.4769	0.4368	0.3804	0.3450	0.3220	0.3063	0.2952	0.2806
y	0.3446	0.4133	0.4137	0.4041	0.3767	0.3516	0.3318	0.3165	0.3048	0.2883
u'	0.4478	0.3050	0.2721	0.2505	0.2251	0.2114	0.2032	0.1981	0.1946	0.1903
v'	0.5320	0.5386	0.5311	0.5214	0.5016	0.4847	0.4712	0.4605	0.4521	0.4399

A5.6 Gas discharge lamps

Relative spectral power distributions

Wave-length, nm	Low pressure sodium	High pressure sodium	Mercury MB	Mercury MBF	Mercury MBTF	Mercury HMI	Xenon
380	0.1	5.55	5.14	4.53	4.73	116.39	93.03
385	0.0	6.11	4.28	3.91	4.01	114.92	94.59
390	0.0	7.25	7.49	6.68	6.40	115.62	96.33
395	0.1	8.02	5.26	4.81	4.85	107.07	100.56
400	0.0	9.22	7.71	7.73	6.77	108.19	102.81
405	0.2	11.37	121.05	107.15	102.94	123.74	100.40
410	0.0	11.84	61.57	54.92	54.53	125.42	100.84
415	0.0	13.33	5.73	6.08	5.97	123.88	101.45
420	0.0	15.39	3.67	4.33	4.19	125.56	102.57
425	0.1	16.95	4.29	4.97	4.82	116.39	101.94
430	0.1	19.30	6.26	7.15	6.75	89.57	101.29
435	0.0	23.70	168.75	146.23	144.91	122.69	101.54
440	0.0	25.45	140.57	123.51	131.86	120.03	103.74
445	0.0	23.36	5.84	6.82	7.00	71.29	103.67
450	0.0	38.22	3.86	4.66	4.96	65.93	110.30
455	0.0	27.95	3.22	3.79	4.28	73.25	112.78
460	0.0	19.25	2.84	3.46	4.08	80.53	116.52
465	0.1	54.71	2.72	3.27	4.05	72.48	129.56
470	0.1	57.56	2.74	3.49	4.34	66.65	141.07
475	0.0	25.63	3.00	3.70	4.68	67.82	126.45
480	0.0	9.94	2.77	3.58	4.68	66.60	115.94
485	0.1	7.81	3.10	3.94	5.20	59.40	118.42
490	0.0	12.68	5.24	6.04	7.27	60.43	111.75
495	0.3	73.42	6.11	6.39	8.03	62.92	113.67
500	0.7	150.21	2.88	3.73	5.59	64.03	105.17
505	0.0	22.10	2.58	3.49	5.55	68.89	103.88
510	0.0	13.02	2.39	3.57	5.73	57.96	102.90
515	0.2	39.88	2.58	3.75	6.16	57.28	102.71
520	0.1	18.38	2.20	3.67	6.24	54.24	102.29
525	0.0	11.27	2.43	3.88	6.70	50.74	101.90
530	0.1	11.97	2.53	4.38	7.29	52.25	101.54
535	0.1	12.91	3.67	6.74	9.43	53.26	101.47
540	0.1	15.01	4.29	9.93	12.21	56.32	101.12
545	0.1	24.05	195.18	198.10	195.59	114.85	101.43
550	0.1	60.75	181.63	171.97	178.84	119.19	100.98
555	0.2	86.90	6.93	10.57	14.46	59.24	100.75
560	0.2	100.00	3.74	7.55	11.60	56.91	100.44
565	2.1	274.28	3.50	6.61	11.26	62.96	100.28
570	8.1	638.70	3.79	6.76	11.92	65.18	100.19
575	1.3	317.64	100.00	100.00	100.00	100.00	100.00
580	1.4	468.03	292.87	273.65	265.84	158.96	99.86
585	131.8	524.92	39.62	42.59	47.39	81.58	100.47
590	1000.0	157.99	4.23	15.05	19.07	82.42	100.33
595	150.6	521.00	3.38	27.64	28.62	72.41	99.04
600	2.0	663.20	2.74	13.65	18.67	74.44	97.17
605	1.3	505.88	2.97	10.75	16.68	62.19	96.65
610	1.3	372.87	2.78	24.34	26.49	61.75	96.84
615	3.5	372.50	3.10	83.81	69.19	61.00	98.50
620	1.9	272.75	2.65	149.80	119.79	60.76	100.21

Wave-length, nm	Low pressure sodium	High pressure sodium	Mercury MB	Mercury MBF	Mercury MBTF	Mercury HMI	Xenon
625	0.6	188.67	3.51	58.07	53.70	62.91	99.91
630	0.6	160.93	2.67	17.99	23.74	55.78	97.80
635	0.8	141.89	2.91	10.51	18.90	50.61	96.84
640	1.8	127.07	2.66	8.09	18.23	52.97	98.92
645	0.6	115.00	2.76	7.86	18.56	49.83	99.18
650	0.7	104.10	2.66	10.98	21.14	46.82	101.99
655	0.4	98.04	2.78	10.76	21.58	47.96	98.78
660	0.3	91.55	2.68	7.25	19.31	56.13	97.14
665	0.2	85.00	2.81	6.29	18.98	53.38	98.43
670	0.5	92.90	2.96	6.06	19.18	67.89	100.30
675	0.2	90.94	3.18	5.91	19.68	73.25	101.29
680	0.0	73.37	2.75	5.18	19.43	47.32	101.97
685	0.0	56.00	2.95	5.48	20.20	50.08	109.55
690	0.1	45.52	6.51	9.83	24.27	44.96	110.88
695	0.5	41.23	5.19	18.35	30.79	56.34	102.08
700	0.2	38.28	2.82	46.48	51.26	69.68	94.49
705	0.6	37.39	3.17	35.63	43.56	36.16	93.05
710	0.1	35.86	4.29	15.82	29.82	28.41	98.10
715	0.0	35.22	3.22	5.92	23.00	27.15	106.20
720	0.1	33.84	2.75	4.61	22.22	28.21	98.24
725	0.2	33.52	2.97	4.48	22.86	34.95	95.69
730	0.0	32.60	2.74	4.11	22.84	34.94	101.73
735	0.0	32.01	2.97	4.23	23.63	27.68	108.38
740	0.2	32.51	2.71	3.92	23.41	24.19	102.62
745	0.0	32.19	2.97	4.08	24.35	22.61	100.28
750	0.4	33.72	2.63	3.84	24.30	21.79	97.33
755	0.2	34.04	3.00	4.13	26.32	30.47	97.45
760	0.1	31.90	2.67	3.67	25.18	33.75	101.71
765	23.6	54.53	3.10	3.99	26.40	36.16	137.21
770	55.8	81.51	2.99	3.79	26.22	32.39	105.22
775	11.1	40.67	3.86	4.61	27.96	26.55	84.41
780	0.0	32.49	2.64	3.41	27.08	24.38	80.55

A5.7 Correlated colour temperatures of D Illuminants

The correlated colour temperatures of Illuminants D_{50}, D_{55}, D_{65}, and D_{75} are as follows:

D_{50}: $T_c = 5000(1.4388/1.4380) = 5003$ K approximately.
D_{55}: $T_c = 5500(1.4388/1.4380) = 5503$ K approximately.
D_{65}: $T_c = 6500(1.4388/1.4380) = 6504$ K approximately.
D_{75}: $T_c = 7500(1.4388/1.4380) = 7504$ K approximately.

Appendix 6

Colorimetric formulae

A6.1 Chromaticity relationships

$$u'=4X/(X+15Y+3Z)$$
$$v'=9Y/(X+15Y+3Z)$$
$$w'=(-3X+6Y+3Z)/(X+15Y+3Z)$$

$$u'=4x/(-2x+12y+3)$$
$$v'=9y/(-2x+12y+3)$$
$$w'=(-6x+3y+3)/(-2x+12y+3)$$

$$x=9u'/(6u'-16v'+12)$$
$$y=4v'/(6u'-16v'+12)$$
$$z=(-3u'-20v'+12)/(6u'-16v'+12)$$

$$u=u'$$
$$v=(2/3)v' \ .$$

A6.2 CIELUV, CIELAB, and U*V*W* relationships

$$u^*=13L^*(u'-u'_n)$$
$$v^*=13L^*(v'-v'_n)$$
$$L^*=116(Y/Y_n)^{1/3}-16$$

$$a^*=500[(X/X_n)^{1/3}-(Y/Y_n)^{1/3}]$$
$$b^*=200[(Y/Y_n)^{1/3}-(Z/Z_n)^{1/3}]$$
$$L^*=116(Y/Y_n)^{1/3}-16$$

$$U^*=13W^*(u-u_n)$$
$$V^*=13W^*(v-v_n)$$
$$W^*=25Y^{1/3}-17 \ .$$

The suffix n indicates that the value is for the reference white. In the formula for W^* it is necessary to express Y as a percentage. If Y/Y_n is equal to or less than 0.008856 the expression for L^* is replaced by:

$$L^*=903.3(Y/Y_n) \ .$$

If any of the ratios X/X_n, Y/Y_n, Z/Z_n, are equal to or less than 0.008856, they are replaced in the above formulae for a^* and b^* by:

$$7.787F+16/116 \ ,$$

where F is X/X_n, Y/Y_n, or Z/Z_n, as the case may be.

$$h_{uv}=\arctan[(v'-v'_n)/(u'-u'_n)]$$
$$s_{uv}=13[(u'-u'_n)^2 + (v'-v'_n)^2]^{1/2}$$

$$h_{uv}=\arctan(v^*/u^*)$$
$$h_{ab}=\arctan(b^*/a^*) \ .$$

h_{uv} lies between 0° and 90° if v^* and u^* are both positive; between 90° and 180° if v^* is positive and u^* is negative; between 180° and 270° if v^* and u^* are both negative; and between 270° and 360° if v^* is negative and u^* is positive (and similarly for h_{ab} and a^* and b^*).

$$C^*_{uv}=(u^{*2}+v^{*2})^{1/2}=L^*s_{uv}$$
$$C^*_{ab}=(a^{*2}+b^{*2})^{1/2}$$

$$\Delta E^*_{uv}=[(\Delta L^*)^2+(\Delta u^*)^2+(\Delta v^*)^2]^{1/2}$$
$$\Delta E^*_{ab}=[(\Delta L^*)^2+(\Delta a^*)^2+(\Delta b^*)^2]^{1/2}$$

$$\Delta E^*_{uv}=[(\Delta L^*)^2+(\Delta H^*_{uv})^2+(\Delta C^*_{uv})^2]^{1/2}$$
$$\Delta E^*_{ab}=[(\Delta L^*)^2+(\Delta H^*_{ab})^2+(\Delta C^*_{ab})^2]^{1/2}$$

where

$$\Delta H^*_{uv}=[(\Delta E^*_{uv})^2-(\Delta L^*)^2-(\Delta C^*_{uv})^2]^{1/2}$$
$$\Delta H^*_{ab}=[(\Delta E^*_{ab})^2-(\Delta L^*)^2-(\Delta C^*_{ab})^2]^{1/2}$$

ΔH^* is positive if indicating an increase in h and negative if indicating a decrease in h.

Appendix 7

Glossary of terms

The following list of terms and their definitions follows broadly the recommendations made in the fourth edition of the *CIE International Lighting Vocabulary* (CIE Publication No. 17.4, 1987). The numbers in brackets after the terms indicate the sections in which these terms are discussed.

Abney's Law (2.3)
An empirical law stating that if two colour stimuli, A and B, are perceived to be of equal brightness, and two other colour stimuli, C and D, are perceived to be of equal brightness, then the additive mixtures of A with C and B with D will also be perceived to be of equal brightness. (The validity of this law depends strongly on the observing conditions and on the colours of the stimuli involved.)

achromatic colour (1.7)
Colour devoid of hue. (The names white, grey, black, neutral, and colourless are commonly used for these colours.)

achromatic signal (1.6 and 12.19)
Visual signal from the retina composed of additions of the cone and rod responses.

achromatic stimulus (3.4)
A stimulus that is chosen to provide a reference that is regarded as achromatic in colorimetry.

action spectra (1.5)
Spectral sensitivity of a visual mechanism in terms of the light incident on the cornea of the eye.

adaptation (1.8)
Visual process whereby approximate compensation is made for changes in the luminances and colours of stimuli, especially in the case of changes in illuminants.

additive mixing (2.3 and 2.4)
Addition of colour stimuli on the retina in such a way that they cannot be perceived individually.

anomalous trichromatism (1.10)
Form of trichromatic vision in which colour discrimination is less than normal.

apostilb, asb (A1.3)
Unit of luminance equal to 0.3183 cd/m^2.

arctan
Abbreviation meaning: angle whose tangent is; sometimes written as: $\tan^{-1}$.

blackness (7.7)
Attribute of a visual sensation according to which an area appears to contain more or less black content.

bright (1.7)
Adjective denoting high brightness.

brightness (1.7 and Table 3.1)
Attribute of a visual sensation according to which an area appears to exhibit more or less light.

candela, cd (A1.3)
Unit of luminous intensity. The candela is the luminous intensity, in a given direction, of a source emitting a monochromatic radiation of frequency 540×10^{12} hertz, the radiant intensity of which in that direction is 1/683 watt per steradian.

candela per square metre, cd/m^2 (A1.3)
Unit of luminance.

chroma (1.9 and Table 3.1)
The colourfulness of an area judged as a proportion of the brightness of a similarly illuminated area that appears to be white or highly transmitting.

chromatic adaptation (3.13 and 12.10)
Visual process whereby approximate compensation is made for changes in the colours of stimuli, especially in the case of changes in illuminants.

chromatic colour (1.7)
Colour exhibiting hue (as distinct from those commonly called white, grey, black, neutral, and colourless).

chromaticity (3.3 and Table 3.1)
Property of a colour stimulus defined by its chromaticity co-ordinates.

chromaticity co-ordinates (3.3)
Ratio of each of a set of tristimulus values to their sum.

chromaticity diagram (3.3)
A two-dimensional diagram in which points specified by chromaticity co-ordinates represent the chromaticities of colour stimuli.

chromaticness (7.7)
1. NCS measure of the chromatic content of a colour.
2. An alternative term for colourfulness.
3. Perceptual colour attribute consisting of the hue and saturation of a colour (obsolete).

CIE (Commission Internationale de l'Éclairage)
The International Commission on Lighting; the body responsible for international recommendations for photometry and colorimetry.

CIE standard photometric observer (2.2 and 2.3)
Ideal observer whose relative spectral sensitivity function conforms to the photopic or scotopic spectral luminous efficiency function.

CIE 1931 standard colorimetric observer (2.6)
Ideal observer whose colour matching properties correspond to the CIE colour-matching functions for the 2° field size.

CIE 1964 supplementary standard colorimetric observer (2.6)
Ideal observer whose colour matching properties correspond to the CIE colour-matching functions for the 10° field size.

CIE 1976 chroma, C^*_{uv}, C^*_{ab} (3.9 and Table 3.1)
Correlate of chroma in the CIELUV and CIELAB colour spaces.

CIE 1976 colour difference, ΔE^*_{uv}, ΔE^*_{ab} (3.10)
Correlate of colour difference in the CIELUV and CIELAB colour spaces.

CIE 1976 hue-angle, h_{uv}, h_{ab} (3.9 and Table 3.1)
Correlate of hue in the CIELUV and CIELAB colour spaces.

CIE 1976 hue-difference ΔH^*_{uv}, ΔH^*_{ab}(3.10)
Correlate of hue difference in the CIELUV and CIELAB colour spaces.

CIE 1976 lightness, L^* (3.9 and Table 3.1)
Correlate of lightness in the CIELUV and CIELAB colour spaces.

CIE 1976 saturation, s_{uv} (3.9 and Table 3.1).
Correlate of saturation in the CIELUV colour space.

CIE 1976 uniform chromaticity scale (UCS) diagram (3.6 and Table 3.1)
Chromaticity diagram in which u' and v' are plotted; equal distances in this diagram represent more nearly than in the x,y diagram equal colour differences for stimuli having the same luminance.

CIE colour rendering index, R_a (4.17 and A5.3)
A CIE method of assessing the degree to which a test illuminant renders colours similar in appearance to their appearance under a reference illuminant.

CIELAB colour space (3.9)
Colour space in which L^*, a^*, b^* are plotted at right angles to one another. Equal distances in the space represent approximately equal colour differences.

CIELUV colour space (3.9)
Colour space in which L^*, u^*, v^* are plotted at right angles to one another. Equal distances in the space represent approximately equal colour differences.

colorimetric purity, p_c (11.3)
Quantity defined by the expression $p_c = L_d/(L_d + L_n)$ where L_d and L_n are the respective luminances of the monochromatic stimulus and of a specified achromatic stimulus that match the colour stimulus considered in an additive mixture. (In the case of purple stimuli, the monochromatic stimulus is replaced by a stimulus whose chromaticity is represented by a point on the purple boundary.)

Coloroid system (7.10)
A colour order system.

colour constancy (1.8 and 12.5)
Effect of visual adaptation whereby the appearance of colours remains approximately constant when the level and colour of the illuminant are changed.

colour difference signal (1.6 and 12.7)
Visual signal from the retina composed of differences between the cone responses.

colour-matching functions (2.4)
The tristimulus values of monochromatic stimuli of equal radiant power per small constant-width wavelength interval throughout the spectrum.

colour stimulus function, $\phi_\lambda(\lambda)$ (2.6)
Description of a colour stimulus by an absolute measure of a radiant quantity per small constant-width wavelength interval throughout the spectrum.

colour rendering index (4.17 and A5.3)
A method of assessing the degree to which a test illuminant renders colours similar in appearance to their appearance under a reference illuminant.

colour temperature, T_c Unit: kelvin, K (4.2)
The temperature of a Planckian radiator whose radiation has the same chromaticity as that of a given stimulus.

colourfulness (1.7 and Table 3.1)
Attribute of a visual sensation according to which the perceived colour of an area appears to exhibit more or less hue.

complementary colour stimulus (3.4)
Two colour stimuli are complementary when it is possible to reproduce the tristimulus values of a specified achromatic stimulus by an additive mixture of these two stimuli.

complementary wavelength, λ_c (3.4)
Wavelength of the monochromatic stimulus that, when additively mixed in suitable proportions with the colour stimulus considered, matches the specified achromatic stimulus.

cones (1.4)
Photoreceptors in the retina that contain light-sensitive pigments capable of initiating the process of photopic vision.

correlated colour temperature, T_{cp} Unit: kelvin, K (4.8)
Temperature of the Planckian radiator whose perceived colour most closely resembles that of a given stimulus seen at the same brightness and under specified viewing conditions. The recommended method of calculating the correlated colour temperature of a stimulus is to determine on the u,v (not the u',v') chromaticity diagram the temperature corresponding to the point on the locus of Planckian radiators that is nearest to the point representing the stimulus.

corresponding colour stimuli (3.13)
Pairs of colour stimuli that look alike when one is seen in one set of adaptation conditions, and the other is seen in a different set.

D illuminants (4.16 and A5.4)
CIE Standard Illuminants having defined relative spectral power distributions that represent phases of daylight with different correlated colour temperatures.

dark (1.9)
Adjective denoting low lightness.

daylight illuminant (4.11, 4.14, and 4.16)
Illuminant having the same, or nearly the same, relative spectral power distribution as a phase of daylight.

daylight locus (4.16)
The locus of points in a chromaticity diagram that represent chromaticities of phases of daylight with different correlated colour temperatures.

defective colour vision (1.10)
Abnormal colour vision in which there is a reduced ability to discriminate some or all colours.

density, (optical) D (A1.6)
Logarithm to the base 10 of the reciprocal of the reflectance factor or the transmittance factor.

deutan (1.10)
Adjective denoting deuteranopia or deuteranomaly.

deuteranomaly (1.10)
Defective colour vision in which discrimination of reddish and greenish colours is reduced, without any colours appearing abnormally dim.

deuteranopia (1.10)
Defective colour vision in which discrimination of reddish and greenish colours is absent, without any colours appearing abnormally dim.

dichromatism (1.10)
Defective colour vision in which all colours can be matched using additive mixtures of only two matching stimuli.

diffuse reflectance, ρ_d (5.9, Table 5.1, and A1.6)
The ratio of the part of the radiant or luminous flux reflected by diffusion to the incident flux, under specified conditions of irradiance.

diffuse transmittance, τ_d (5.9, Table 5.2, and A1.6)
The ratio of the part of the radiant or luminous flux transmitted by diffusion to the incident flux, under specified conditions of irradiance.

dim (1.7)
Adjective denoting low brightness.

DIN system (7.9)
A colour order system.

distribution temperature T_d Unit: kelvin, K (4.8)
Temperature of a Planckian radiator whose relative spectral power distribution is the same as that of the radiation considered.

dominant wavelength, λ_d (3.4 and Table 3.1)
Wavelength of the monochromatic stimulus that, when additively mixed in suitable proportions with the specified achromatic stimulus, matches the colour stimulus considered.

doubly diffuse transmittance, τ_{dd} (5.9, Table 5.2, and A1.6)
The ratio of the transmitted to the incident radiant or luminous flux when the sample is irradiated and viewed diffusely.

emission
The process of emitting radiation.

emissivity (4.8)
Ratio of the radiant exitance of a radiator to that of a Planckian radiator at the same temperature.

equi-energy stimulus, S_E (2.4)
Stimulus consisting of equal amounts of power per small constant-width wavelength interval throughout the spectrum.

erg
Unit of energy or work equal to 10^{-7} joules.

excitation purity, p_e (3.4 and Table 3.1)
Quantity defined by the ratio NC/ND of two collinear distances on the x,y, or on the x_{10},y_{10}, chromaticity diagram. NC is the distance between the point C representing the colour stimulus considered and the point N representing the specified achromatic stimulus; ND is the distance between the point N and the point D on the spectral locus at the dominant wavelength of the colour stimulus considered. In the case of purple stimuli, the point on the spectral locus is replaced by a point on the purple boundary.

exitance
At a point on a surface, the flux leaving the surface per unit area.

fluorescence (9.1)
Process whereby colours absorb radiant power at one wavelength and immediately re-emit it at another (usually longer) wavelength.

flux
Rate of flow per unit cross-section normal to the direction of flow.

foot-candle, fc (A1.3)
Unit of illuminance equal to 10.76 lux.

foot-lambert, fL (A1.3)
Unit of luminance equal to 3.426 cd/m^2.

fovea (1.3)
Central part of the retina that contains almost exclusively cones, and forming the site of most distinct vision. It subtends an angle of about 0.026 radians (1.5°) in the visual field.

foveola (1.3)
Central region of the fovea that contains only cones and is limited to a diameter of about 0.3 mm; it subtends an angle of about 0.017 radians (1°) in the visual field.

gloss trap (5.9 and Tables 5.1 and 5.2)
Device used in spectrophotometry to eliminate specular components from measurements.

Grassmann's Laws (2.4)
Three empirical laws that describe the colour-matching properties of additive mixtures of colour stimuli:

1. To specify a colour match, three independent variables are necessary and sufficient.
2. For an additive mixture of colour stimuli, only their tristimulus values are relevant, not their spectral compositions.
3. In additive mixtures of colour stimuli, if one or more components of the mixture are gradually changed, the resulting tristimulus values also change gradually.

Helmholtz–Kohlrausch effect (2.3, 3.2, 11.4, and 12.2)
Change in brightness of perceived colours produced by increasing the purity of a colour stimulus while keeping its luminance constant (within the range of photopic vision).

Helson–Judd effect (12.11)
Tendency, in coloured illumination, for light colours to be tinged with the hue of the illuminant, and for dark colours to be tinged with the complementary hue.

hertz, hz
Unit of frequency denoting the number of events per second.

hue (1.7 and Table 3.1)
Attribute of a visual sensation according to which an area appears to be similar to one, or to proportions of two, of the perceived colours, red, yellow, green, and blue.

illuminance, E. Unit: lux, lx (A1.3 and Table A1.1)
Luminous flux per unit area incident on a surface.

intensity, I. Unit: candela, cd (A1.3 and Table A1.1)
1. Flux per unit solid angle.
2. General term used to indicate the magnitude of a variable.

irradiance, E_e. Unit: Watt per square metre, $W.m^{-2}$ (Table A1.1)
Radiant flux per unit area incident on a surface.

isotropic (5.7)
Independent of direction.

joule, J
Unit of power expended or consumed, equal to 1 watt second.

Judd correction (2.8 and A2)
Modification of the $V(\lambda)$ function to correct the low values at wavelengths below 460 nm, called the *CIE 1988 2° spectral luminous efficiency function for photopic vision,* $V_M(\lambda)$.

kelvin, K (4.8)
Unit of temperature used for expressing colour temperatures. The temperature in kelvin is equal to that in Celsius plus 273.

lambert, L (A1.3)
Unit of luminance equal to 3183 cd/m^2.

light (1.9)
Adjective denoting high lightness.

light exposure, H. Unit: lux seconds, lx.s. (Table A1.1)
Quantity of light received per unit area.

lightness (1.9 and Table 3.1)
The brightness of an area judged relative to the brightness of a similarly illuminated area that appears to be white or highly transmitting.

lumen, lm (A1.3 and Table A1.1)
Unit of luminous flux. The lumen is the luminous flux emitted within unit solid angle (1 steradian) by a point source having an isotropic luminous intensity of 1 candela. It is the luminous flux of a beam of monochromatic radiation whose frequency is 540×10^{12} hertz and whose radiant flux is 1/683 watt.

lumen per square foot, $lm.ft^{-2}$ (A1.3)
Unit of illuminance equal to 10.76 lux.

lumen per square metre, lm.m^{-2} (Table A1.1)
1. Unit of illuminance called lux, lx.
2. Unit of luminous exitance.

luminance, L. Unit: cd.m^{-2} (Table 3.1, A1.3 and Table A1.1)
In a given direction, at a point in the path of a beam, the luminous intensity per unit projected area (the projected area being at right angles to the given direction).

luminance factor, β (2.6, 3.2, Table 3.1, and A1.6)
Ratio of the luminance to that of the perfect diffuser identically illuminated.

luminescence
Emission of radiation in excess of that caused by thermal radiation.

luminous (2.6, A1.2, and A1.6)
1. Adjective denoting measures evaluated in terms of spectral power weighted by the $V(\lambda)$ function.
2. Adjective denoting stimuli that produce light or appear to do so.

luminous efficacy. Unit: lm.W^{-1}(A1.3)
1. Of a source: ratio of luminous flux emitted to power consumed.
2. Of a radiation: ratio of luminous flux to radiant flux.

luminous efficiency (A1.3)
Ratio of radiant flux weighted according to the $V(\lambda)$ function, to the radiant flux.

luminous exitance, M. Unit: lm.m^{-2} (Table A1.1)
At a point on a surface, the radiant flux leaving the surface per unit area.

luminous flux, F. Unit: lumen, lm (A1.3 and Table A1.1)
Radiant flux weighted by the V(λ) function.

luminous intensity, I. Unit: candela, cd (A1.3 and Table A1.1)
Luminous flux per unit solid angle.

lux, lx (A1.3 and Table A1.1)
Unit of illumination equal to 1 lumen per square metre.

macula lutea (1.3)
Layer of photostable pigment covering parts of the retina in the foveal region.

mesopic vision (1.4)
Vision intermediate between photopic and scotopic vision.

metameric colour stimuli (2.7 and 6.1)
Spectrally different colour stimuli that have the same tristimulus values.

metamerism index (6.6, 6.7, and 6.8)
Measure of the extent to which two stimuli that match one another become different when the illuminant or the observer is changed.

millilambert, ml (A1.3)
Unit of luminance equal to 3.183 cd/m^2.

monochromat (1.10)
Observer who is completely unable to discriminate stimuli by their colours.

monochromatic stimulus (2.4)
A stimulus consisting of a very small range of wavelengths which can be adequately described by stating a single wavelength.

monochromatism (1.10)
Defective colour vision in which all colours can be matched using only a single matching stimulus.

Munsell system (7.4 and 7.5)
A colour order system.

nanometre, nm (1.2)
Very small unit of length equal to 10^{-9} metre commonly used for identifying wavelengths of the spectrum.

Natural Colour System (NCS) (7.7 and 7.8)
A colour order system.

NCS (7.7 and 7.8)
Abbreviation for Natural Colour System.

neuron (1.6)
Nerve cell.

nit, nt (A1.3)
Obsolete name for the unit of luminance, cd/m^2.

non-self-luminous colour stimulus (5.6)
Stimulus that consists of an illuminated object.

observer metamerism (6.7)
Variations of colour matches (of spectrally different stimuli) amongst different observers.

optical axis (of the eye) (1.3)
The direction defined by a line passing normally through the optical elements of the eye.

optical brightening agent (5.7)
Compound added to coloured objects to increase their luminance factors by fluorescence, particularly in the case of whites.

Optical Society of America system (OSA system) (7.11)
A colour order system.

optimal colour stimuli (7.3)
Colour stimuli whose luminance factors have maximum possible values for each chromaticity when their spectral radiance factors do not exceed 1 for any wavelength.

OSA system (7.11)
Abbreviation for Optical Society of America system.

paramers (6.11)
Spectrally different colour stimuli that have nearly the same tristimulus values. The corresponding property is called *paramerism.*

pearlescent (5.8)
Adjective to denote reflecting colours that contain metallic or other particles which impart reflective properties similar to those of pearls.

perfect diffuser (2.6, 5.7, and A1.6)
An ideal isotropic diffuser with a reflectance (or transmittance) equal to unity.

phosphorescence
Photoluminescence that continues appreciably after the exciting radiation is removed.

photoluminescence
Luminescence cause by ultraviolet, visible, or infrared radiation.

photon
A quantum of light or of other electromagnetic radiation.

photopic vision (1.4)
Vision by the normal eye when it is adapted to levels of luminance of at least several candelas per square metre. (The cones are the principal photoreceptors that are active in photopic vision.)

Planckian locus (4.8, A1.5, and A5.5)
The locus of points in a chromaticity diagram that represent the chromaticities of the radiation of Planckian radiators at different temperatures.

Planckian radiator (4.8, A1.5, and A5.5)
A body that emits radiation, because of its temperature, according to Planck's Law.

point brilliance. Unit: lux, lx (A1.3)
The illuminance produced by a source on a plane at the observer's eye, when the apparent diameter of the source is inappreciable.

point source (A1.3)
Source of radiation the dimensions of which are small enough, compared with the distance between source and detector, for them to be neglected in calculations.

power, P. Unit: watt, W
Energy per unit time.

protan (1.10)
Adjective denoting protanopia or protanomaly.

protanomaly (1.10)
Defective colour vision in which discrimination of reddish and greenish colours is reduced, with reddish colours appearing abnormally dim.

protanopia (1.10)
Defective colour vision in which discrimination of the reddish and greenish contents of colours is absent, with reddish colours appearing abnormally dim.

purity (3.4)
A measure of the proportions of the amounts of the monochromatic stimulus and of the specified achromatic stimulus that, when additively mixed, match the colour stimulus considered. (In the case of purple stimuli, the monochromatic stimulus is replaced by a stimulus whose chromaticity is represented by a point on the purple boundary.)

Purkinje phenomenon (1.4 and 2.3)
Reduction in the brightness of a predominantly long-wavelength colour stimulus relative to that of a predominantly short-wavelength colour stimulus, when the luminances are reduced in the same proportion from photopic to mesopic or scotopic levels, without changing the respective relative spectral distributions of the stimuli involved.

purple boundary (3.3)
The line in a chromaticity diagram, or the surface in a colour space, that represents additive mixtures of monochromatic stimuli of wavelengths of approximately 380 and 780 nm.

purple stimulus (3.3)
Stimulus that is represented in any chromaticity diagram by a point lying within the triangle defined by the point representing the specified achromatic stimulus and the two ends of the spectral locus, which correspond approximately to the wavelengths 380 and 780 nm.

quantum (A1.2, A1.6, and Table A1.1)
1. Extremely small indivisible unit of energy. (In the case of light or of other electromagnetic radiation, the term *photon* may be used.)
2. Adjective denoting measures evaluated in terms of number of quanta.

quantum exitance, M_q. Unit: $s^{-1}.m^{-2}$ (Table A1.1)
At a point on a surface, the quantum flux leaving the surface per unit area.

quantum exposure, H_q. Unit: m^{-2} (Table A1.1)
Number of quanta received.

quantum flux, F_q. Unit: s^{-1} (Table A1.1)
Number of quanta emitted, transferred or received per unit of time.

quantum intensity, I_q. Unit: $s^{-1}.sr^{-1}$ (Table A1.1)
Quantum flux per unit solid angle.

quantum irradiance, E_q. Unit: $s^{-1}.m^{-2}$ (Table A1.1)
Quantum flux per unit area incident on a surface.

quantum radiance, L_q. Unit: $s^{-1}.\ sr^{-1}.\ m^{-2}$ (Table A1.1)
In a given direction, at a point in the path of a beam, the quantum intensity per unit projected area (the projected area being at right angles to the given direction).

radiance, L_e Unit: $W.sr^{-1}.m^{-2}$ (Table A1.1)
In a given direction, at a point in the path of a beam, the radiant intensity per unit projected area (the projected area being at right angles to the given direction).

radiance factor, β_e (2.6, 5.9, Tables 5.1 and 5.2, and A1.1)
Ratio of the radiance to that of the perfect diffuser identically irradiated.

radiant (2.6, A1.2, and A1.6)
Adjective denoting measures evaluated in terms of power.

radiant efficiency (A1.3)
Ratio of the radiant flux emitted to the power consumed by a source.

radiant exitance, M_e. Unit: W.m^{-2} (Table A1.1)
At a point on a surface, the radiant flux leaving the surface per unit area.

radiant exposure, H_e. Unit: J.m^{-2} (Table A1.1)
Quantity of radiant energy received per unit area.

radiant flux, F_e. Unit: W (Table A1.1)
Power (energy per unit time) emitted, transferred, or received in the form of radiation.

radiant intensity, I_e. Unit: W.sr^{-1} (Table A1.1)
Radiant flux per unit solid angle.

reflectance, ρ (2.6, 5.9, Table 5.1, and A1.6))
Ratio of the reflected radiant or luminous flux to the incident flux under specified conditions of irradiation.

reflectance factor, R (2.6, 5.9, Table 5.1, and A1.6)
Ratio of the radiant or luminous flux reflected in a given cone, whose apex is on the surface considered, to that reflected in the same directions by the perfect diffuser identically irradiated.

reflection (A1.6)
Return of radiation by a medium without change of frequency (that is, without fluorescence).

regular reflectance, ρ_r (5.9, Table 5.1, and A1.6)
Ratio of the regularly reflected radiant or luminous flux to the incident flux, under specified conditions of irradiance.

regular reflection (5.9, Table 5.1, and A1.6)
Reflection as in a mirror without deviation by scattering, diffraction, or diffusion.

regular transmission (Table 5.2, and A1.6)
Transmission without deviation by scattering, diffraction, or diffusion.

regular transmittance, τ_r (Table 5.2, and A1.6)
Ratio of the regularly transmitted radiant or luminous flux to the incident flux, under specified conditions of irradiance.

related colour (1.9, 12.1, 12.25, and 12.27)
Colour perceived to belong to an area seen in relation to other colours.

relative colour stimulus function, $\phi(\lambda)$ (2.6)
Relative spectral power distribution of a colour stimulus.

relative spectral power distribution, $S(\lambda)$ (2.6)
Spectral power per small constant-width wavelength interval throughout the spectrum relative to a fixed reference value.

retina (1.3)
Light sensitive layer on the inside of the back of the eye; it contains the photoreceptors (cones and rods) and nerve cells that transmit the visual signals to the optic nerve.

rhodopsin (1.5)
Visual pigment present in the rods of the retina.

rods (1.4)
Photoreceptors in the retina that contain a light-sensitive pigment capable of initiating the process of scotopic vision.

saturation (1.9 and Table 3.1)
The colourfulness of an area judged in proportion to its brightness.

scotopic vision (1.4)
Vision by the normal eye when it is adapted to levels of luminance less than some hundredths of a candela per square metre.

self-luminous colour stimulus (5.5)
Stimulus that produces its own light.

shadow series (7.9)
A series of colours of constant chromaticity but varying luminance factor.

solid angle (2.2)
The part of space that is bounded by lines radiating from a point and passing through a closed curve that does not contain the point.

spectral (2.6 and A1.6)
Adjective denoting that monochromatic concepts are being considered.

spectral conventional reflectometer value, $\rho_C(\lambda)$ (9.2)
The apparent spectral reflectance factor obtained when a fluorescent sample is measured relative to a non-fluorescent white sample, using monochromatic illumination and heterochromatic detection.

spectral luminescent radiance factor, $\beta_L(\lambda)$ (9.2)
Ratio, at a given wavelength, of the radiance produced by fluorescence by a sample to that produced by the perfect reflecting diffuser identically irradiated.

spectral locus (3.3)
Locus in a chromaticity diagram or colour space of the points that represent monochromatic stimuli throughout the spectrum.

spectral luminous efficiency, $V(\lambda)$, $V'(\lambda)$ (2.2 and 2.3)
Weighting functions used to derive photometric measures from radiometric measures in photometry. Normally the $V(\lambda)$ function is used, but, if the conditions are such that the vision is scotopic, the $V'(\lambda)$ function is used and the measures are then distinguished by the adjective 'scotopic' and the symbols by the superscript $'$. (See also *Judd correction.*)

spectral power distribution, $S_\lambda(\lambda)$ (2.6)
Spectral power per small constant-width wavelength interval throughout the spectrum.

spectral radiance factor, $\beta_e(\lambda)$ (9.2)
Ratio of the spectral radiance to that of the perfect diffuser identically irradiated.

spectral reflectance, $\rho(\lambda)$ (5.9, Table 5.1, and A1.6)
Ratio of the spectral reflected radiant or luminous flux to the incident flux under specified conditions of irradiation.

spectral reflectance factor, $R(\lambda)$ (5.9, Table 5.1, and A1.6)
Ratio of the spectral radiant or luminous flux reflected in a given cone, whose apex is on the surface considered, to that reflected in the same directions by the perfect diffuser identically irradiated.

spectral reflected radiance factor, $\beta_S(\lambda)$ (9.2)
Ratio of the spectral radiance produced by reflection by a sample to that produced by the perfect diffuser identically irradiated.

spectral total radiance factor, $\beta_T(\lambda)$ (9.2)
The sum of the spectral reflected and spectral luminescent radiance factors.

spectral transmittance $\tau(\lambda)$ (5.9, Table 5.2, and A1.6)
Ratio of the spectral transmitted radiant or luminous flux to the incident flux under specified conditions of irradiation.

spectral transmittance factor, $T(\lambda)$ (5.9, Table 5.2, and A1.6)
Ratio of the spectral radiant or luminous flux transmitted in a given cone, whose apex is on the surface considered, to that transmitted in the same directions by the perfect transmitting diffuser identically irradiated.

spectrophotometry (5.3)
Measurement of the relative amounts of radiant flux at each wavelength of the spectrum.

spectroradiometry (5.3)
(1) Measurement of the absolute amounts of radiant flux at each wavelength of the spectrum.
(2) Measurement of the relative amounts of radiant flux at each wavelength of the spectrum for self-luminous colour stimuli.

specular reflection (5.8, 5.9, Table 5.1, and A1.6)
Reflection as in a mirror without deviation by scattering, diffraction, or diffusion.

SPEX (5.9, Tables 5.1 and 5.2, and A1.6)
Abbreviation for 'specular excluded' in spectrophotometry.

SPINC (5.9, Tables 5.1 and 5.2, and A1.6)
Abbreviation for 'specular included' in spectrophotometry.

Standard Illuminants (4.12, 4.13. 4.14. and 4.16)
Relative spectral power distributions defining illuminants for use in colorimetric computations.

Standard Sources (4.12 and 4.15)
Light sources specified for use in colorimetry and defined by relative spectral power distributions.

steradian, sr (2.2)
Unit of solid angle defined as : the solid angle that, having its vertex in the middle of a sphere, cuts off an area on the surface of the sphere equal to that of a square with side of length equal to that of the radius of the sphere.

Stiles–Crawford effect (1.3)
Decrease of the brightness of a light stimulus with increasing eccentricity of the entry of the light through the pupil of the eye. If the variation is in hue and saturation instead of in brightness, it is called the Stiles–Crawford effect of the second kind.

strong (1.9)
Adjective denoting high chroma.

subtractive mixing of colorants (2.3)
Production of colours by mixing colorants in such a way that they each subtract light from some parts of the spectral power distribution used to illuminate them.

tele-spectroradiometry (5.4)
Spectroradiometry carried out remotely from a sample with the aid of a telescopic optical system.

tint (of whites) (11.2)
Reddish or greenish hue of whites.

total spectral radiance factor, $\beta_T(\lambda)$ (9.2)
The sum of the spectral reflected and spectral luminescent radiance factors.

transformation equations (2.5)
Set of three simultaneous equations used to transform a colour specification from one set of matching stimuli to another.

transmission (A1.6)
Passage of radiation through a medium without change of frequency (that is, without fluorescence).

transmittance, τ (2.6, Table 5.2, and A1.6)
Ratio of the transmitted radiant or luminous flux to the incident flux under specified conditions of irradiation.

transmittance factor, T (2.6, Table 5.2, and A1.6)
Ratio of the radiant or luminous flux transmitted in a given cone, whose apex is on the surface considered, to that transmitted in the same directions by the perfect diffuser identically irradiated.

trichromatic matching (2.4)
Action of making a colour stimulus appear the same colour as a given stimulus by adjusting three components of an additive colour mixture.

tristimulus values (2.4)
Amounts of the three matching stimuli, in a given trichromatic system, required to match the stimulus considered.

tritan (1.10)
Adjective denoting tritanopia or tritanomaly.

tritanomaly (1.10)
Defective colour vision in which discrimination of the bluish and yellowish content of colours is reduced.

tritanopia (1.10)
Defective colour vision in which discrimination of the bluish and yellowish content of colours is absent.

troland, td (A1.3)
Unit used to express a quantity proportional to retinal illuminance produced by a light stimulus. When the eye views a surface of uniform luminance, the number of trolands is equal to the product of the area in square millimetres of the limiting pupil, natural or artificial, times the luminance of the surface in candelas per square metre.

u,v diagram (3.6)
Uniform chromaticity diagram introduced by the CIE in 1960, but now superseded by the u',v' diagram.

u',v' diagram (3.6)
Uniform chromaticity diagram introduced by the CIE in 1976.

uniform chromaticity diagram (3.6)
Chromaticity diagram in which equal distances approximately represent equal colour differences for stimuli having the same luminance.

uniform colour space (3.9)
Colour space in which equal distances approximately represent equal colour differences.

unique hue (7.6)
Perceived hue that cannot be further described by the use of hue names other than its own; there are four unique hues: red, green, yellow, and blue.

unrelated colour (1.9, 12.26, and 12.28)
Colour perceived to belong to an area seen in isolation from other colours.

$V(\lambda)$ function (2.3)
Weighting function used to derive photometric measures from radiometric measures in photometry.

$V'(\lambda)$ function (2.2)
Weighting function used to derive scotopic photometric measures from radiometric measures in photometry. Normally the $V(\lambda)$ function is used, but if the conditions are such that the vision is scotopic the $V'(\lambda)$ is used and the measures are then distinguished by the adjective 'scotopic' and the symbols by the superscript $'$.

visual axis (of the eye) (1.3)
The direction defined by a line joining the centre of the fovea to the centre of the pupil of the eye; the visual axis is offset from the optical axis of the eye by about 4°.

Von Kries transformation (3.13)
Algebraic transformation whereby changes in adaptation are represented as adjustments of the sensitivities of the three cone systems such as to compensate fully for changes in the colour of illuminants.

watt, W
Unit of power equal to 1 joule per second or 10^7 ergs per second.

weak (1.9)
Adjective denoting low chroma.

whiteness (7.7 and 11.2)

1. Attribute of a visual sensation according to which an area appears to contain more or less white content.
2. Attribute that enables whites of different colours to be ranked in order of increasing similarity to some ideal white.

x,y diagram (3.3)
Chromaticity diagram in which the *x,y* chromaticity co-ordinates of the CIE XYZ system are used.

yellow spot (1.3)
Layer of photostable pigment covering parts of the retina in the foveal region (also called the macula lutea).

REFERENCE

CIE Publication No. 17.4, *CIE International Lighting Vocabulary* (1987).

Index

ELLIS HORWOOD SERIES IN APPLIED SCIENCE AND INDUSTRIAL TECHNOLOGY

Series Editor: Dr D. H. SHARP, OBE, former General Secretary, Society of Chemical Industry; formerly General Secretary, Institution of Chemical Engineers; and former Technical Director, Confederation of British Industry.

MECHANICS OF WOOL STRUCTURES
R. POSTLE, University of New South Wales, Sydney, Australia, G. A. CARNABY, Wool Research Organization of New Zealand, Lincoln, New Zealand, and S. de JONG, CSIRO, New South Wales, Australia
MICROCOMPUTERS IN THE PROCESS INDUSTRY
E. R. ROBINSON, Head of Chemical Engineering, North East London Polytechnic
BIOPROTEIN MANUFACTURE: A Critical Assessment
D. H. SHARP, OBE, former General Secretary, Society of Chemical Industry; formerly General Secretary, Institution of Chemical Engineers; and former Technical Director, Confederation of British Industry
QUALITY ASSURANCE: The Route to Efficiency and Competitiveness, Second Edition
L. STEBBING, Quality Management Consultant
QUALITY MANAGEMENT IN THE SERVICE INDUSTRY
L. STEBBING, Quality Management Consultant
INDUSTRIAL CHEMISTRY
E. STOCCHI, Milan, with additions by K. A. K. LOTT and E. L. SHORT, Brunel
REFRACTORIES TECHNOLOGY
C. STOREY, Consultant, Durham; former General Manager, Refractories, British Steel Corporation
COATINGS AND SURFACE TREATMENT FOR CORROSION AND WEAR RESISTANCE
K. N. STRAFFORD and P. K. DATTA, School of Material Engineering, Newcastle upon Tyne Polytechnic, and C. G. GOOGAN, Global Corrosion Consultants Limited, Telford
TEXTILE OBJECTIVE MEASUREMENT AND AUTOMATION IN GARMENT MANUFACTURE
G. STYLIOS, Department of Industrial Technology, University of Bradford
PREPARATIVE AND PROCESS-SCALE LIQUID CHROMATOGRAPHY
G. SUBRAMANIAN, Department of Chemical Engineering, Loughborough University of Technology
INDUSTRIAL PAINT FINISHING TECHNIQUES AND PROCESSES
G. F. TANK, Educational Services, Graco Robotics Inc., Michigan, USA
MODERN BATTERY TECHNOLOGY
Editor: C. D. S. TUCK, Alcan International Ltd, Oxon
FIRE AND EXPLOSION PROTECTION: A Systems Approach
D. TUHTAR, Institute of Fire and Explosion Protection, Yugoslavia
PERFUMERY TECHNOLOGY 2nd Edition
F. V. WELLS, Consultant Perfumer and former Editor of *Soap, Perfumery and Cosmetics,* and M. BILLOT, former Chief Perfumer to Houbigant-Cheramy, Paris, Président d'Honneur de la Société Technique des Parfumeurs de la France
THE MANUFACTURE OF SOAPS, OTHER DETERGENTS AND GLYCERINE
E. WOOLLATT, Consultant, formerly Unilever plc